WAVES, TIDES AND SHALLOW-WATER PROCESSES

THE OCEANOGRAPHY COURSE TEAM

Authors
Evelyn Brown (*Waves, Tides, etc.; Ocean Chemistry*)
Angela Colling (*Ocean Circulation; Seawater (2nd edn); Waves, Tides, etc. (2nd edn); Case Studies*)
Dave Park (*Waves, Tides, etc.*)
John Phillips (*Case Studies*)
Dave Rothery (*Ocean Basins*)
John Wright (*Ocean Basins; Seawater; Waves, Tides, etc. (2nd edn); Ocean Chemistry; Case Studies*)

Course Manager
Fiona McGibbon

Designers
Jane Sheppard
Liz Yeomans

Graphic Artist
Sue Dobson

Cartographer
Ray Munns

Editor
Gerry Bearman

This Volume forms part of an Open University course. For general availability of all the Volumes in the Oceanography Series, please contact your regular supplier, or in case of difficulty the appropriate Butterworth–Heinemann office.

Further information on Open University courses may be obtained from: The Admissions Office, The Open University, P.O. Box 48, Walton Hall, Milton Keynes MK7 6AA, UK or from the Open University website: http://www.open.ac.uk

Cover illustration: Satellite photograph showing distribution of phytoplankton pigments in the North Atlantic off the US coast in the region of the Gulf Stream and the Labrador Current. (*NASA, and O. Brown and R. Evans, University of Miami.*)

WAVES, TIDES AND SHALLOW-WATER PROCESSES

PREPARED BY AN OPEN UNIVERSITY COURSE TEAM
SECOND EDITION REVISED FOR THE COURSE TEAM BY JOHN WRIGHT,
ANGELA COLLING AND DAVE PARK

in association with

THE OPEN UNIVERSITY, WALTON HALL,
MILTON KEYNES, MK7 6AA, ENGLAND

Butterworth–Heinemann
Linacre House, Jordan Hill, Oxford OX2 8DP
A division of Reed Educational and Professional Publishing Ltd

A member of the Reed Elsevier plc group

BOSTON JOHANNESBURG
MELBOURNE NEW DELHI OXFORD

First edition 1989. Reprinted with corrections 1991, 1995, 1997

Second edition 1999. Reprinted with corrections 2001, 2002, 2005

British Library Cataloguing in Publication Data
A catalogue record for this book is available from the British Library

ISBN 0 7506 4281 5

Library of Congress Cataloguing in Publication Data
A catalogue record for this book is available from the Library of Congress

Jointly published by the Open University, Walton Hall, Milton Keynes MK7 6AA and Butterworth–Heinemann

Edited, typeset, illustrated and designed by The Open University
Printed in Singapore by Kyodo under the supervision of MRM Graphics Ltd., UK

s330v4i2.4

CONTENTS

ABOUT THIS VOLUME

This is one of a Series of Volumes on Oceanography. It is designed so that it can be read on its own, like any other textbook, or studied as part of S330 *Oceanography*, a third level course for Open University students. The science of oceanography as a whole is multidisciplinary. However, different aspects fall naturally within the scope of one or other of the major 'traditional' disciplines. Thus, you will get the most out of this Volume if you have some previous experience of studying physics and/or geology. Other Volumes in this Series lie more within the fields of chemistry or biology (and their associated sub-branches) according to subject matter.

Chapter 1 describes the qualitative aspects of water waves, briefly reviews modern methods of wave measurement, and explores some of the simple relationships of wave dimensions and characteristics. It also examines the concept of wave energy, the behaviour of waves as they approach the shore and expend that energy in breaking, and the features and causes of unusual waves.

Tides are a special type of wave, and Chapter 2 outlines the mechanism whereby tides are generated by the gravitational attractions of the Sun and Moon, but constrained by the configuration of the ocean basins. The effects of tides are most evident in shallow water and at the shoreline, and this Chapter also deals with the behaviour of tidal currents in shallow seas along coasts and in estuaries, and with the prediction of tides.

In coastal and shallow marine areas, waves and tidal currents are responsible for sediment movement and deposition. Chapter 3 introduces the nature of shallow marine sediments and the types of environment in which they are deposited. Chapter 4 goes on to consider, in general terms, the physical conditions that lead to the erosion, transport and deposition of sediments by flowing water. Some of the problems of applying the theory of fluid flow to the natural marine environment are discussed.

Chapter 5 examines the factors controlling beach slopes, the conditions under which sediment is moved by waves, the rate at which it is moved, and the way in which waves affect current patterns.

Chapter 6 examines coastal areas where tidal processes are more important than wave-related processes: estuaries, lagoons and tidal flats. Estuaries vary considerably in character because of variations in tidal range and river discharge which affect patterns of water circulation and sedimentation, and the extent to which seawater and river water mix. Their influence is shown to extend for significant distances offshore. Patterns of sedimentation in lagoons and on tidal flats are controlled mainly by the flood and ebb of the tides.

Where the sediment discharge from a river is so high that waves and tidal currents are unable to disperse it at the river mouth, a delta accumulates seawards of the mouth. The estuarine mixing processes discussed in Chapter 6 also apply to processes at the distributary mouths of deltas, and Chapter 7 explains how differences in relative influences of river flow, tidal currents and wave energy lead to differences in sediment dispersal and give deltas their characteristic shapes.

Finally, Chapter 8 outlines how the wave and current regimes of shelf seas affect both the seasonally varying structure of the water column (including development of fronts) and the movement of sediments. There is also a summary of mineral resources on continental shelves.

You will find questions designed to help you to develop arguments and/or test your own understanding as you read, with answers provided at the back of the Volume. Important technical terms are printed in **bold** type where they are first introduced or defined.

Note: The terms 'speed' and 'velocity' are used frequently throughout this Volume. Strictly speaking, *speed* is the rate at which a particular distance is covered and the units are metres per second ($m\,s^{-1}$). *Velocity* is a quantity that specifies both speed and the direction of motion, and its units are also $m\,s^{-1}$. We have attempted to maintain the distinction in this Volume.

ABOUT THIS SERIES

The Volumes in this Series are all presented in the same style and format, and together provide a comprehensive introduction to marine science. *Ocean Basins* deals with structure and formation of oceanic crust, hydrothermal circulation, and factors affecting sea-level. *Seawater* considers the seawater solution and leads naturally into *Ocean Circulation*, which is the 'core' of the Series. It provides a largely non-mathematical treatment of ocean atmosphere interaction and the dynamics of wind-driven surface current systems, and of density-driven circulation in the deep oceans. *Waves, Tides and Shallow-Water Processes* introduces the physical processes which control water movement and sediment transport in the nearshore environment (beaches, estuaries, deltas, shelf seas). *Ocean Chemistry and Deep-Sea Sediments* is concerned with biogeochemical cycling of elements within the seawater solution and with water–sediment interaction at the ocean floor. *Case Studies in Oceanography and Marine Affairs* examines the effect of human intervention in the marine environment and introduces the essentials of Law of the Sea. The two case studies respectively review marine affairs in the Arctic from an historical standpoint, and outline the causes and effects of the tropical climatic phenomenon known as El Niño.

Biological Oceanography: An Introduction (by C. M. Lalli and T. R. Parsons) is a companion Volume to the Series, and is in the same style and format. It describes and explains interactions between marine plants and animals in relation to the physical/chemical properties and dynamic behaviour of the seawater in which they live.

CHAPTER 1 WAVES

'... the chidden billow seems to pelt the clouds ...'

Othello, Act II, Scene I.

Sea waves have attracted attention and comment throughout recorded history. Aristotle (384–322 BC) observed the existence of a relationship between wind and waves, and the nature of this relationship has been a subject of study ever since. However, at the present day, understanding of the mechanism of wave formation and the way that waves travel across the oceans is by no means complete. This is partly because observations of wave characteristics at sea are difficult, and partly because mathematical models of wave behaviour are based upon the dynamics of idealized fluids, and ocean waters do not conform precisely with those ideals. Nevertheless, some facts about waves are well established, at least to a first approximation, and the purpose of this Chapter is to outline the qualitative aspects of water waves and to explore some of the simple relationships of wave dimensions and characteristics.

We start by examining the dimensions of an idealized water wave, and the terminology used for describing waves (Figure 1.1).

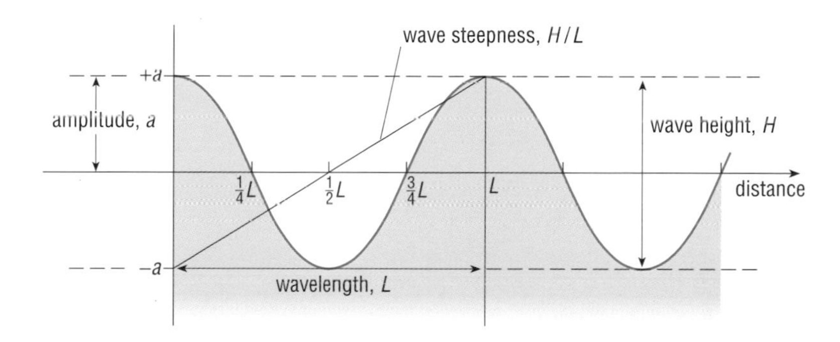

Figure 1.1 Vertical profile of two successive idealized ocean waves, showing their linear dimensions and sinusoidal shape.

Wave height (H) refers to the overall vertical change in height between the wave crest (or peak) and the wave trough. The wave height is twice the wave **amplitude** (a). **Wavelength** (L) is the distance between two successive peaks (or two successive troughs). **Steepness** is defined as wave height divided by wavelength (H/L) and, as can be seen in Figure 1.1, is not the same thing as the slope of the sea-surface between a wave crest and its adjacent trough. The time interval between two successive peaks (or two successive troughs) passing a fixed point is known as the **period** (T), and is generally measured in seconds. The number of peaks (or the number of troughs) which pass a fixed point per second is known as the **frequency** (f).

QUESTION 1.1 If a wave has a frequency of 0.2 s^{-1}, what is its period?

As the answer to Question 1.1 shows, period is the reciprocal of frequency. We will return to this concept in Section 1.2.

1.1 WHAT ARE WAVES?

Waves are a common occurrence in everyday life, and are manifested as, for example, sound, the motion of a plucked guitar string, ripples on a pond, or the billows on the ocean. It is not easy to define a wave. Before attempting to do so, let us consider some of the characteristics of wave motion:

1 A wave transfers a disturbance from one part of a material to another. (The disturbance caused by dropping a stone into a pond is transmitted across the pond by ripples.)

2 The disturbance is propagated through the material without any substantial overall motion of the material itself. (A floating cork merely bobs up and down on the ripples, but experiences very little overall movement in the direction of travel of the ripples.)

3 The disturbance is propagated without any significant distortion of the wave form. (A ripple shows very little change in shape as it travels across a pond.)

4 The disturbance appears to be propagated with constant speed.

If the material itself is not being transported by wave propagation, then what *is* being transported?

The answer, 'energy', provides a reasonable working definition of wave motion – a means whereby energy is transported across or through a material without any significant overall transport of the material itself.

So, if energy, and not material, is being transported, what is the nature of the movement observed when ripples cross a pond?

There are two aspects to be considered: first, the progress of the waves (which we have already noted), and secondly, the movement of the water particles themselves. Superficial observation of the effect of ripples on a floating cork suggests that the water particles move 'up and down', but closer observation will reveal that, provided the water is very much deeper than the ripple height, the cork is describing a nearly circular path in a vertical plane, parallel with the direction of wave movement. In a more general sense, the particles are displaced from an equilibrium position, and a wave motion is the propagation of regular oscillations about that equilibrium position. Thus, the particles experience a displacing force and a restoring force. The nature of these forces is often used in the descriptions of various types of waves.

1.1.1 TYPES OF WAVES

All waves can be regarded as **progressive waves**, in that energy is moving through, or across the surface of, the material. The so-called **standing wave**, of which the plucked guitar string is an example, can be considered as the sum of two progressive waves of equal dimensions, but travelling in opposite directions. We examine this in more detail in Section 1.6.4.

Waves which travel through the material are called body waves. Examples of body waves are sound waves and seismic P- and S-waves, but our main concern in this Volume is with *surface waves* (Figure 1.2). The most familiar surface waves are those which occur at the interface between atmosphere and ocean, caused by the wind blowing over the sea. Other external forces acting on the fluid can also generate waves. Examples range from raindrops falling into tidal pools, through diving gannets and ocean-going liners to earthquakes (see Section 1.6.3).

The tides are also waves (Figure 1.2), caused by the gravitational influence of the Sun and Moon and having periods corresponding to the causative forces. This aspect is considered in more detail in Chapter 2. Most other waves, however, result from a non-periodic disturbance of the water. The water particles are displaced from an equilibrium position, and to regain that position they require a restoring force, as mentioned above. The restoring force causes a particle to 'overshoot' on either side of the equilibrium position. Such alternate displacements and restorations establish a characteristic oscillatory 'wave motion', which in its simplest form has sinusoidal characteristics (Figures 1.1 and 1.6), and is sometimes referred to as simple harmonic motion. In the case of surface waves on water, there are two such restoring forces which maintain wave motion:

1 The gravitational force exerted by the Earth.

2 Surface tension, which is the tendency of water molecules to stick together and present the smallest possible surface to the air. So far as the effect on water waves is concerned, it is as if a weak elastic skin were stretched over the water surface.

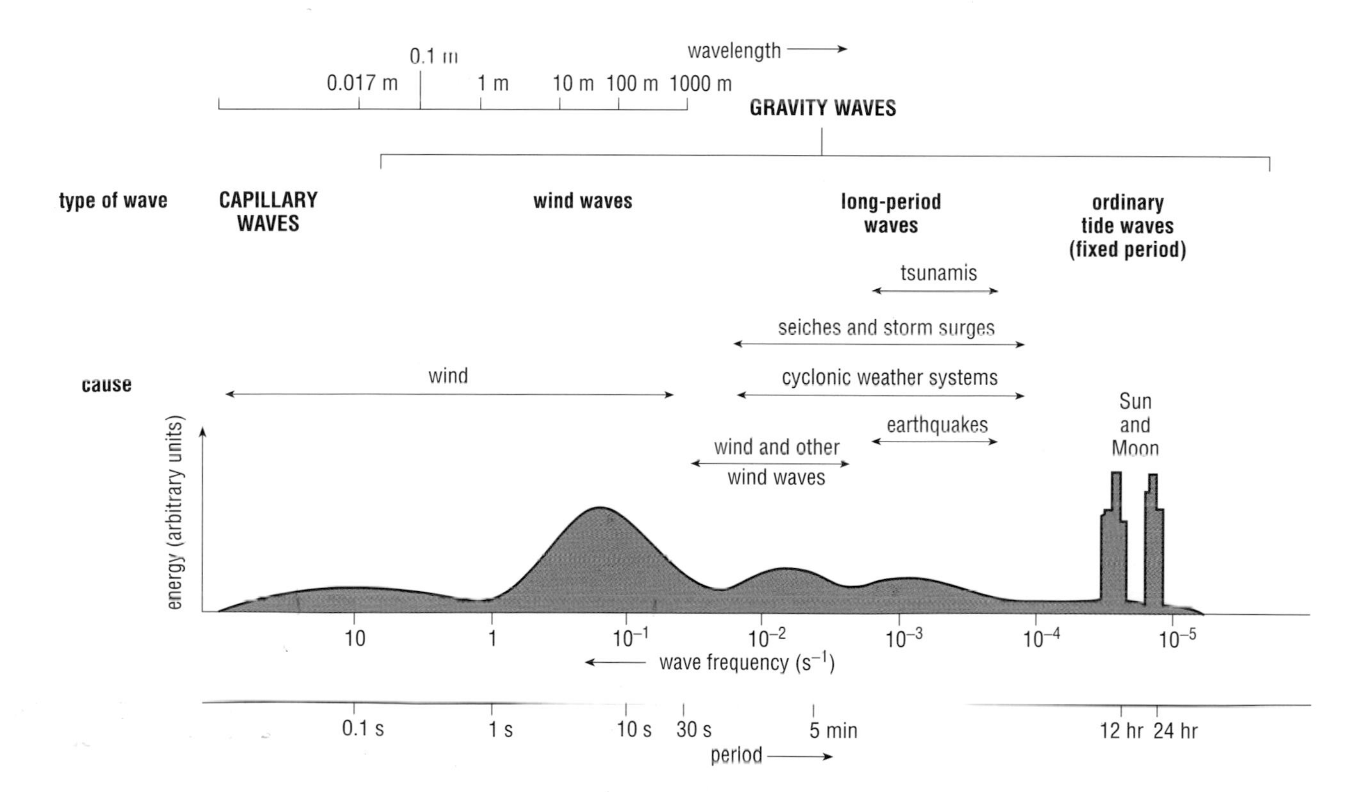

Figure 1.2 Types of surface waves, showing the relationships between wavelength, wave frequency and period, the nature of the forces that cause them, and the relative amounts of energy in each type of wave. Unfamiliar terms will be explained later. *Note:* Waves caused by 'other wind waves' are waves resulting from interactions between waves of higher frequency as they move away from storm areas – see Section 1.4.2.

Water waves are affected by both of these forces. In the case of waves with wavelengths less than about 1.7 cm, the principal restoring force is surface tension, and such waves are known as **capillary waves**. They are important in the context of remote sensing of the oceans (Section 1.7.1). However, the main interest of oceanographers lies with surface waves of wavelengths greater than 1.7 cm, and the principal restoring force for such waves is gravity; hence they are known as **gravity waves** (Figure 1.2).

Gravity waves can also be generated at an interface between two layers of ocean water of differing densities. Because the interface is a surface, such waves are, strictly speaking, surface waves, but oceanographers usually refer to them as **internal waves**. These occur most commonly where there is a rapid increase of density with depth, i.e. a steep density gradient, or **pycnocline**. Pycnoclines themselves result from steep gradients of temperature and/or salinity, the two properties which together govern the density of seawater. Because the difference in density between two water layers is much smaller than that between water and air, less energy is required to displace the interface from its equilibrium position, and oscillations are more easily set up at an internal interface than at the sea-surface. Internal waves travel considerably more slowly than most surface waves. They have greater amplitudes than all but the largest surface waves (up to a few tens of metres), as well as longer periods (minutes or hours rather than seconds, cf. Figure 1.2) and longer wavelengths (hundreds rather than tens of metres). Internal waves are of considerable importance in the context of vertical mixing processes in the oceans, especially when they break.

Not all waves in the oceans are displaced primarily in a vertical plane. For example, because atmosphere and oceans are on a rotating Earth, variation of **planetary vorticity** with latitude (i.e. variation in the angular velocity of the Earth's surface and hence in the effect of the Earth's rotation on horizontal motions) causes horizontal deflection of atmospheric and oceanic currents, and provides restoring forces which establish oscillations mainly in a horizontal plane, so that easterly or westerly currents tend to swing back and forth about an equilibrium latitude. These large-scale horizontal oscillations are known as **planetary** (or **Rossby**) **waves**, and may occur as surface or as internal waves. They are not gravity waves (i.e. the restoring force is not gravity) and so do not appear in Figure 1.2.

1.1.2 WIND-GENERATED WAVES IN THE OCEAN

In 1774, Benjamin Franklin said: 'Air in motion, which is wind, in passing over the smooth surface of the water, may rub, as it were, upon that surface, and raise it into wrinkles, which, if the wind continues, are the elements of future waves'.

In other words, if two fluid layers having differing speeds are in contact, there is frictional stress between them and there is a transfer of momentum and energy. The frictional stress exerted by a moving fluid is proportional to the *square* of the speed of the fluid, so the **wind stress** exerted upon a water surface is proportional to the square of the wind speed. At the sea-surface, most of the transferred energy results in waves, although a small proportion is manifest as wind-driven currents. In 1925, Harold Jeffreys suggested that waves obtain energy from the wind by virtue of pressure differences caused by the sheltering effect provided by wave crests (Figure 1.3).

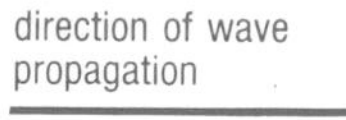

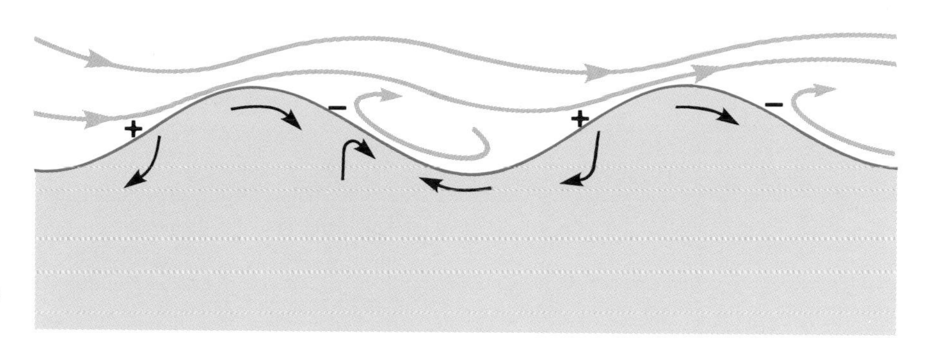

Figure 1.3 Jeffreys' 'sheltering' model of wave generation. Curved grey lines indicate air flow; shorter, black arrows show water movement. The rear face of the wave against which the wind blows experiences a higher pressure than the front face, which is sheltered from the force of the wind. Air eddies are formed in front of each wave, leading to excesses and deficiencies of pressure (shown by plus and minus signs respectively), and the pressure difference pushes the wave along.

Although Jeffreys' hypothesis fails to explain the formation of very small waves, it does seem to work if:

1 Wind speed exceeds wave speed.

2 Wind speed exceeds 1 m s^{-1}.

3 The waves are steep enough to provide a sheltering effect.

Empirically, it can be shown that the sheltering effect is at a maximum when wind speed is approximately three times the wave speed. In general, the greater the amount by which wind speed exceeds wave speed, the steeper the wave. In the open oceans, most wind-generated waves have steepness (H/L) of about 0.03 to 0.06. However, as we shall see later, wave speed in deep water is not related to wave steepness, but to wavelength – the greater the wavelength, the faster the wave travels.

QUESTION 1.2 Two waves have the same height, but differing steepness. Which of the two waves will travel the faster? the less steep wave

Consider the sequence of events that occurs if, after a period of calm weather, a wind starts to blow, rapidly increases to a gale, and continues to blow at constant gale force for a considerable time. No significant wave growth occurs until wind speed exceeds 1 m s^{-1}. Then, small steep waves form as the wind speed increases. Even after the wind has reached a constant gale force, the waves continue to grow with increasing rapidity until they reach a size and wavelength appropriate to a speed which corresponds to one-third of the wind speed. Beyond this point, the waves continue to grow in size, wavelength and speed, but at an ever-diminishing rate. On the face of it, one might expect that wave growth would continue until wave speed was the same as wind speed. However, in practice wave growth ceases whilst wave speed is still at some value below wind speed. This is because:

1 Some of the wind energy is transferred to the ocean surface via a tangential force, producing a surface current.

2 Some wind energy is dissipated by friction, and is converted to heat and sound.

3 Energy is lost from larger waves as a result of **white-capping**, i.e. breaking of the tip of the wave crest because it is being driven forward by the wind faster than the wave itself is travelling. Much of the energy dissipated during white-capping is converted into forward momentum of the water itself, reinforcing the surface current initiated by process 1 above.

1.1.3 THE FULLY DEVELOPED SEA

We have already seen that the size of waves in deep water is governed not only by the actual wind speed, but also by the length of time the wind has been blowing at that speed. Wave size also depends upon the unobstructed distance of sea, known as the **fetch**, over which the wind blows.

Provided the fetch is extensive enough and the wind blows at constant speed for long enough, an equilibrium is eventually reached, in which energy is being dissipated by the waves at the same rate as the waves receive energy from the wind. Such an equilibrium results in a sea state called a **fully developed sea**, in which the size and characteristics of the waves are not changing. However, the wind speed is usually variable, so the ideal fully developed sea, with waves of uniform size, rarely occurs. Variation in wind speed produces variation in wave size, so, in practice, a fully developed sea consists of a range of wave sizes known as a **wave field**. Waves coming into the area from elsewhere will also contribute to the range of wave sizes, as will interaction between waves – a process we explain in Section 1.4.2.

Oceanographers find it convenient to consider a wave field as a spectrum of wave energies (Figure 1.4). The energy contained in an individual wave is proportional to the square of the wave height (see Section 1.4).

QUESTION 1.3 Examine Figure 1.4. Does the energy contained in a wave field increase or decrease as the average frequency of the constituent waves increases?

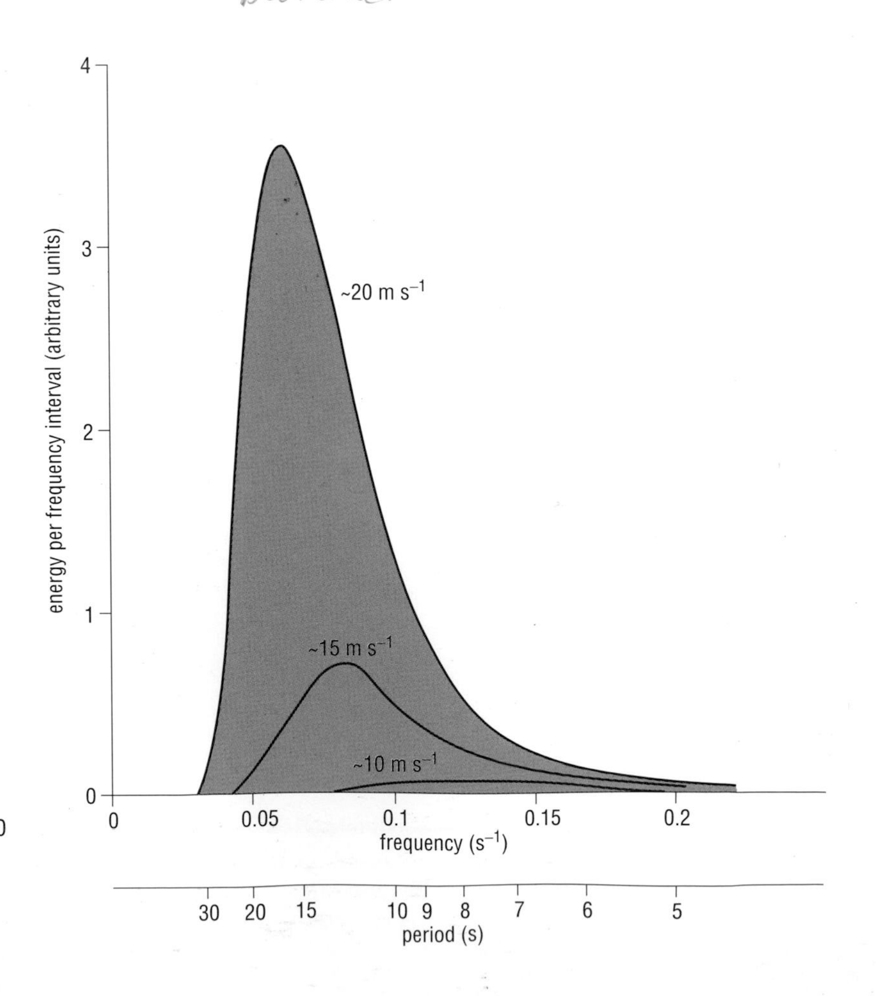

Figure 1.4 Wave energy spectra for three fully developed seas, related to wind speeds of 20, 30 and 40 knots (about 10, 15 and 20 m s^{-1} respectively). The area under each curve is a measure of the total energy in that particular wave field.

1.1.4 WAVE HEIGHT AND WAVE STEEPNESS

As was hinted in Section 1.1.3, the height of any real wave is determined by many component waves, of different frequencies and amplitudes, which move into and out of phase with, and across each other ('in phase' means that peaks and troughs coincide). In theory, if the heights and frequencies of all the contributing waves were known, it would be possible to predict the heights and frequencies of the real waves accurately. In practice, this is rarely possible. Figure 1.5 illustrates the range of wave heights occurring over a short time at one location – there is no obvious pattern to the variation of wave height.

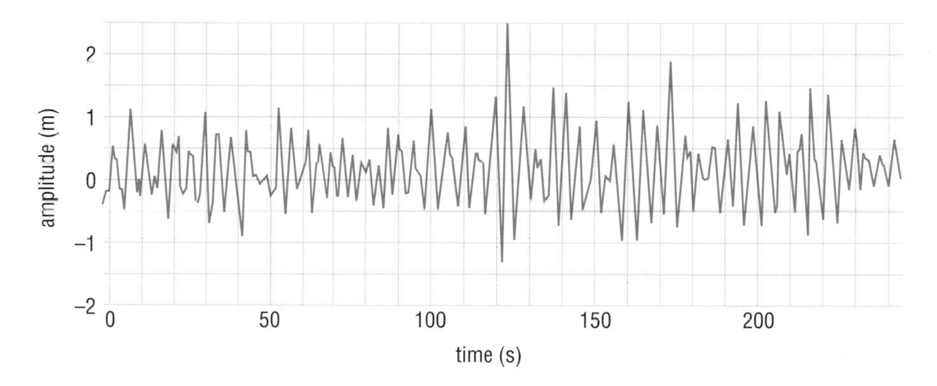

Figure 1.5 A typical wave record, i.e. a record of variation in water level (displacement from equilibrium) with time at one position.

For many applications of wave research, it is necessary to choose a single wave height which characterizes a particular sea state. Many oceanographers use the **significant wave height**, $H_{1/3}$, which is the average height of the highest one-third of all waves occurring in a particular time period. In any wave record, there will also be a maximum wave height, H_{max}. Prediction of H_{max} for a given period of time has great value in the design of structures such as flood barriers, harbour installations and drilling platforms. To build these structures with too great a margin of safety would be unnecessarily expensive, but to underestimate H_{max} could have tragic consequences. However, it is necessary to emphasize the essentially random nature of H_{max}. Although the wave $H_{max(25\text{ years})}$, will occur *on average* once every 25 years, this does not mean such a wave will automatically occur every 25 years – there may be periods much longer than that without one. On the other hand, two such waves might appear next week.

As wind speed increases, so $H_{1/3}$ in the fully developed sea increases. The relationship between sea state, $H_{1/3}$ and wind speed is expressed by the **Beaufort Scale** (Table 1.1, overleaf). The Beaufort Scale can be used to estimate wind speed at sea, but is valid only for waves generated within the local weather system, and assumes that there has been sufficient time for a fully developed sea to have become established (cf. Figure 1.4).

The absolute height of a wave is less important to sailors than is its steepness (H/L). As mentioned in Section 1.1.2, most wind-generated waves have a steepness in the order of 0.03 to 0.06. Waves steeper than this can present problems for shipping, but fortunately it is very rare for wave steepness to exceed 0.1. In general, wave steepness diminishes with increasing wavelength. The short choppy seas rapidly generated by local squalls are particularly unpleasant to small boats because the waves are steep, even though not particularly high. On the open ocean, very high waves can usually be ridden with little discomfort because of their relatively long wavelengths.

Table 1.1 A selection of information from the Beaufort Wind Scale.

Beaufort No.	Name	Wind speed (mean) knots	Wind speed (mean) m s^{-1}	State of the sea-surface	Significant wave height, $H_{1/3}$ (m)
0	Calm	<1	0.0–0.2	Sea like a mirror	0
1	Light air	1–3	0.3–1.5	Ripples with appearance of scales; no foam crests	0.1–0.2
2	Light breeze	4–6	1.6–3.3	Small wavelets; crests have glassy appearance but do not break	0.3–0.5
3	Gentle breeze	7–10	3.4–5.4	Large wavelets; crests begin to break; scattered white horses	0.6–1.0
4	Moderate breeze	11–16	5.5–7.9	Small waves, becoming longer; fairly frequent white horses	1.5
5	Fresh breeze	17–21	8.0–10.7	Moderate waves taking longer form; many white horses and chance of some spray	2.0
6	Strong breeze	22–27	10.8–13.8	Large waves forming; white foam crests extensive everywhere and spray probable	3.5
7	Near gale	28–33	13.9–17.1	Sea heaps up and white foam from breaking waves begins to be blown in streaks; spindrift begins to be seen	5.0
8	Gale	34–40	17.2–20.7	Moderately high waves of greater length; edges of crests break into spindrift; foam is blown in well-marked streaks	7.5
9	Strong gale	41–47	20.8–24.4	High waves; dense streaks of foam; sea begins to roll; spray may affect visibility	9.5
10	Storm	48–55	24.5–28.4	Very high waves with overhanging crests; sea-surface takes on white appearance as foam in great patches is blown in very dense streaks; rolling of sea is heavy and visibility reduced	12.0
11	Violent storm	56–64	28.5–32.7	Exceptionally high waves; sea covered with long white patches of foam; small and medium-sized ships might be lost to view behind waves for long times; visibility further reduced	15.0
12	Hurricane	>64	>32.7	Air filled with foam and spray; sea completely white with driving spray; visibility greatly reduced	>15

1.2 SURFACE WAVE THEORY

To simplify the theory of surface waves, we assume here that the wave-form is sinusoidal and can be represented by the curves shown in Figures 1.1 and 1.6. This assumption allows us to consider wave **displacement** (η) as simple harmonic motion, i.e. a sinusoidal variation in water level caused by the wave's passage. Figure 1.1 shows how the displacement varies over distance at a fixed instant in time – a 'snapshot' of the passing waves – whereas Figure 1.6 shows how wave displacement varies with time at a fixed point.

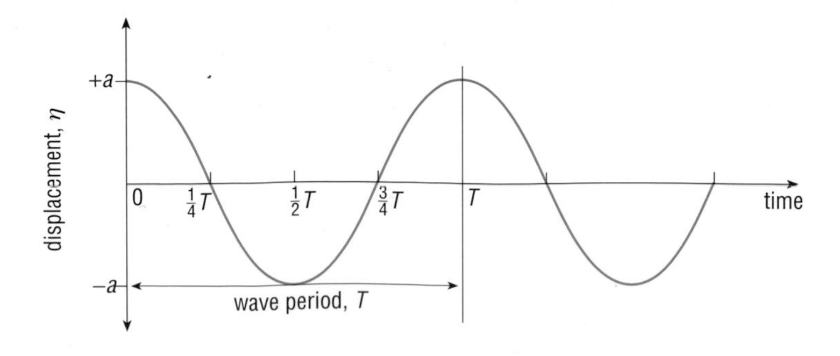

Figure 1.6 The displacement of an idealized wave at a fixed point, plotted against time. Maximum and minimum displacements are recorded in fractions of the period, T.

Before examining displacement, let us remind ourselves of the relationship between period and frequency.

QUESTION 1.4 If 16 successive wave troughs pass a fixed point during a time interval of one minute and four seconds, what is the frequency of the waves?

The displacement (η) of a wave at a fixed instant in time, or at a fixed point in space, varies between $+a$ (at the peak) and $-a$ (in the trough). Displacement is zero where $L = \frac{1}{4}L$ on Figure 1.1 (and at intervals of $L/2$ along the distance axis). Displacement is also zero at $T = \frac{1}{4}T$ on Figure 1.6 (and at intervals of $T/2$ along the time axis).

QUESTION 1.5 Use Figure 1.6 to help you answer the following questions. The peak, or crest, of a wave having a wavelength of 624 m, a frequency of 0.05 s^{-1}, and travelling in deep water, passes a fixed point P. What is the displacement at P (in terms of the amplitude, a):

(a) 30 seconds after the peak has passed?

(b) 80 seconds after the peak has passed?

(c) 85 seconds after the peak has passed?

What is the displacement at a second point, Q, which is 312 m away from P in the direction of wave propagation:

(d) when the displacement at P is zero?

(e) when the displacement at P is $-a$?

(f) 5 seconds after a trough has passed P?

The curves shown in Figures 1.1 and 1.6 are both sinusoidal. However, most wind-generated waves do not have simple sinusoidal forms. The steeper the wave, the further it departs from a simple sine curve. Very steep waves resemble a trochoidal curve, which is illustrated in Figure 1.7.

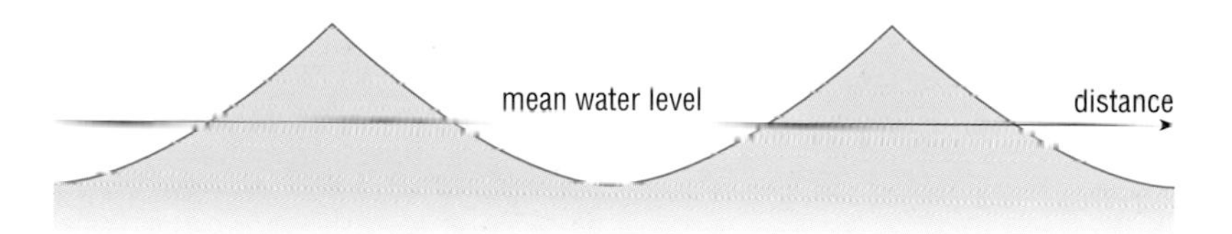

Figure 1.7 Profile of trochoidal waves.

A point marked on the rim of a car tyre will appear to trace out a trochoidal curve as the car is driven past an observer. Invert that pattern and you have the profile of a trochoidal water wave. We do not need to delve into the mathematical complexities of trochoidal wave forms here, because the sinusoidal model is sufficient for our purposes.

1.2.1 MOTION OF WATER PARTICLES

Water particles in a wave in deep water move in an almost closed circular path. At wave crests, the particles are moving in the same direction as wave propagation, whereas in the troughs they are moving in the opposite direction. At the surface, the orbital diameter corresponds to wave height, but the diameters decrease exponentially with increasing depth, until at a depth roughly equal to half the wavelength, the orbital diameter is negligible, and there is virtually no displacement of the water particles (Figure 1.8(a)). This has some important practical applications. For example, a submarine only has to submerge about 150 m to avoid the effects of even the most severe storm at sea, and knowledge of the exponential decrease of wave influence with depth has implications for the design of stable floating oil rigs.

It is important to realize that the orbits are only approximately circular. There is a small net component of forward motion, particularly in waves of large amplitude, so that the orbits are not quite closed, and the water, whilst in the crests, moves slightly further forward than it moves backward whilst in the troughs. This small net forward displacement of water in the direction of wave travel is termed **wave drift** (see Figure 1.8(b)). In shallow water, where depth is less than half the wavelength and the waves 'feel' the sea-bed, the orbits become progressively flattened with depth (Figure 1.8(c) and (d)). The significance of these changes will be seen in Section 1.5, and in the Chapters on sediment movement.

The orbital motions relevant to internal waves (Section 1.1.1) are shown in Figure 1.8(e): they are in opposite directions on either side of the interface. The passage of internal waves can often be detected visually by secondary effects at the surface, especially if the upper layer (above the pycnocline) is not very thick and the sea is relatively calm. As the internal waves travel along, the upper layer becomes alternately thinner (over the internal wave crests) and thicker (over the troughs). The result is that there are **convergences** and **divergences** of water at the surface, as water is displaced back and forth between the thinner and thicker parts of the upper layer.

A combination of these to-and-fro motions and the opposing particle orbits on either side of the interface may, under certain conditions, influence the movement of vessels with a draught comparable to the depth to the interface, sometimes causing an unexpected drag on the hull, thus making the vessels sluggish and difficult to handle.

Sometimes the convergences compress short wavelength surface waves, making them more visible, but commonly they bring together organic material (especially oils released by marine organisms), which increases the surface tension and tends to suppress ripples, so that the water is smoother than elsewhere. Alternating bands of smooth and rippled surface water at intervals of a few hundred metres may thus indicate the passage of an internal wave, but whether ripples represent convergences or divergences depends upon local conditions.

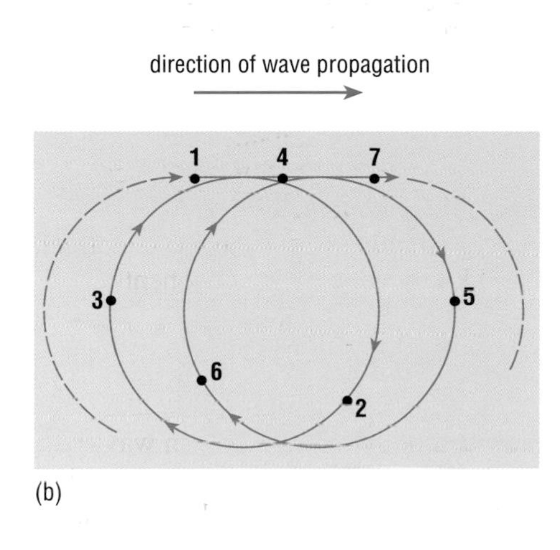

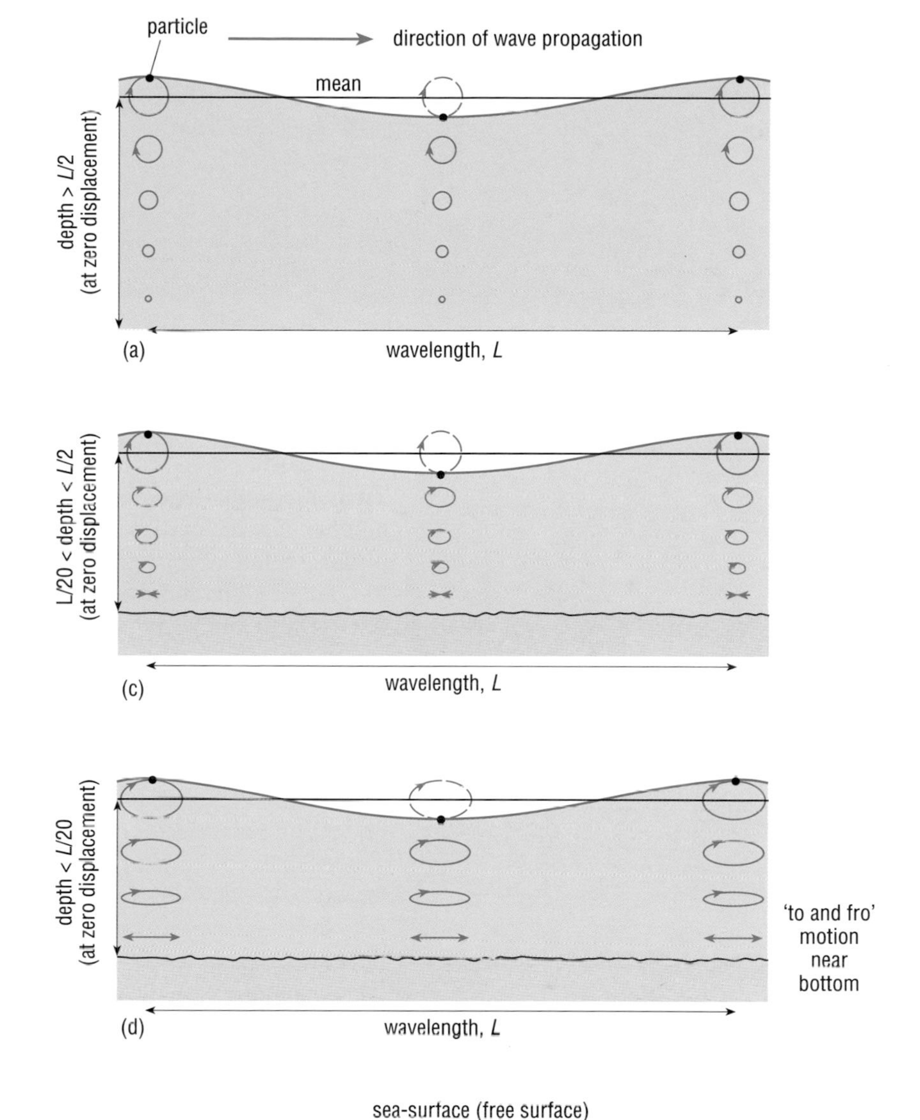

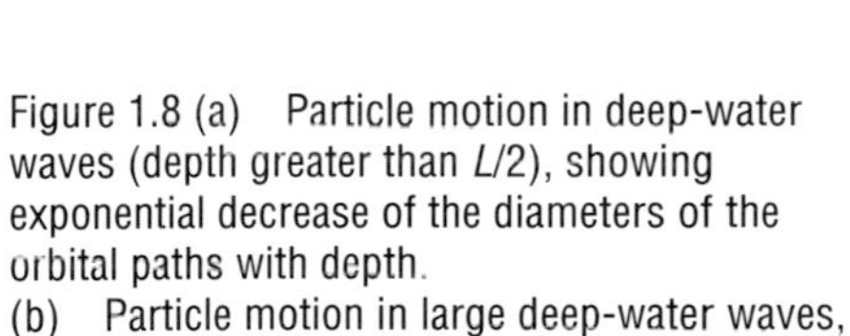

sea-surface (free surface)

convergence

divergence

convergence

divergence

direction of wave propagation

(e)

Figure 1.8 (a) Particle motion in deep-water waves (depth greater than $L/2$), showing exponential decrease of the diameters of the orbital paths with depth.
(b) Particle motion in large deep-water waves, showing wave drift.
(c) Particle motion in waves where water depth is less than $L/2$ but greater than $L/20$ (see Section 1.2.3), showing both decrease in horizontal orbital diameter and progressive flattening of the orbits near the sea-bed.
(d) Particle motion in shallow-water waves (depth less than $L/20$, see Section 1.2.3), showing progressive flattening of the orbits (but no decrease in horizontal diameter) near the sea-bed.
(e) Particle motion in internal waves (Section 1.1.1), and the convergences and divergences associated with their passage. The orbits will only be truly circular if the layers are thick enough (i.e. greater than half the wavelength). The orbital diameters decrease linearly (i.e. not exponentially) with distance from the interface but may not reach zero at the free surface above the interface (where any undulations do not necessarily reflect those of the internal waves). Orbital motions are in *opposite* directions above and below the interface.

1.2.2 WAVE SPEED

As we have already hinted, there are mathematical relationships linking the characteristics of wavelength (L), wave period (T) and wave height (H) to wave speed in deep water and to wave energy. First, let us consider **wave speed** (c) (the c stands for '*celerity of propagation*').

The speed of a wave can be ascertained from the time taken for one wavelength to pass a fixed point. As one wavelength (L) takes one wave period (T) to pass a fixed point, then:

$$c = L/T \tag{1.1}$$

which is simply a form of the well-known expression: speed = distance/time.

So, if we know any two of the variables in Equation 1.1, we can calculate the third.

Two other terms you may meet in oceanographic literature are the *wave number, k*, which is $2\pi/L$, and the *angular frequency* σ, which is $2\pi/T$; both of these relate to the sinusoidal nature of the idealized wave form. The units of k are m^{-1} (i.e. number of waves per metre), and the units of σ are s^{-1} (i.e. number of cycles (waves) per second).

QUESTION 1.6 How would c be expressed in terms of k and σ?

1.2.3 WAVE SPEED IN DEEP AND IN SHALLOW WATER

You may have noticed that when wave speeds have been mentioned we have been careful to state that the waves described were travelling in deep water. Thus you might have suspected that in shallow water, water depth has an effect on wave speed, because of interaction with the sea-bed. If so, you were quite right. Wave speed in any water depth can be represented by the general equation:

$$c = \sqrt{\frac{gL}{2\pi}\tanh\left(\frac{2\pi d}{L}\right)} \tag{1.2}$$

where the acceleration due to gravity $g = 9.8\ m\ s^{-2}$, L = wavelength (m), and d = water depth (m). Tanh is a mathematical function known as the hyperbolic tangent. All you need to know about it in this context is that if x is small, say less than 0.05, then $\tanh x \approx x$, and if x is larger than π, then $\tanh x \approx 1$.

QUESTION 1.7 Armed with Equation 1.2, and the information given above about the tanh function, work out the answers to the following questions:

(a) What does Equation 1.2 become if the water depth exceeds half the wavelength?

(b) What does Equation 1.2 become if the water depth is very much smaller than L?

In summary, the implications of your answers to Question 1.7 in terms of factors affecting wave speed are as follows (cf. Figure 1.8):

1 In water deeper than half the wavelength, wave speed depends upon the wavelength, and Equation 1.2 approximates to:

$$c = \sqrt{\frac{gL}{2\pi}} \tag{1.3}$$

Diameters of circular particle orbits decrease exponentially downwards to near zero at depth = $L/2$ (Figure 1.8(a)).

2 In water very much shallower than the wavelength (in practice, when $d < L/20$), wave speed is determined by water depth, and Equation 1.2 approximates to:

$$c = \sqrt{gd} \tag{1.4}$$

Horizontal diameters of particle orbits remain constant in size with depth, but ellipticity increases downwards (Figure 1.8(d)).

3 When d lies between $L/20$ and $L/2$, the full form of Equation 1.2 is required. Hence, to calculate wave speed you would need to know wavelength and depth, and have access to a set of hyperbolic tangent tables, or a calculator with hyperbolic functions on its keyboard. Particle orbits decrease in size downwards and become progressively more elliptical (Figure 1.8(c)).

The answer to Question 1.7(a) (i.e. Equation 1.3) allows us to explore further the relationships between T and L. We saw in Equation 1.1 that $c = L/T$, so it is possible to combine Equations 1.1 and 1.3.

QUESTION 1.8 Derive an equation for wavelength (L) in terms of period (T), using Equations 1.1 and 1.3.

The answer to Question 1.8 provides an equation expressing L in terms of T, i.e.

$$L = \frac{gT^2}{2\pi} \tag{1.5}$$

A similar exercise, substituting the expression obtained for L from Equation 1.5 into Equation 1.1, will give c in terms of T:

$$c = \frac{gT}{2\pi} \tag{1.6}$$

Thus, it is possible, given only one of the wave characteristics c, T or L, to calculate either of the other two. Moreover, by substituting the numerical values of the constants involved, the equations can be simplified as follows:

Equation 1.3 becomes $c = \sqrt{1.56L}$ (1.7)

Equation 1.5 becomes $L = 1.56T^2$ (1.8)

Equation 1.6 becomes $c = 1.56T$ (1.9)

QUESTION 1.9 Show (a) how the numerical factor 1.56 in each of Equations 1.7 to 1.9 is derived, and (b) that the units in those equations work out correctly.

QUESTION 1.10

(a) The period of a wave is 20 s. At what speed will it travel in deep water?

(b) At what speed will a wave of wavelength 312 m travel in deep water?

(c) At what speeds will each of the waves referred to in (a) and (b) above travel in water of 12 m depth?

The answer to Question 1.10(c) highlights an important conclusion about wave speed in shallow water. *Provided that depth is less than 1/20 of their wavelengths*, all waves will travel at the same speed in water of a given depth.

1.2.4 ASSUMPTIONS MADE IN SURFACE WAVE THEORY

The simple wave theory introduced above is a first-order approximation, and makes the following assumptions:

1 The wave shapes are sinusoidal.

2 The wave amplitudes are very small compared with wavelengths and depths.

3 Viscosity and surface tension can be ignored.

4 The Coriolis force (see Section 2.3) and vorticity (Section 1.1.1), which result from the Earth's rotation, can be ignored.

5 The depth is uniform, and the bottom has no bumps or hummocks.

6 The waves are not constrained or deflected by land masses, or by any other obstruction.

7 That real three-dimensional waves behave in a way that is analogous to a two-dimensional model.

None of the above assumptions is valid in the strictest sense, but predictions based on simple models of surface wave behaviour approximate closely to how wind-generated waves behave in practice.

1.3 WAVE DISPERSION AND GROUP SPEED

Those deep-water waves that have the greatest wavelengths and longest periods travel fastest, and thus are first to arrive in regions distant from the storm which generated them. This separation of waves by virtue of their differing rates of travel is known as **dispersion**, and Equation 1.3 ($c = \sqrt{gL / 2\pi}$) is sometimes known as the *dispersion equation*, because it shows that waves of longer wavelength (L) travel faster than shorter wavelength waves.

The simple experiment of tossing a stone into a still pond shows that a band of ripples is created, which gets wider with increasing distance from the original disturbance. Ripples of longer wavelength progressively out-distance shorter ones – an example of dispersion in action. There is a second feature of the ripple band, which is not obvious at first sight. Each individual ripple travels faster than the band of ripples as a whole. A ripple appears at the back of the band, travels through it, and disappears out of the front. The speed of the band, called the **group speed**, is about half the wave speed of the individual ripples which travel through that band.

To understand the relationship between wave speed and group speed, the additive effect of two sets of waves (or *wave trains*) needs to be examined. If the difference between the wavelengths of two sets of waves is relatively small, the two sets will 'interfere' and produce a single set of resultant waves. Figure 1.9 shows a simplified and idealized example of interference. Where the crests of the two wave trains coincide (i.e. they are 'in phase'), the wave amplitudes are added, and the resultant wave has about twice the amplitude of the two original waves. Where the two wave trains are 'out of phase', such that the crests of one wave train coincide with the troughs of the other, the amplitudes cancel out, and the water surface has minimal displacement.

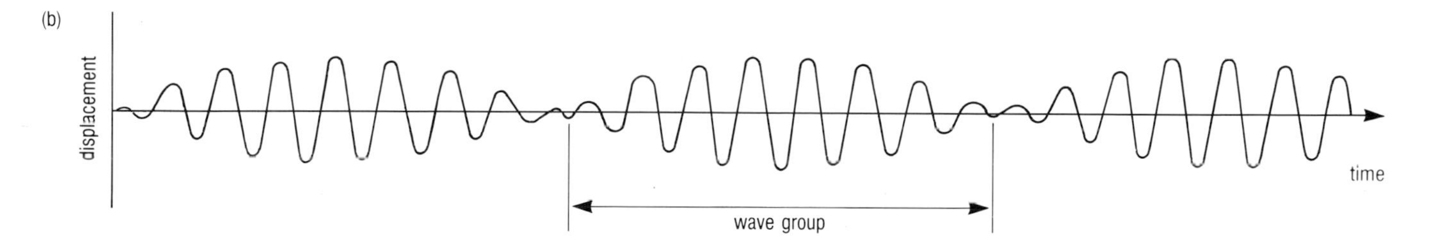

Figure 1.9 (a) The merging of two wave trains (shown in red and blue) of slightly different wavelengths (but the same amplitudes), to form wave groups (b).

The two component wave trains thus interact, each losing its individual identity, and combine to form a series of wave groups, separated by regions almost free from waves. The wave group advances more slowly than individual waves in the group, and thus in terms of the occurrence and propagation of waves, group speed is more significant than speeds of the individual waves. Individual waves do not persist for long in the open ocean, only as long as they take to pass through the group. Figure 1.10 shows the relationship between wave speed (sometimes called *phase speed*) and group speed in the open ocean.

The group speed is *half* the average speed of the two wave trains, and for your interest we present below an abbreviated form of how this relationship is derived. It is not necessary to follow all the steps that lead to Equation 1.10 (overleaf), still less to memorize them.

If two sets of waves are interfering to produce a succession of wave groups, the group speed (c_g) is the difference between the two angular frequencies (σ_1 and σ_2) divided by the difference between the two wave numbers (k_1 and k_2 respectively), i.e.

$$c_g = \frac{\sigma_1 - \sigma_2}{k_1 - k_2}$$

Figure 1.10 The relationship between wave speed (phase speed) and group speed. As the wave advances from left to right, each wave moves through the group to die out at the front (e.g. wave 1), as new waves form at the rear (e.g. wave 6). In this process, the distance travelled by each individual wave as it moves from the rear to front of the group is twice that travelled by the group as a whole. Hence, the wave speed is twice that of the group speed.

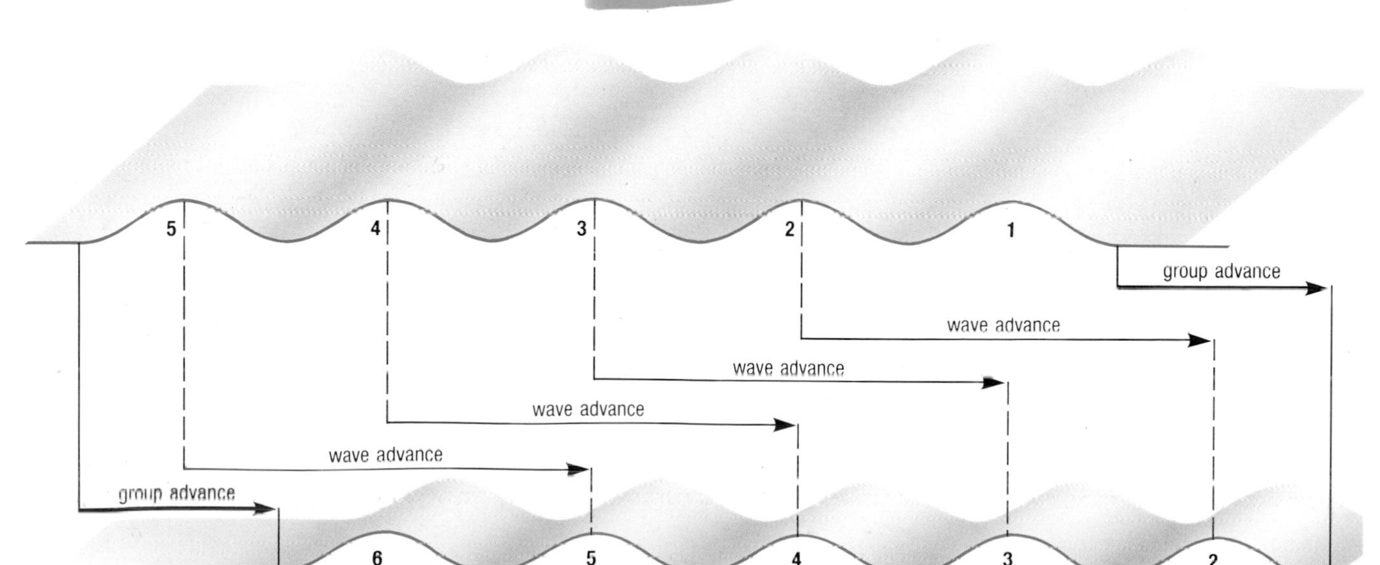

We have seen (Question 1.6) that $c = \sigma/k$, and we know (Section 1.2.2) that angular frequency, σ, can be expressed in terms of T, also that wave number k, can be expressed in terms of L. In addition, we know that Equations 1.6 and 1.3, respectively, enable us to express both T and L in terms of c (Section 1.2.3). Hence, c_g can be expressed in terms of the respective speeds, c_1 and c_2, of the two wave trains. The equation obtained is:

$$c_g = \frac{c_1 \times c_2}{c_1 + c_2}$$

If c_1 is nearly equal to c_2, this equation simplifies to:

$$c_g \approx c^2/2c$$

$$\text{or } c_g \approx c/2 \tag{1.10}$$

where c is the average speed of the two wave trains.

What happens to group speed when waves enter shallow water?

Equation 1.2 shows that as the water becomes shallower, wavelength becomes less important, and depth more important, in determining wave speed. As a result, in shoaling water, wave (phase) speed decreases, becoming closer to group speed. Eventually, at depths less than $L/20$, all waves travel at the same depth-determined speed, there will be no wave–wave interference, and therefore in effect each wave will represent its own 'group'. Thus, in shallow water, group speed can be regarded as equal to wave (phase) speed.

1.4 WAVE ENERGY

The energy possessed by a wave is in two forms:

1 kinetic energy, which is the energy inherent in the orbital motion of the water particles; and

2 potential energy possessed by the particles as a result of being displaced from their mean (equilibrium) position.

For a water particle in a given wave, energy is continually being converted from potential energy (at crest and trough) to kinetic energy (as it passes through the equilibrium position), and back again.

The total energy (E) *per unit area* of a wave is given by:

$$E = \frac{1}{8}(\rho g H^2) \tag{1.11}$$

where ρ is the density of the water (in kg m^{-3}), g is 9.8 m s^{-2}, and H is the wave height (m). The energy (E) is then in joules per square metre (J m^{-2}). Equation 1.11 shows that wave energy is proportional to the square of the wave height.

QUESTION 1.11 Would the total energy of a wave be doubled if its amplitude were doubled? No it would increase by x4.

1.4.1 PROPAGATION OF WAVE ENERGY

Figures 1.9 and 1.10 show that waves travel in groups in deep water, with areas of minimal disturbance between groups. Individual waves die out at the front of each group. It is obvious that no energy is being transmitted across regions where there are no waves, i.e. between the groups. It follows that the energy is contained within the wave group, and travels at the group speed. The *rate* at which energy is supplied at a particular location (e.g. a beach) is called **wave power**, and is the product of group speed (c_g) and wave energy per unit area (E), expressed *per unit length of wave crest.*

QUESTION 1.12

(a) In the case of waves in deep water, what is the energy per square metre of a wave field made up of waves with an average amplitude of 1.3 m? (Use $\rho = 1.03 \times 10^3$ kg m^{-3}.)

(b) What would be the wave power (in kW per metre of crest length) if the waves had a steepness of 0.04? (1 watt = 1 J s^{-1}, and one kilowatt (kW) = 10^3 W.)

1.4.2 ATTENUATION OF WAVE ENERGY

Wave **attenuation** involves loss or dissipation of wave energy, resulting in a reduction of wave height. Energy is dissipated in four main ways:

1 White-capping, which involves transfer of wave energy to the kinetic energy of moving water, thus reinforcing the wind-driven surface current (Section 1.1.2).

2 Viscous attenuation, which is only important for very high frequency capillary waves, and involves dissipation of energy into heat by friction between water molecules.

3 Air resistance, which applies to large steep waves soon after they leave the area in which they were generated and enter regions of calm or contrary winds.

4 **Non-linear wave–wave interaction**, which is more complicated than the simple (linear) combination of frequencies to produce wave groups as outlined in Section 1.3.

Non-linear interaction appears to be most important in the frequency range 0.2 to 0.3 s^{-1}. Groups of three or four frequencies can interact in complex non-linear ways, to transfer energy to waves of both higher and lower frequencies. A rough but useful analogy is that of the collision of two drops of water. A linear combination would simply involve the two drops coalescing (adding together) into one big drop, whereas a non-linear combination is akin to a collision between the drops so that they split into a number of drops of differing sizes. The total amount of water in the drops (analogous to the total amount of energy in the waves) is the same before and after the collision.

Thus, non-linear wave–wave interaction involves no loss of energy in itself, because energy is simply 'swapped' between different frequencies. However, the total amount of energy available for such 'swapping' will gradually decrease, because higher frequency waves are more likely to dissipate energy in the ways described under 1 and 2 above. For example, higher frequency waves are likely to be steep, and thus more prone to white-capping. Wave attenuation is greatest in the storm-generating area, where there are waves of many frequencies, and hence more opportunities for energy exchange between them.

1.4.3 SWELL

The sea-surface is rarely still. Even when there is no wind, and the sea 'looks like a mirror', a careful observer will notice waves of very long wavelength (say 300 to 600 m) and only a few centimetres amplitude. At other times, a sea may include locally generated waves of small wavelength, and travelling through these waves, possibly at a large angle to the wind, other waves of much greater wavelength. Such long waves are known as **swell**, which is simply defined as waves that have been generated elsewhere and have travelled far from their place of origin. If you look out to sea on a calm day, the waves that you see will be swell waves from a distant storm.

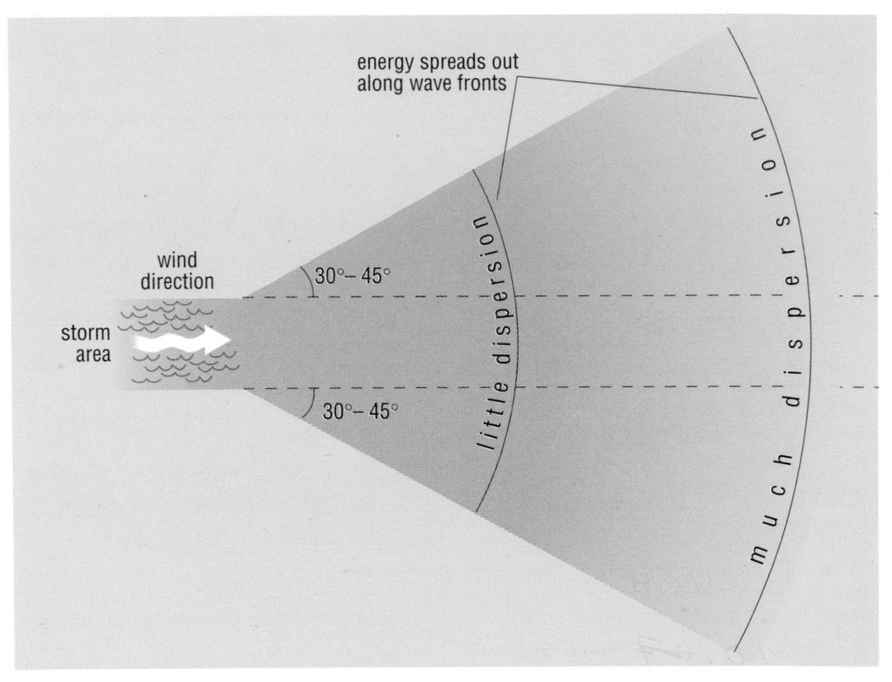

(a)

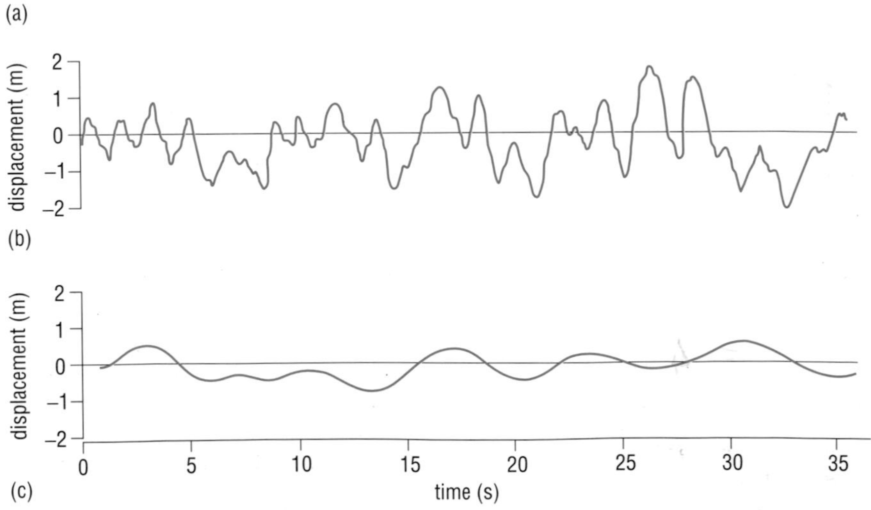

Figure 1.11 (a) The spreading of swell from a storm centre, showing the area in which swell might be expected. As distance from the storm increases, the length of the wave crest increases, with a corresponding decrease in wave height and energy per unit length of wave – spreading loss, discussed opposite.
(b) Wave record near a storm centre.
(c) Wave record of swell, well away from the storm centre.

Systematic observations show that local winds and waves have very little effect on the size and progress of swell waves, and swell seems able to pass through locally generated seas without hindrance or interaction. Once swell waves have left the storm area, their wave height gradually diminishes, chiefly because the wave crests lengthen over a progressively wider front (Figure 1.11(a)) but also because of energy loss caused by air resistance to steep waves (item 3 of Section 1.4.2). Once wave height has diminished to a few tens of centimetres, swell waves are not steep enough to be significantly influenced by local winds.

In the ocean, we find waves travelling in many directions, resulting in a confused sea. To achieve a complete description of such a sea-surface, the amplitude, frequency and direction of travel of each component are needed. The energy distribution of the sea-surface (cf. Figure 1.4) can then be calculated, but, as you might imagine, such a complex process requires expensive equipment to measure the wave characteristics, and computer facilities to perform the necessary calculations.

One or more components of a confused sea may be long waves or swell resulting from distant storms. In practice, about 90% of the sea-surface energy generated by the storm propagates within an angle of 30° to 45° either side of the wind direction. Consequently, waves generated by a storm in a localized region of a large ocean radiate outwards as a segment of a circle (Figure 1.11(a)). As the circumference of the circle increases, the energy per unit length of wave crest must decrease (and so must wave height), so that the total energy of the wave front remains the same. This decrease in energy per unit length of wave crest is known as **spreading loss** (of wave energy), and in the case of established swell waves there is very little loss of wave energy apart from that caused by spreading over a progressively wider front.

The waves with the longest periods travel fastest, and progressively out-distance waves of higher frequencies (shorter periods). Near to the storm, dispersion is likely to be minimal (Figure 1.11(b)), but the further one moves from the storm location, the more clearly separated waves of differing frequencies become, resulting in the regular wave motions we know as swell (Figure 1.11(c)).

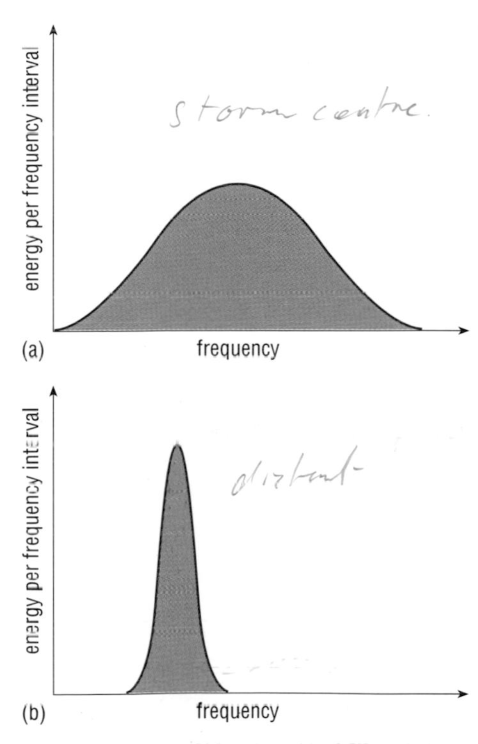

Figure 1.12 Wave-energy spectra, each determined from wave heights measured over a short time interval, for two areas (a) and (b) in the same ocean (not to scale). One area is a storm centre, and the other is far away from the storm. (For use with Question 1.13.)

QUESTION 1.13 Figure 1.12 shows two wave-energy spectra, (a) and (b) (cf. Figure 1.4). One represents the wave field energy in a storm-generating area; and the other represents the energy of the wave field in an area far away from the storm, but receiving swell from it. Which of the two spectra represents which situation?

If you recorded the waves arriving from a storm a great distance (over 1000 km) away, you would, as time progressed, see the peak in the wave-energy spectrum move progressively towards higher frequencies (i.e. shorter periods). By recording the frequencies of each of a series of swell waves arriving at a point, it would be possible to calculate each of their speeds. From the set of speeds, a graph could be plotted to estimate the time and place of their origin. Before the days of meteorological satellites, this method was often used to pinpoint where and when storms had occurred in remote parts of the oceans.

1.4.4 USES OF WAVE ENERGY

Wave energy is a potential source of pollution-free 'alternative energy', and has been used for some time on a small scale, e.g. to recharge batteries on buoys carrying navigation lights (see below). Harnessing wave power on a large scale presents a number of problems:

1 Prevailing sea conditions must ensure a supply of waves with amplitudes sufficient to make conversion worthwhile.

2 Installations must not be a hazard to navigation, or to marine ecosystems. The nature of wave energy is such that rows of converters many kilometres in length are needed to generate amounts of electricity comparable with conventional power stations. These would form offshore barrages which might interfere with shipping routes, although sea conditions would be made calmer on the shoreward side. Calmer conditions, however, lead to reduced water circulation, less sediment transport, and increased growth of quiet-water plants and animals. Pollutants are less easily flushed away from such an environment.

Some of these problems can be avoided by use of another type of wave-energy converter, the circular 'clam' or 'atoll', designed so that waves from any direction are guided by a system of radiating vanes to a central vertical channel where the water spirals downwards to drive a turbine and generator. Such converters would be deployed in arrays rather than in long rows, but they are big, and many of them would be required to generate useful amounts of power commercially. For example, a 60 m diameter 'clam' could generate an average of about 600 kW, so many hundreds would be needed to equal the power provided by even one conventional fossil-fuel or nuclear power station.

3 The capital cost of such floating power stations and their related energy transmission and storage systems is enormous. Installations need to be robust enough to withstand storm conditions, yet sensitive enough to be able to generate power from a wide range of wave sizes. Such conditions are expensive to meet, and make it difficult for large-scale wave-energy schemes to be as cost effective as conventional energy sources.

A large (2 MW) oscillating water column converter (OSPREY = *O*cean *S*well *P*owered *R*enewable *E*nerg*Y*) was installed off northern Scotland in 1995. Soon afterwards it was extensively damaged by storms and had to be taken ashore for repairs, which set back the UK wave-energy programme by some years. Relatively small-scale utilization of wave power is more feasible. For example, oscillating water-column converters provide power for both buoys and lighthouses in several parts of the world.

Other approaches have been demonstrated by the Norwegians, who in 1985 brought into operation a wave-powered generator of 850 kW. This machine was sited on the west coast of Norway, where waves were funnelled into a narrow bay and increased substantially in height and thus in energy. Unfortunately, this installation also was wrecked by storms a few years later, but similar schemes have been successfully developed elsewhere.

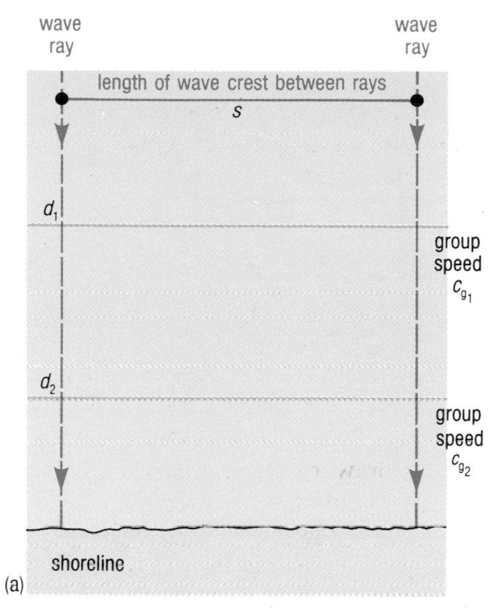

Figure 1.13 Plan view illustrating changes in the speed of waves approaching the shore. Grey lines represent wave crests at depths d_1 and d_2. Wave rays (dashed blue lines) are at right angles (i.e. normal) to the wave crests. For further explanation, see text.

1.5 WAVES APPROACHING THE SHORE

It is a matter of common observation that waves coming onto a beach increase in height and steepness and eventually break. Figure 1.13 shows a length of wave crest, *s*, which is directly approaching a beach. As the water is shoaling, the wave crest passes a first point where the water depth (d_1) is greater than at a second point nearer the shore (where the depth is d_2). We assume that the amount of energy within this length of wave crest remains constant, the wave is not yet ready to break, and that water depth is less than 1/20 of the wavelength (i.e. Equation 1.4 applies: $c = \sqrt{gd}$). Because wave speed in shallow water is related to depth, the speed c_1 at depth d_1 is greater than the speed c_2 at depth d_2. If energy remains constant per unit length of wave crest, then

$$E_1c_1s = E_2c_2s$$

$$\text{or } \frac{E_2}{E_1} = \frac{c_1}{c_2} \tag{1.12}$$

and because energy is proportional to the square of the wave height (Equation 1.11) then we can write

$$\frac{E_2}{E_1} = \frac{c_1}{c_2} = \frac{H_2^2}{H_1^2} \tag{1.13}$$

Thus, both the square of the wave height and wave energy are inversely proportional to wave speed in shallow water.

This relationship is straightforward once the wave has entered shallow water. But what happens during the transition from deep to shallow water?

This is quite a difficult question, best answered by considering the highly simplified case illustrated in Figure 1.14. Imagine waves travelling shoreward over deep water (depth greater than half the wavelength). Wave speed is then governed solely by wavelength (Equation 1.3, $c = \sqrt{gL/2\pi}$. The energy is being propagated at the group speed (c_g) which is approximately half the wave speed (c), Section 1.3. As the waves move into shallower water, wave speed becomes governed by both depth and wavelength (Equation 1.2), but once the waves have moved into shallow water, where $d < L/20$, wave speed becomes governed solely by depth (Equation 1.4) and is much reduced. Remember from Section 1.3 that in shallow water group speed is equal to wave speed. The rate at which energy arrives from offshore (Figure 1.14, overleaf) must be equal to the rate at which energy moves inshore; so if the group speed in shallow water is less than half the original wave speed (and hence less than the original group speed) in deep water, the waves will show corresponding increases in height and in energy per unit area.

However, it is essential to realize that while the energy and height of individual waves will increase as they enter shallow water, *the rate of supply of wave energy* (wave power, Section 1.4.1) *must remain constant* (ignoring frictional losses).

As mentioned earlier, when waves move into shallow water, the waves begin to 'feel the bottom', the circular orbits of the water particles become flattened (Figure 1.8(c) and (d)), and wave energy will be dissipated by friction at the sea-bed, resulting in to-and-fro movement of sediments. The gentler the slope of the immediate offshore region, the sooner the incoming waves will 'feel' the bottom, the greater will be the friction with the sea-bed and the greater the energy loss before the waves finally break (see Section 1.5.2).

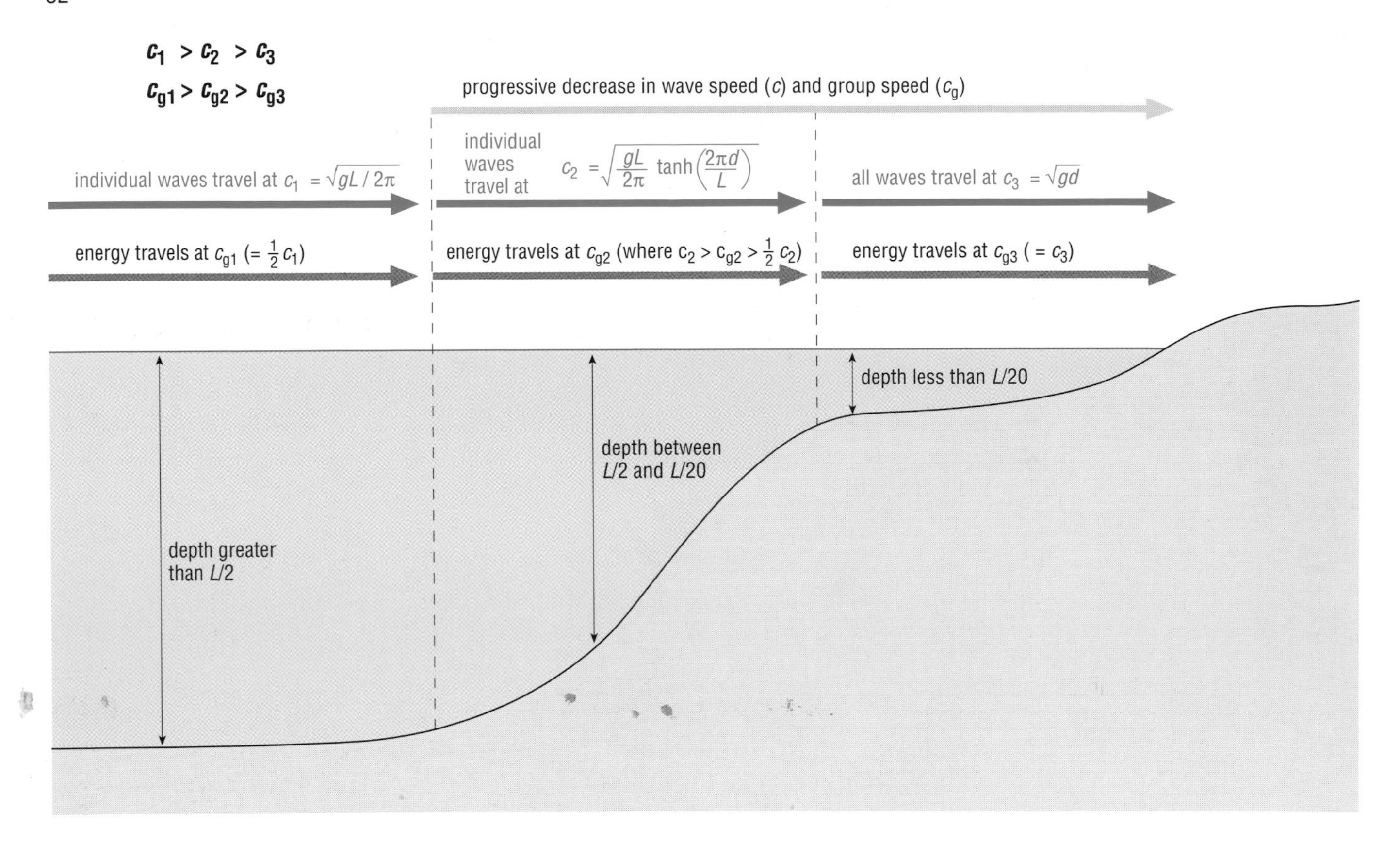

Figure 1.14 Vertical section (not to scale) illustrating changes in the relationship between wave speed and group speed, and how this affects the speed of energy propagation as waves move from deep to shallow water. The energy is being brought in from offshore at the same rate as it is being removed as the waves break at the shore (see Section 1.5.2). As group speed (c_g) in shallow water is less than in deep water, then the waves in shallow water must have more energy per unit length of wave crest, and a greater wave height than the waves in deep water.

1.5.1 WAVE REFRACTION

Figure 1.15 shows an idealized linear wave crest (length s_1, between A and B) approaching a shoreline at an angle. Because the waves are travelling in shallow water, their speed is depth-determined (Equation 1.4, $c = \sqrt{gd}$). The depth at A exceeds the depth at B, so the wave at A will travel faster than the wave at B, leading to the phenomenon known as **refraction**, which will tend to 'swing' the wave crest to an alignment parallel with the depth contours.

Can the extent of refraction be quantified?

Refraction of waves in progressively shallowing (shoaling) water can be described by a relationship equivalent to Snell's law, which describes refraction of light rays through materials of different refractive indices. Rays can be drawn perpendicular to the wave crests, and will indicate the direction of wave movement. The angle between these wave rays and lines drawn perpendicular to the depth contours can be related to wave speeds at various depths. In Figure 1.15, a wave ray approaching shoaling water at an angle θ_1, where water depth is d_1, will be at an angle θ_2 when it reaches depth d_2. Angles θ_1 and θ_2 are related to wave speed by:

$$\frac{\sin\theta_1}{\sin\theta_2} = \frac{c_1}{c_2} = \frac{\sqrt{gd_1}}{\sqrt{gd_2}} = \frac{\sqrt{d_1}}{\sqrt{d_2}} = \sqrt{\frac{d_1}{d_2}} \quad (1.14)$$

where c_1 and c_2 are the respective wave speeds at depths d_1 and d_2.

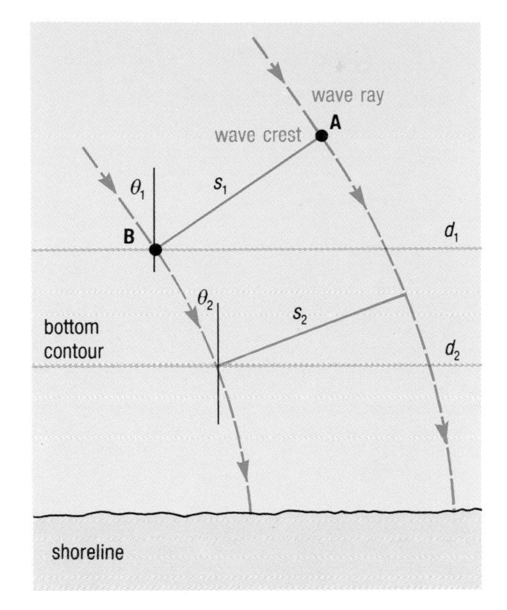

Figure 1.15 Plan view illustrating the relationship between wave approach angle (θ), water depth (d), and length of wave crest (s). The wave rays (broken blue lines), are normal to the wave crests, and are the paths followed by points on the wave crests. For further explanation, see text.

You might ask: why go to the trouble of drawing perpendiculars? Why not simply use the angles between wave crests and bottom contours?

Well, of course, one could do that, and obtain analogous relationships between the relevant angles, depths and wave speeds. However, as you will see presently, wave rays are often more useful than wave crests in determining regions that are likely to experience waves that are higher or lower than normal because of refraction.

Consider a length, s_1, of ideal wave crest, with energy per unit length E_1, which is bounded by two wave rays, as in Figure 1.15. To a first approximation, we may assume that the total energy of the wave crest between these two rays will remain constant as the wave progresses. Therefore, if the two rays converge, the same amount of energy is contained within a shorter length of wave crest, so that, for the total wave energy to remain constant, the wave height will have to increase (Equation 1.11). Conversely, if the wave rays were to diverge, then the wave height would decrease.

If, as they finally approach the shore, the two wave rays are separated by a length s_2, as in Figure 1.15, and if the wave energy is conserved, then the final wave energy must equal the initial wave energy, i.e. $E_1s_1 = E_2s_2$, or in terms of wave heights (remember Equation 1.11):

$$H_1^2s_1 = H_2^2s_2 \tag{1.15}$$

For simplicity, s_2 in Figure 1.15 is the same length as s_1, but it is common for wave rays to converge or diverge, and in general they converge (focus) on headlands and diverge in bays (Figure 1.16).

In more complicated situations, wave refraction diagrams can be plotted for a region by using the wave of most common period and the most common direction of approach, and in this way areas in which wave rays are focused or defocused can be identified.

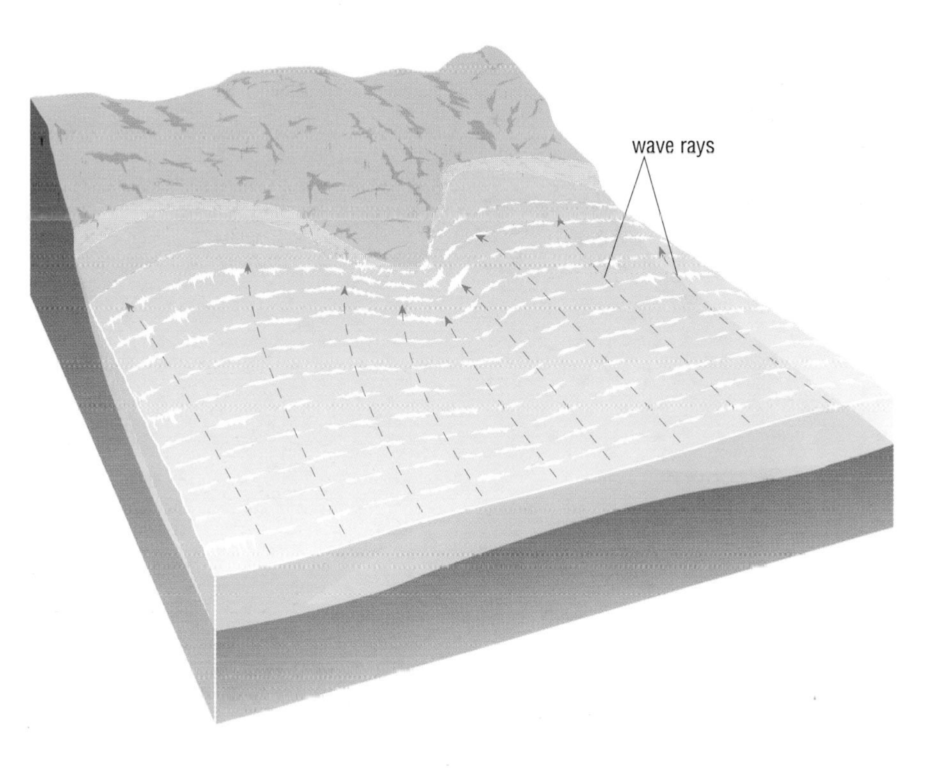

Figure 1.16 Waves are refracted and the wave rays show how wave energy is focused on headlands, where erosion is active, while deposition occurs in the bays, where the wave rays diverge and wave energy is less. Waves 'feel the bottom' and are slowed first in the shallow areas off the headland. The parts of the wave fronts that move through the deeper water leading into the bays are not slowed until they are well into the bays.

QUESTION 1.14 Figure 1.17 is a bathymetric map and storm wave refraction diagram for the Hudson River submarine canyon on the Atlantic coast of the USA. In what area covered by the refraction diagram would you advise fishermen to leave their boats to minimize the likelihood of major damage, and why?

Long Branch the waves are divergent; so there energy is less.

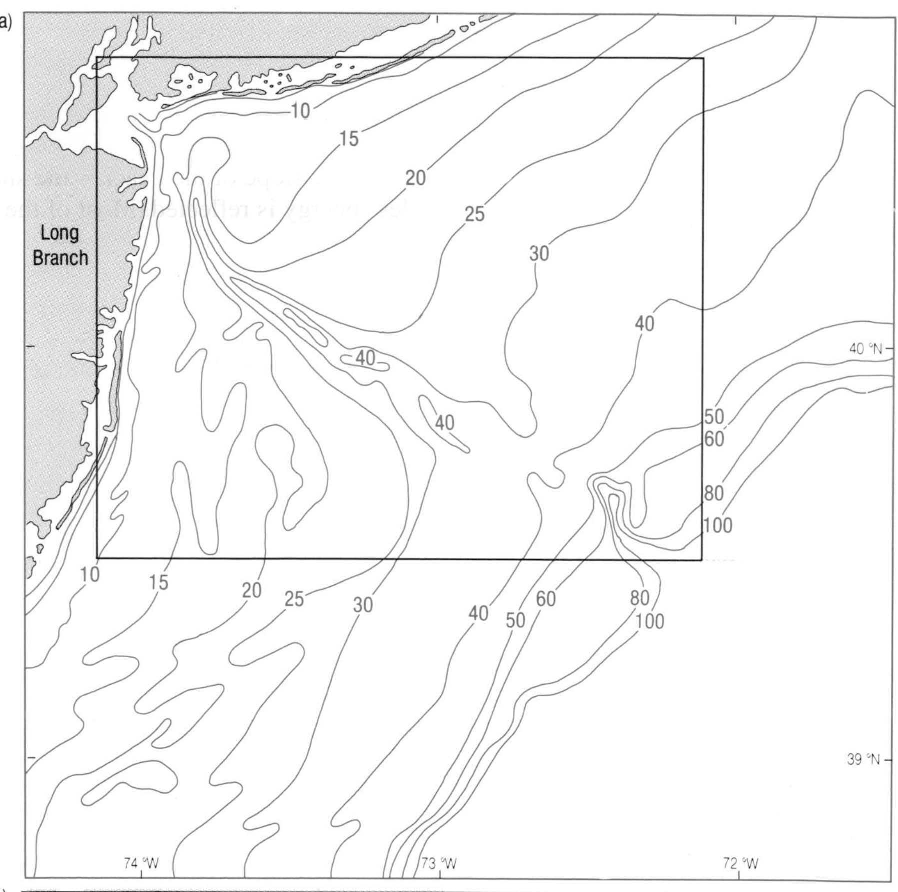

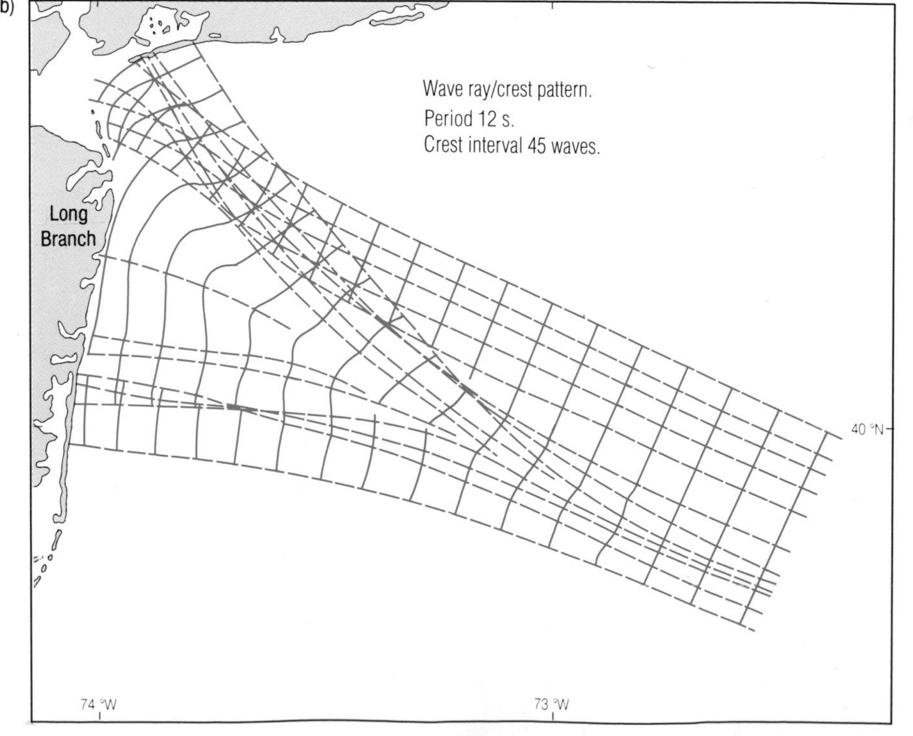

Figure 1.17 (a) Bathymetric map of the continental shelf off New York harbour at the mouth of the Hudson River on the Atlantic coast of the USA. The rectangle shows the area of map (b). The position of the Hudson Canyon can be deduced from the submarine contours (in fathoms).
(b) Pattern of wave crests (wave fronts) and wave rays off part of Long Branch beach. The wave fronts are drawn at intervals of 45 waves of period 12 s.

We can estimate increase or decrease in wave size by measuring the distances between wave rays, and applying Equation 1.15. This method is quite useful provided wave rays neither approach each other too closely nor cross over, as in these cases the waves become high, steep and unstable, and simple wave theory becomes inadequate.

1.5.2 WAVES BREAKING UPON THE SHORE

As a wave breaks upon the shore, the energy it received from the wind is dissipated. Some energy is reflected back out to sea, the amount depending upon the slope of the beach – the shallower the angle of the beach slope, the less energy is reflected. Most of the energy is dissipated as heat and sound (the 'roar' of the surf) in the final small-scale mixing of foaming water, sand and shingle. Some energy is used in fracturing large rock or mineral particles into smaller ones, and yet more may be used to move sediments and increase the height and hence the potential energy of the beach form. This last aspect depends upon the type of waves. Small gentle waves and swell tend to build up beaches, whereas storm waves tear them down (see also Chapter 5).

A breaking wave is a highly complex system. Even some distance before the wave breaks, its shape is substantially distorted from a simple sinusoidal wave. Hence the mathematical model of such a wave is more complicated than we have assumed in this Chapter.

Four major types of breaker can be identified, though you may often see breakers of intermediate character and/or of more than one type on the same beach at the same time.

1 Spilling breakers are characterized by foam and turbulence at the wave crest. Spilling usually starts some distance from shore and is caused when a layer of water at the crest moves forward faster than the wave as a whole. Foam eventually covers the leading face of the wave, and such waves are characteristic of a gently sloping shoreline. A tidal bore (Section 2.4.3) is an extreme form of a spilling breaker. Breakers seen on beaches *during* a storm, when the waves are steep and short, are of the spilling type. They dissipate their energy gradually as the top of the wave spills down the front of the crest, which gives a violent and formidable aspect to the sea because of the more extended period of breaking.

2 Plunging breakers are the most spectacular type. The classical form, much beloved by surf-riders, is arched, with a convex back and concave front. The crest curls over and plunges downwards with considerable force, dissipating its energy over a short distance. Plunging breakers on beaches of relatively gentle slope are usually associated with the long swells generated by distant storms. Locally generated storm waves seldom develop into plunging breakers on gently sloping beaches, but may do so on steeper ones. The energy dissipated by plunging breakers is concentrated at the *plunge point* (i.e. where the water hits the bed) and can have great erosive effect.

3 Collapsing breakers are similar to plunging breakers, except that the waves may be less steep and instead of the crest curling over, the front face collapses. Such breakers occur on beaches with moderately steep slopes, and under moderate wind conditions, and represent a transition from plunging to surging breakers.

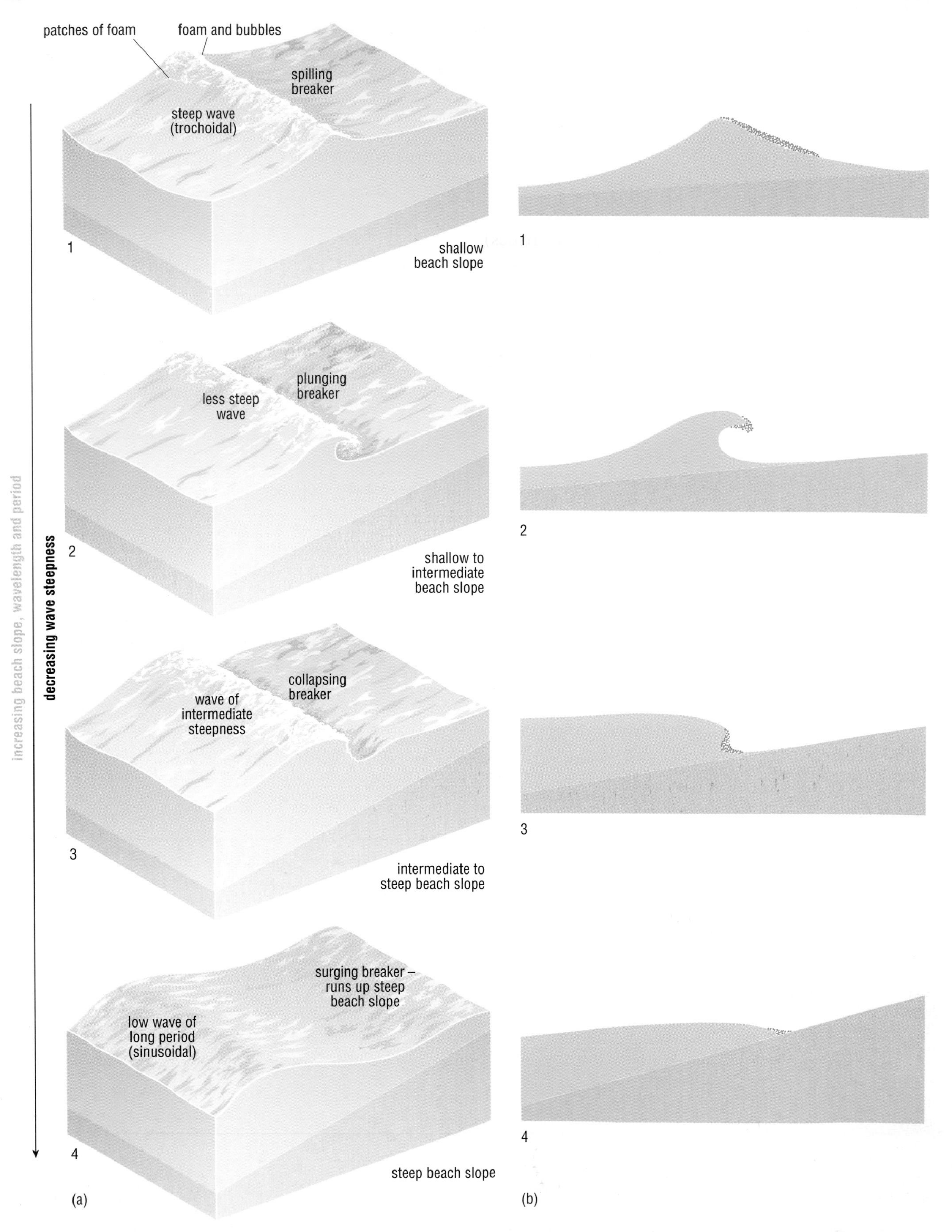
patches of foam
foam and bubbles
spilling breaker
steep wave (trochoidal)
1
shallow beach slope
less steep wave
plunging breaker
2
shallow to intermediate beach slope
wave of intermediate steepness
collapsing breaker
3
intermediate to steep beach slope
surging breaker – runs up steep beach slope
low wave of long period (sinusoidal)
4
steep beach slope
increasing beach slope, wavelength and period
decreasing wave steepness
(a)
1
2
3
4
(b)

4 Surging breakers are found on the very steepest beaches. Surging breakers are typically formed from long, low waves, and the front faces and crests thus remain relatively unbroken as the waves slide up the beach.

Figure 1.18 illustrates the relationship between wave steepness, beach steepness and breaker type.

The way breaker shape changes from top to bottom of the picture depends upon:

1 Increasing beach slope (if considered independently from wave characteristics).

2 Increasing wavelength and period and correspondingly decreasing wave steepness, if these characteristics are considered independently of beach slope.

It is not always possible to consider 1 and 2 separately, because as you will see in later Chapters, beach slope is partly influenced by prevailing wave type and partly by the particle sizes of the beach sediments, which in turn depend upon the energy of the waves which erode, transport and deposit them.

QUESTION 1.15 If you observed plunging breakers on a beach and walked along towards a region where the beach became steeper, what different types of breaker might you expect to see?

From the descriptions, Figure 1.18, and the answer to Question 1.15, it can be seen that the four types of breaker form a continuous series. The spilling breaker, characteristic of shallow beaches and steep waves (i.e. with short periods and large amplitudes), forms one end of the series. At the other end of the series is the surging breaker, characteristic of steep beaches and of waves with long periods and small amplitudes. For a given beach, the arrival of waves steeper than usual will tend to give a type of breaker nearer the 'spilling' end of the series, whereas calmer weather favours the surging type. The dynamics of collapsing (3) and surging (4) breakers are affected by bottom slope more than those of spilling (1) and plunging (2) breakers. Spilling and plunging breakers can also occur in deep water, partly because the sea-bed is far below and does not affect wave dynamics. Collapsing and surging breakers do not occur in deep water.

(c)

Figure 1.18 (a) The four types of breaker seen in perspective view from top to bottom (1–4): spilling, plunging, collapsing, surging. The vertical arrow shows their relationships to beach slope, wave period, length and steepness.
(b) Cross-sections through the four breaker types.
(c) Photograph of a breaker, part spilling, part plunging. See text for further discussion.

IMPORTANT: When examining Figure 1.18, you need to be aware that the four types of breaker illustrated are just stages in a continuous spectrum; changes from one to another are gradual, not instantaneous.

1.6 WAVES OF UNUSUAL CHARACTER

Waves of unusual character may result from any one of a number of conditions, such as a particular combination of wave frequencies; the constraining effect of nearby land masses; interaction between waves and ocean currents; or a submarine earthquake. The destructive effects of abnormally large waves are well known, and prediction of where and when they will occur is of extreme importance to all who live or work beside or upon the sea.

1.6.1 WAVES AND CURRENTS

Anyone who regularly sails a small boat into and out of estuaries will be well aware that at certain states of the tide the waves can become abnormally large and uncomfortable. Such large waves are usually associated with waves propagating against an ebbing tide. Because the strength of the tidal current varies with position as well as with time, waves propagating into an estuary during an ebb tide often advance into progressively stronger counter-currents. The only ocean waves that can disturb the estuary as a whole during an ebbing tide are those that have speeds sufficiently high to overcome the effects of the counter-current.

Consider a simple system of deep-water waves, moving from a region with little or no current (A) into another region (B) where there is a current flowing parallel to the direction of wave propagation. Imagine two points, one in region A and one in region B, each of which are fixed with respect to the sea-bed. The number of waves passing each point in a given time must be the same, otherwise waves would either have to disappear, or be generated, between the two points. In other words, the wave frequency (and period) must be the same at each point.

How will wavelength and wave height be affected if the current is flowing (a) with or (b) against the direction of wave propagation?

Clearly, a current flowing with the waves will have the effect of increasing the speed of the waves, although the wave period (T) must remain constant, i.e.

$$T = \frac{L_0}{c_0} = \frac{L}{c+u} \tag{1.16}$$

where L_0 = wavelength when current is zero;

c_0 = wave speed when current is zero;

L = wavelength in the current;

c = wave speed in the current;

u = speed of the current

Because $c + u$ is greater than c_0, then if T is to remain constant, L must be greater than L_0, i.e. the waves get longer. Moreover, the waves get correspondingly lower, because the rate of energy transfer depends upon group speed (half wave speed) and wave height. If the rate of energy transfer is to remain constant, then as wave speed increases, wave height must decrease. However, in practice, not all of the wave energy is retained in the wave system: some is transferred to the current, causing wave height to decrease still further.

Conversely, if the current flows counter to the direction of wave propagation, then L will decrease and the waves will get shorter and higher. Wave height will be further increased as a result of energy gained from the current. In theory, a point could be reached where wave speed is reduced to zero, so that a giant wave builds up to an infinite height (it can be shown mathematically that this occurs when the counter-current exceeds half the group speed of the wave in still water). However, in practice, as waves propagate against a counter-current of ever-increasing strength, they become shorter, steeper and higher until they become unstable and break; so waves do not propagate against a counter-current of more than half their group speed.

As currents affect wave speed, they can also refract waves, i.e. change their direction of propagation. In such situations, the refraction diagram of the wave rays can be plotted in a similar way to that outlined in Section 1.5.1 (Figure 1.19(b); see also Section 1.6.2).

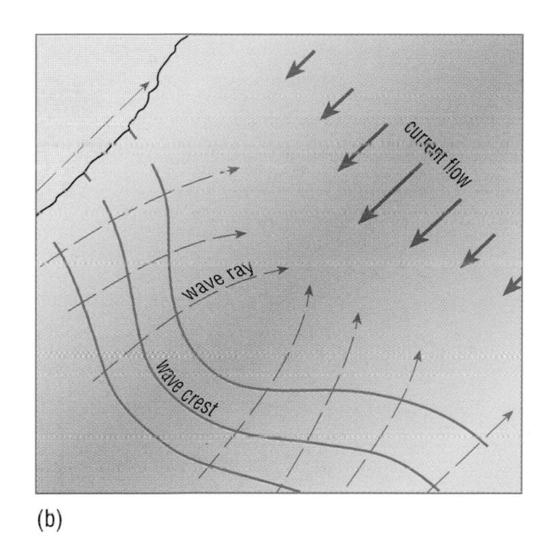

(b)

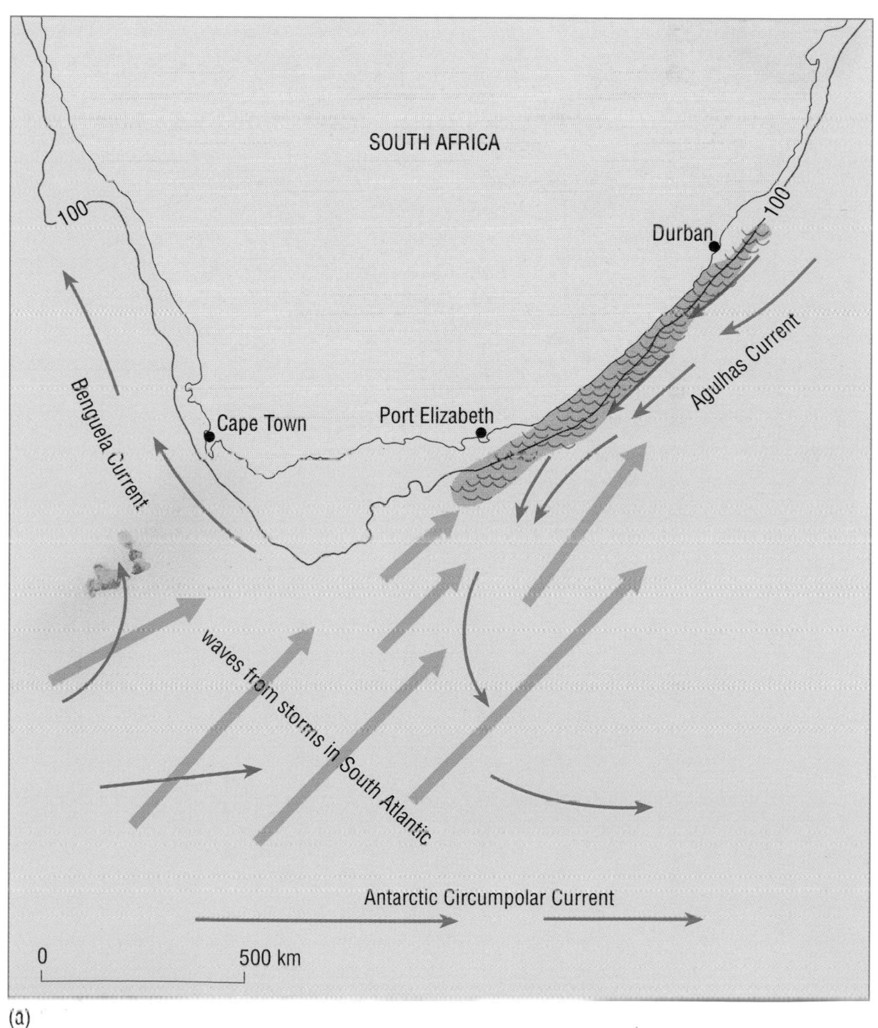

(a)

Figure 1.19 (a) (right) Giant waves have been recorded in the vicinity of the Agulhas Current (area shaded in mauve). Thinner, darker blue arrows show current direction, and thicker, lighter blue arrows indicate the direction of wave movement. The irregular black line is the 100 fathom contour.

(b) Diagram to illustrate how waves might be refracted and wave energy focused by lateral shear in a counter-current (i.e. flowing against the direction of wave propagation). The lengths of the solid blue arrows are proportional to the current velocity, curved solid lines are wave crests, and curved broken arrows are wave rays. Wave rays are focused by counter-currents in much the same way as they are at headlands.

1.6.2 GIANT WAVES

The cultures of all seafaring nations abound with legends of ships being swamped by gigantic waves, and of sightings of waves of almost unbelievable size.

An early objective method for estimating the height of large waves was to send a seaman to climb the rigging until he could just see the horizon over the top of the highest waves when the ship was in a wave trough. This technique was used by the *Venus* during her circumnavigation of the world from 1836 to 1839. She did not meet any particularly high waves during her voyage – the highest estimated by this method was about 8 m high, off Cape Horn. The highest reliably measured wave was encountered by the US tanker *Ramapo*, *en route* from Manila to San Diego across the North Pacific in 1933, when she was overtaken by waves having heights up to 34 m.

A region of the ocean that is infamous for encounters with giant waves is the Agulhas Current off the east coast of South Africa (Figure 1.19(a)). Waves travelling north-east from the southern Atlantic Ocean tend to be refracted and focused by the current, and wave rays can be plotted, as outlined in Section 1.5.1, and illustrated diagrammatically in Figure 1.19(b)); and they are further steepened and shortened by the counter-current effect outlined in Section 1.6.1. Wave periods of 14 s, with correspondingly long wavelengths of about 300 m, are quite common. Wave heights in this region can be of the order of 30 m which would result in a very steep wave (steepness = 0.1). Such high and steep waves are sometimes preceded by correspondingly deep troughs, which are particularly dangerous as they can only be seen by vessels that are on the crest of the preceding wave.

Look again at Figure 1.5. The high wave occurring 122 s into the wave record is preceded by a particularly deep trough, but that is not the case for the high wave occurring at 173 s. As we saw in Section 1.1.4, ocean waves are rarely regular, and it is usually not possible to predict the heights of individual waves, nor the depths of individual troughs.

QUESTION 1.16 An elderly ex-seaman, in his cups, claims to have seen gigantic waves in the Southern Ocean, successive peaks of which took 30 seconds to pass, and which had wavelengths twice as long as his ship. Can you believe him?

Before dismissing the sailor's claim in Question 1.16 as a tall story, let us examine it more closely. Let us suppose his ship was travelling in the same direction as the waves and was being *overtaken* by them, and he had made the simple mistake of not taking account of the ship's velocity with respect to the waves when timing the intervals between successive peaks.

QUESTION 1.17 Further conversation with the seaman established that his ship was the cruiser HMS *Exeter* (1929–42), which at the time of the incident was steaming at 23 knots (11.8 m s^{-1}) in the same direction as the waves. Given that the *Exeter* was 575 feet in length (175 m), can you believe him now?

1.6.3 TSUNAMIS

Tsunami is a Japanese word for ocean waves of very great wavelength. Such waves are caused chiefly by seismic disturbances (earthquakes), but also by slumping of unstable masses of submarine sediment or rock (see Chapter 3), and by 'splashdown' of comets or asteroids in the sea. Among the most famous tsunamis is the one triggered by the Lisbon earthquake of 1755, which caused at least as much destruction as the earthquake itself. Although commonly referred to as 'tidal waves', *tsunamis are not related to the tides.* Tsunamis commonly have wavelengths of the order of hundreds of kilometres, so even in the open ocean, the ratio of wavelength to depth is such that a tsunami travels as a shallow-water wave, i.e. its speed is *always* governed by the depth of ocean over which it is passing.

QUESTION 1.18 What would be the speed of a tsunami across the open ocean above an abyssal plain? (Assume the average depth is 5.5 km.)

Although a tsunami travels at great speed in the open ocean, its wave height is small, usually in the order of one metre, and so it often remains undetected. On reaching shallow coastal waters, however, the speed diminishes, while the *power* of the wave remains the same (Section 1.4.1). Hence the wave energy (and therefore the wave height, Equations 1.11 and 1.13) must increase.

Great destruction can be wreaked by a tsunami. It is not unknown for people on board ships at anchor offshore to be unaware of a tsunami passing beneath them, but to witness the adjacent shoreline being pounded by large waves only a few seconds later. Tsunamis occur most frequently in the Pacific, because that ocean experiences frequent seismic activity, particularly around its margins. Accurate earthquake detection can give warning of the approach of tsunamis to coasts some distance from the earthquakes. Around and across the Pacific Ocean is a long-established system of warning stations, of which Honolulu is the administrative and geographical centre.

1.6.4 SEICHES

A **seiche** is a standing wave, which can be considered as the sum of two progressive waves travelling in opposite directions (Section 1.1). Seiches can occur in lakes, and also in bays, estuaries or harbours which are open to the sea at one end. A seiche can be approximately modelled by sliding up and down in the bath, or – perhaps more accurately – with a flat dish containing water, gently moving it to and fro and setting the water into oscillatory motion, as often happens inadvertently when the tray below the freezer compartment of a defrosted refrigerator is removed, and the water starts sloshing back and forth.

In most bays and estuaries, the water is relatively shallow compared with the seiche wavelength (L), and the period of the seiche is determined by the length of the basin and the depth of water in it. Figure 1.20(a, b) (overleaf) shows the idealized water motions in a seiche in a closed basin (e.g. a lake). At either end of the container, water level is alternately high and low, whereas in the middle the water level remains constant. The length of the container (l) corresponds to half the wavelength (L) of the seiche.

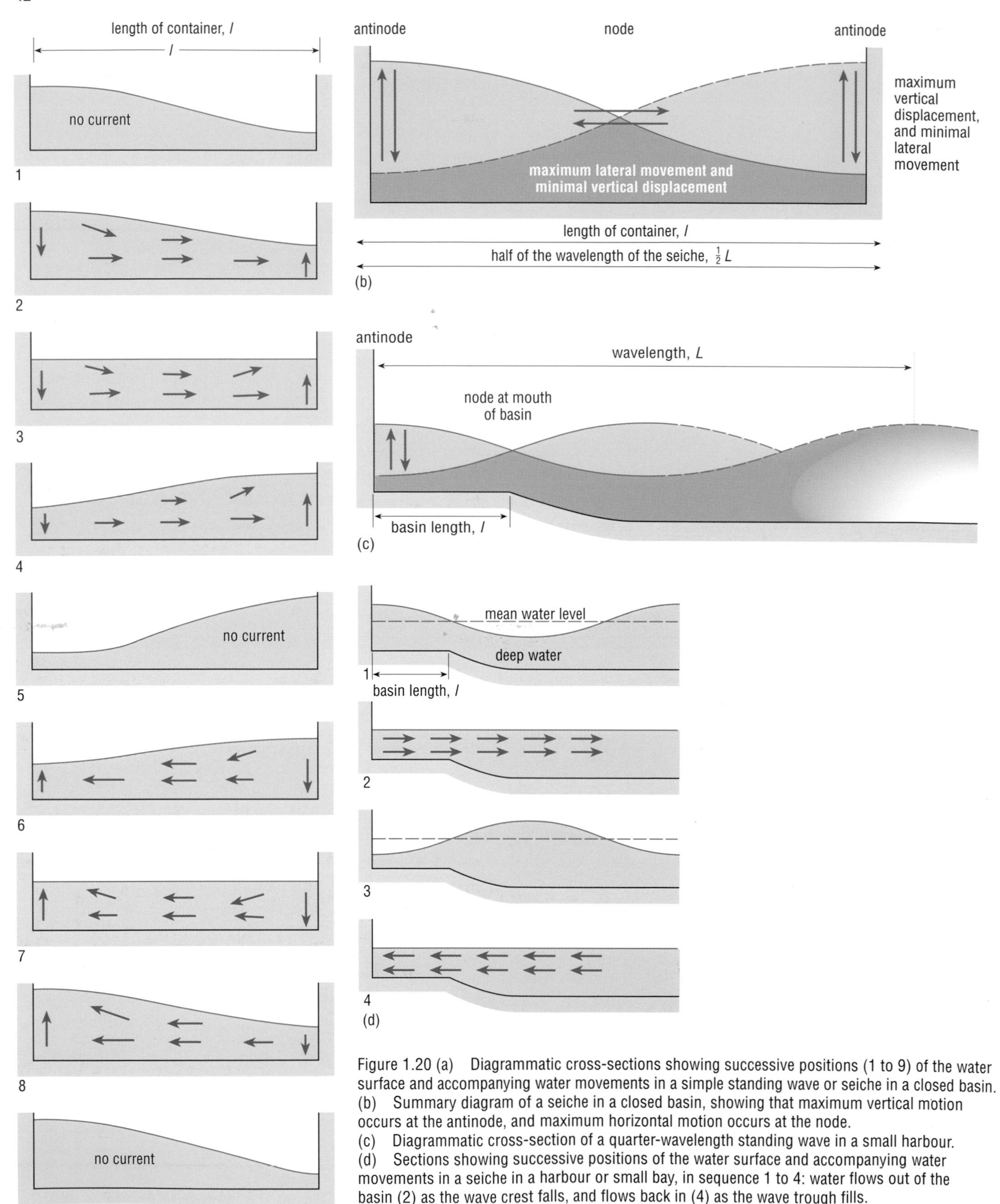

Figure 1.20 (a) Diagrammatic cross-sections showing successive positions (1 to 9) of the water surface and accompanying water movements in a simple standing wave or seiche in a closed basin. (b) Summary diagram of a seiche in a closed basin, showing that maximum vertical motion occurs at the antinode, and maximum horizontal motion occurs at the node. (c) Diagrammatic cross-section of a quarter-wavelength standing wave in a small harbour. (d) Sections showing successive positions of the water surface and accompanying water movements in a seiche in a harbour or small bay, in sequence 1 to 4: water flows out of the basin (2) as the wave crest falls, and flows back in (4) as the wave trough fills.

Where the water level is constant (the **node**), the horizontal flow of water from one end of the container to the other is greatest, as the crest and trough of the seiche alternate at either end of the basin. Where the fluctuation of water level is greatest (the **antinodes**, at either end of the basin), there is minimal horizontal movement of the water. These oscillatory water movements in a seiche in a closed basin are summarized in the idealized cross-section, Figure 1.20(b).

If the water depth divided by the length of the container is less than 0.1, then the waves can be considered to behave as shallow-water waves, with speed $= \sqrt{gd}$, and the period of oscillation, T (in seconds), is given by:

$$T = \frac{2l}{\sqrt{gd}} \tag{1.17}$$

where l = length of container (in metres); d = depth (in metres); and $g = 9.8 \text{ m s}^{-2}$.

In some basins open to the sea at one end, i.e. in some bays and estuaries, it is possible for a node to occur at the entrance to the basin and an antinode at the landward end. Figure 1.20(c) is an idealized vertical section of such a situation, and Figure 1.20(d) shows the corresponding oscillatory water movements. In this case, the length of the basin (l) corresponds to a quarter of the wavelength of the seiche (L). The corresponding equation for the period is therefore:

$$T = \frac{4l}{\sqrt{gd}} \tag{1.18}$$

T is here known as the **resonant period**.

For standing waves to develop, the resonant period of the basin must be equal to the period of the wave motion or to a small whole number of multiples of that period.

QUESTION 1.19 A small harbour, open to the sea at one end, is 90 m long and 10 m deep at high water. What would be the effect of swell waves of period 18 s arriving at the harbour mouth?

Your answer to Question 1.19 is an example of how the arrival of waves of a certain frequency can create problems for moored vessels in small harbours by setting up a standing wave. Just as the seiche in a flat dish will ‘slop over’ if your standing wave gets too big, so a standing wave in a harbour may dash vessels against the harbour wall, or even throw them ashore. When the standing wave is at the low point of the antinode, there is also the danger of vessels being grounded, thus suffering damage to their hulls.

1.7 MEASUREMENT OF WAVES

Figure 1.21 Solar-powered directional wave buoy (which in use is moored to the sea-bed). The buoy contains sensors that measure roll, pitch and yaw, as well as wave displacement, velocity and acceleration.

Various instruments have been devised to measure wave characteristics. For example, a pressure gauge can be placed upon the sea-floor and will detect changes in the pressure (and therefore height) of the water column above it, and hence the frequency and size of waves at the surface (see Figure 4.16(a)). Pressure gauges with a sensitivity of about one part in a million are available and such gauges can detect pressure changes corresponding to a water level difference of less than a centimetre in a depth of several thousand metres of water. Another method is to place instruments such as accelerometers and pitch-and-roll recorders on moored buoys to detect and record the rate of rise and fall, and the direction of slope, of the sea-surface in response to waves (Figure 1.21).

1.7.1 SATELLITE OBSERVATIONS OF WAVES

The observation of ocean surface waves is particularly amenable to the use of satellite-based remote-sensing. Satellite remote-sensing techniques provide a unique elevated and synoptic (simultaneous) view of wide expanses of ocean, and offer the only practical method of obtaining repeated observations with the wide geographic coverage needed for reliable prediction. Four main techniques are considered here, but the technical details are not important in the context of this Volume.

1 Radar altimetry

A radar altimeter, mounted in a satellite, emits radar pulses at a rate of about 1000 pulses per second directly towards the sea-surface, and each reflected pulse is picked up by a sensor. Radar pulses of the frequency used in satellites pass through (i.e. are not reflected by) water droplets (clouds) in the atmosphere, but are reflected by the sea-surface. (Shorter wavelengths characterize the radar used in weather forecasting which 'sees' clouds and precipitation.) The average time which the pulses take to travel to the sea-surface and back again enables the mean satellite-to-surface distance to be calculated. The accuracy with which the distance can be measured depends upon the sea state. For values of $H_{1/3}$ up to 8 m, the accuracy is ± 8 cm, and for values of $H_{1/3}$ over 8 m, the accuracy is about $\pm x$ cm, where x is the significant wave height in metres. Changes in the shape and amplitude of the returning pulses can be used to give a measure of wave heights, from which the significant wave height ($H_{1/3}$) can be calculated. The accuracy achieved is about ± 10% of $H_{1/3}$ for waves of significant wave height greater than 3 m, and about ± 0.3 m for smaller values of $H_{1/3}$.

QUESTION 1.20 Estimate the uncertainties which would be expected in (i) the mean satellite-to-surface distance, and (ii) the significant wave height, if they were measured by radar altimeter under the following conditions:

(a) Force 1 wind on the Beaufort Scale, no cloud;

(b) Force 4 wind on the Beaufort Scale, dense cloud cover;

(c) Force 10 wind on the Beaufort Scale, light cloud cover.

Use Table 1.1 on p. 18, and assume a steady sea state has been established in each case.

2 Synthetic aperture radar (SAR)

This technique maps variations in the proportion of the radar signal that is back-scattered to the radar antenna by the sea-surface, and gives a measure of its 'roughness'. Short radar pulses are transmitted at an oblique angle to the sea-surface, and variations in the Doppler shift (i.e. apparent changes in signal frequency caused by the relative movement between target and detector) of the scattered return signal are analysed to produce an image.

SAR images often show a wave-like pattern (Figure 1.22), but this pattern is not a direct indication of wavelength. The regular variations in reflectivity can, however, be related to the characteristics of larger (swell) waves from which an idea of their wavelength and direction of propagation can be obtained. The precise interpretation of SAR images is hampered by such difficulties as:

(i) SAR assumes a moving satellite and a stationary target, whereas both waves and water are moving. This motion complicates the return signal, so that elaborate procedures have to be applied to render the image into an interpretable form.

(ii) Image 'roughness' is determined by the entire spectrum of waves present, not just by the swell waves. Even capillary waves of a few millimetres wavelength influence the image, and hence complicate its interpretation.

(iii) White-capping tends to scatter incident radar randomly, thus obscuring the more regular back-scatter obtained from smoother surfaces. If waves actually break, the foam tends to absorb the radar signals, reducing both clarity and contrast in the final image.

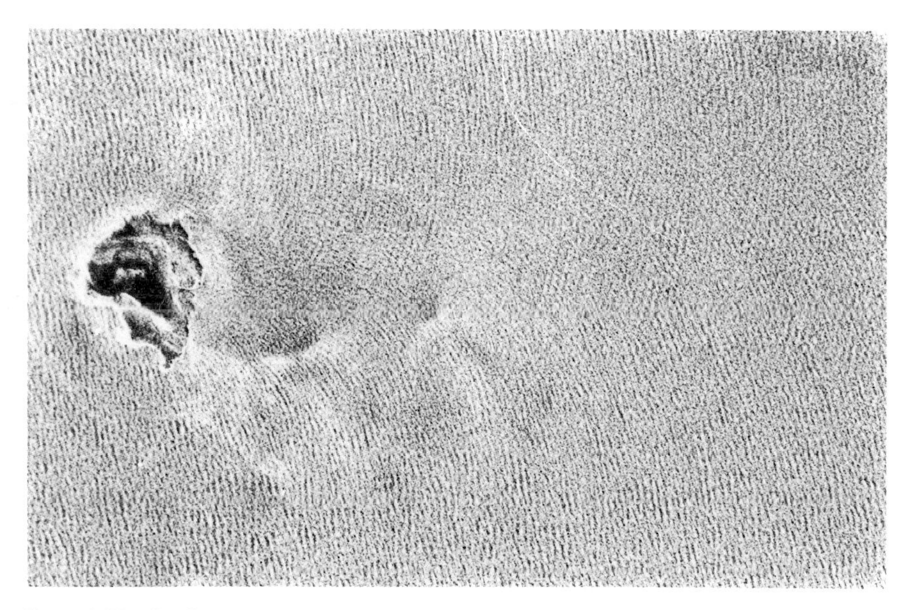

Figure 1.22 Swell-wave pattern around the island of Foula (70 km NW of Fair Isle, NE Atlantic) obtained by *S*ynthetic *A*perture *R*adar (SAR). Area ≈ 30 km × 19 km.

3 Scatterometry

Radar scatterometers are used to analyse the strength and polarization of radar echoes, from which surface wind strength and direction can be determined. The method is complex and relies on the different back-scattering properties of wind-generated surface waves when viewed obliquely in upwind, downwind and crosswind directions. (A related land-based technique is known as OSCR (*O*cean *S*urface *C*urrent *R*adar), which exploits the fact that reflections of radio waves of suitable wavelength from surface waves produce specific interference patterns (analogous to Bragg reflections in crystallography). While originally devised as a way of measuring currents from the pattern of waves interacting with them, OSCR provides much useful information about the waves themselves.)

4 Photography

Changes in the amount of reflected sunlight correlate with local roughness and wave steepness, thus revealing wave patterns on photographs. Even internal waves can be detected by photography because of their effect on surface roughness (Section 1.2.1). The photographs in Figure 1.23 were taken from a manned spacecraft with a hand-held camera, and show the surface manifestations of some internal waves in the South China Sea and in the Mediterranean.

(a)

(b)

Figure 1.23 (a) Internal waves in the South China Sea (Hainan Island visible beneath clouds on lower left). Four wave packets are visible.
(b) Tidally generated internal waves propagating into the Mediterranean from the Straits of Gibraltar. The internal waves (which have amplitudes of the order of 15 m) are visible because of variations in the roughness of the sea-surface (see p. 20). Area = 74 km × 74 km.

1.8 SUMMARY OF CHAPTER 1

1 Idealized waves of sinusoidal form have wavelength (length between successive crests), height (vertical difference between trough and crest), steepness (ratio of height to length), amplitude (half the wave height), period (length of time between successive waves passing a fixed point) and frequency (reciprocal of period). Water waves show cyclical variations in water level (displacement), from $-a$ (amplitude) in the trough to $+a$ at the crest. Displacement varies not only in space (one wavelength between successive crests) but also in time (one period between crests at one location). Steeper waves depart from the simple sinusoidal model, and more closely resemble a trochoidal wave form.

2 Waves transfer energy across/through material without significant *overall* motion of the material itself, but individual particles are displaced from, and return to, equilibrium positions as each wave passes. Surface waves occur at interfaces between fluids, either because of relative movement between the fluids, or because the fluids are disturbed by an external force (e.g. wind). Waves occurring at interfaces between oceanic water layers are called internal waves. Wind-generated waves, once initiated, are maintained by surface tension and gravity, although only the latter is significant for water waves over 1.7 cm wavelength.

3 Most sea-surface waves are wind-generated. The stronger the wind, the larger the wave, so variable winds produce a range of wave sizes. A constant wind speed produces a fully developed sea, with waves of $H_{1/3}$ (average height of highest 33% of the waves) characteristic of that wind speed. The Beaufort Scale relates sea state and $H_{1/3}$ to the causative wind speed.

4 Water particles in waves in deep water follow almost circular paths, but with a small net forward drift. Path diameters at the surface correspond to wave heights, but decrease exponentially with depth. In shallow water, the orbits become flattened near the sea-bed. For waves in water deeper than 1/2 wavelength, wave speed equals wavelength/period ($c = L/T$) and is proportional to the square root of the wavelength ($c = \sqrt{gL/2\pi}$); it is unaffected by depth. For waves in water shallower than 1/20 wavelength, wave speed is proportional to the square root of the depth ($c = \sqrt{gd}$) and does not depend upon the wavelength. For idealized water waves, the three characteristics, c, L and T, are related by the equation $c = L/T$. In addition, each can be expressed in terms of each of the other two. For example, $c = 1.56T$ and $L = 1.56T^2$.

5 Waves of different wavelengths become dispersed, because those with greater wavelengths and longer periods travel faster than smaller waves. If two wave trains of similar wavelength and amplitude travel over the same sea area, they interact. Where they are in phase, displacement is doubled, whereas where they are out of phase, displacement is zero. A single wave train results, travelling as a series of wave groups, each separated from adjacent groups by an almost wave-free region. Wave group speed in deep water is half the wave (phase) speed. In shallowing water, wave speed approaches group speed, until the two coincide at depths less than 1/20 of the wavelength, where $c = \sqrt{gd}$.

6 Wave energy is proportional to the square of the wave height, and travels at the group speed. Wave power is rate of supply of wave energy, and so it is wave energy multiplied by wave (or group) speed, i.e. it is wave energy propagated per second per unit length of wave crest (or wave speed multiplied by wave energy per unit area). Total wave power is conserved, so waves entering shallowing water and/or funnelled into a bay or estuary (see also 7 below) increase in height as their group speed falls. Wave energy has been successfully harnessed on a small scale, but large-scale utilization involves environmental and navigational problems, and huge capital outlay.

7 Dissipation of wave energy (attenuation of waves) results from white-capping, friction between water molecules, air resistance, and non-linear wave–wave interaction (exchange of energy between waves of differing frequencies). Most attenuation takes place in and near the storm area. Swell waves are storm-generated waves that have travelled far from their place of origin, and are little affected by wind or by shorter, high-frequency waves. The wave energy associated with a given length of wave crest decreases with increasing distance from the storm, as the wave energy is spread over an ever-increasing length of wave front.

8 Waves in shallow water may be refracted. Variations in depth cause variations in speed of different parts of the wave crest; the resulting refraction causes wave crests to become increasingly parallel with bottom contours. The energy of refracted waves is conserved, so converging waves tend to increase, and diverging waves to diminish, in height. Waves in shallow water dissipate energy by frictional interaction with the sea-bed, and by breaking. In general, the steeper the wave and the shallower the beach, the further offshore dissipation begins. Breakers form a continuous series from steep spilling types to long-period surging breakers.

9 Waves propagating with a current have diminished heights, whereas a counter-current increases wave height, unless current speed exceeds half the wave group speed. If so, waves no longer propagate, but increase in height until they become unstable and break. Tsunamis are caused by earthquakes or by slumping of sediments, and their great wavelength means their speed is always governed by the ocean depth. Wave height is small in the open ocean, but can become destructively large near the shore. Seiches (standing waves) are oscillations of water bodies, such that at antinodes there are great variations of water level but little lateral water movement, whereas at nodes the converse is true. The period of oscillation is proportional to basin length and inversely proportional to the square root of the depth. A seiche is readily established when the wavelength of incoming waves is four times the length of the basin.

10 Waves are measured by a variety of methods, e.g. pressure gauges on the sea-floor, accelerometers in buoys on the sea-surface, and via remote-sensing from satellites.

Now try the following questions to consolidate your understanding of this Chapter.

QUESTION 1.21 A wave of period 10 s approaching the shore has a height of 1 m in deep water. Calculate:

(a) the wave speed and group speed in deep water;

(b) the wave steepness in deep water;

(c) the wave power per metre length of crest in deep water;

(d) the wave power per metre length of crest in water 2.5 m deep.

QUESTION 1.22 A wave system consisting of short waves (wavelength 6 m), together with a swell of period 22 s, propagates through a narrow inlet, in which a current of 3 knots (1.54 m s^{-1}) runs counter to the direction of wave propagation. With the help of Equations 1.7 to 1.9 as well as Equation 1.16, work out what might happen to the two sets of waves and how they are likely to behave in relation to the current:

(a) in the narrow inlet;

(b) at a point where the waves have passed beyond the narrow inlet into a region where the current is negligible.

(Assume the water is very deep at all the locations described.)

QUESTION 1.23 The *Ramapo* (refer back to report of waves 34 m in height, Section 1.6.2) was a tanker 146 m long. Assume the ship was steaming at a reduced speed of 10 knots (5.14 m s^{-1}), and that the wave crests took 6.3 s to pass the ship from stern to bow.

(a) What was the wave speed?

(b) What was the wave steepness?

(c) How does the wave period that is consistent with the answers to (a) and (b) compare with the wave period of 14.8 s reported by the *Ramapo*?

(d) What would be the wave steepness if the period of these waves really was 14.8 s?

QUESTION 1.24 What sort of waves would you expect to see on a beach of intermediate slope:

(a) after a prolonged spell of calm weather?

(b) during a severe gale, with a Force 9 wind blowing onshore?

CHAPTER 2 TIDES

' ... being governed by the watery Moon ... '

Richard III, Act II, Scene II.

The longest oceanic waves are those associated with the tides, and are characterized by the rhythmic rise and fall of sea-level over a period of half a day or a day (Figure 1.2). The rise and fall result from horizontal movements of water (tidal currents) in the tidal wave. The rising tide is usually referred to as the flow (or **flood**), whereas the falling tide is called the **ebb**. The tides are commonly regarded as a coastal phenomenon, and those who see tidal fluctuations only on beaches and in estuaries tend to think (and speak) of the tide as 'coming in' and 'going out'. However, it is important to realize that the ebb and flow of the tide at the coast is a manifestation of the general rise and fall in sea-level caused by a long-wavelength wave motion that affects the oceans as well as shallow coastal waters. Nonetheless, because of their long period and wavelength (Figure 1.2), tidal waves behave as *shallow-water waves*. Do bear in mind also, from Section 1.6.3, that the destructive waves generated by earthquakes are not 'tidal waves' as so often reported in the press – they are *tsunamis*, which also behave as shallow-water waves because of their long wavelength.

From the earliest times, it has been realized that there is some connection between the tides and the Moon. High tides are highest and low tides are lowest when the Moon is full or new, and the times of high tide at any given location can be approximately (but not exactly) related to the position of the Moon in the sky; and, as we shall see, the Sun also influences the tides.

Before discussing these relationships, we shall first describe some principal features of tidal wave motions. Figure 2.1 is a tidal record, showing regular vertical movements of the water surface relative to a mean level, over a period of about a month.

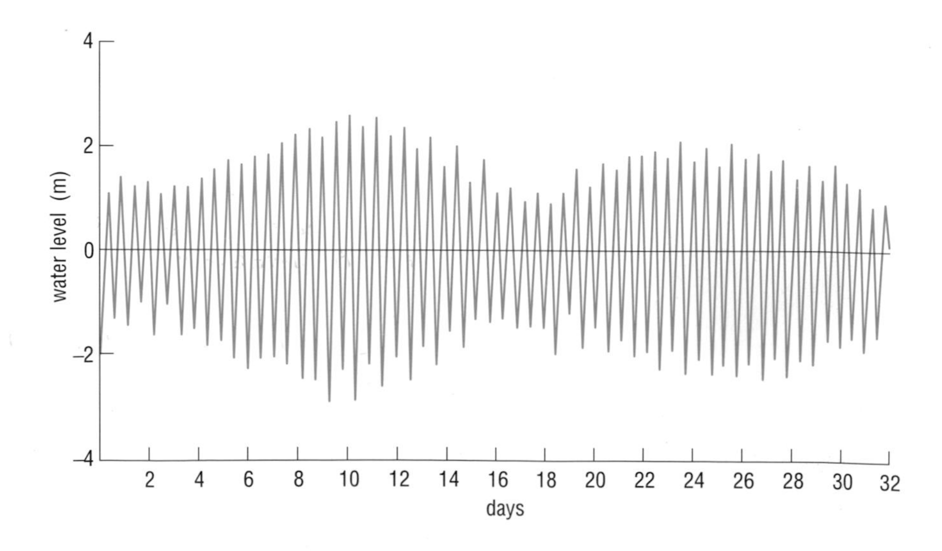

Figure 2.1 A typical 30-day tidal record showing oscillations in water level with a period of about 12.5 hours, at a station in the Tay estuary, Scotland.

If you compare Figures 2.1 and 1.5, you will see two important differences between wave motions resulting from the tides and those associated with wind-generated waves. These are:

1 The period of the oscillations of wind-generated waves (Figure 1.5) is typically in the order of seconds to a few tens of seconds, and both period and amplitude of the oscillations can be quite irregular. In contrast, Figure 2.1 shows the period of the tides to be about 12.5 hours, i.e. high and low tides occur twice a day, and both period and amplitude vary in a systematic way. (Figure 2.1 illustrates a semi-diurnal tide; we shall consider the different types of tide later.)

2 Although the amplitude (and height) of tidal and wind-generated wave motions is of the same order in both Figures 1.5 and 2.1, we have seen that the heights of wind waves can range from virtually zero to 30 m or more (Section 1.6.2). By contrast, in most places the **tidal range** is typically of the order of a few metres, and tidal ranges of more than about 10 m are known only at a few locations. Tidal range nearly always varies within the same limits at any particular location (Figure 2.1), and because the cause of tidal wave motion is both continuous and regular, so that the periodicities that result are pre-determined and fixed (as you will see shortly), tidal range can be very reliably predicted. Wind-generated waves, on the other hand, are much less predictable, because of the inherent variability of the winds. Tidal waves are what are known as 'forced waves' because they are generated by regular (periodic) external forces, and therefore do not behave exactly like the gravity waves considered in Chapter 1. For practical purposes, however, they can be treated as gravity waves, especially in the deep oceans.

In addition to the ~12-hour period of oscillations in Figure 2.1, can you discern another periodic variation?

A 7–8-day periodicity can also be seen: around days 9 to 11 and 23 to 26, the tidal range is more than twice what it is around days 0 to 2 and 16 to 18. This 7–8-day alternation of high and low tidal range (*spring* and *neap* tides, respectively) can also be predicted with great accuracy and characterizes tides all over the world (see Section 2.2.1).

What are the ranges of the spring and neap tides on Figure 2.1?

Spring tides have an amplitude of nearly 3 m (i.e. above and below the mean water level), so the spring tidal *range* is close to 6 m. In contrast, the neap tides have a range of little more than 2 m.

Where there is urban or industrial development in coastal areas, it is common for high and low tidal levels to be quite rigorously identified, because along gently sloping shorelines a tidal range of even a couple of metres results in substantial areas of ground being alternately covered and exposed by the flooding and ebbing tides. In coastal areas, maps and plans commonly indicate Mean High Water and Mean Low Water, as well as the Mean Tide Level. The Mean Tide Level is often used as a datum or baseline for topographic survey work, i.e. it is the baseline for all measurements of elevation and depth on maps and charts. For example, in Britain, this baseline (known as the Ordnance Datum) is the Mean Tide Level at a specific location at Newlyn in Cornwall.

This discussion of tidal levels raises an important general point about people's perception of the tides. As mentioned earlier, those who see tidal fluctuations only on beaches or in estuaries tend to perceive the tide as 'coming in' and 'going out'. In fact, the sea advances over and retreats from the land *only* because the water level is rising and falling with the passage of tidal waves like those illustrated in Figure 2.1.

So much for some basic descriptions of the tides. We must now consider the forces that cause them. The relative motions of the Earth, Sun and Moon are complicated, and so their influence on tidal events results in an equally complex pattern. Nevertheless, as we have just seen, the actual motions of the tides are quite regular, and the magnitudes of the tide-generating forces can be precisely formulated. Although the response of the oceans to these forces is modified by topography and by the transient effects of weather patterns, it is possible to make reliable predictions of the tides for centuries ahead (and indeed to relate specific historical events to tidal states many centuries in the past).

2.1 TIDE-PRODUCING FORCES – THE EARTH–MOON SYSTEM

The Earth and the Moon behave as a single system, rotating about a common centre of mass, with a period of 27.3 days. The orbits are in fact elliptical, but to simplify matters we will treat them as circular for the time being. The Earth rotates eccentrically about the common centre of mass (centre of gravity), which is within the Earth and lies about 4700 km from its centre. Figure 2.2 illustrates the motions that result. The principal consequence of the eccentric motion about the Earth–Moon centre of mass is this: All points on and within the Earth must also rotate about the common centre of mass and so they must all follow the same elliptical path. So each point must have the same angular velocity (2π/27.3 days), and hence will experience the same centrifugal force (which is proportional to acceleration towards the centre, i.e. to the product of the radius and the square of the angular velocity).

The eccentric motion described above *has nothing whatsoever to do with the Earth's rotation (spin) upon its own axis, and should not be confused with it* (we have shown the Earth's rotation axis on Figure 2.2 for the situation where the Moon is directly above the Equator, which happens only twice every 27.3 days – see Section 2.1.1 and Figure 2.8). Nor should the centrifugal force resulting from the eccentric motion (which is equal at all points on Earth) be confused with the centrifugal force caused by the Earth's spin (which increases with distance from the rotation axis).

If you find these concepts difficult, the following simple analogy may help. Imagine you are whirling a small bunch of keys on a short length (say 25 cm) of chain. The keys represent the Moon, and your hand represents the Earth. You are rotating your hand eccentrically (but unlike the Earth it is not spinning as well), and all points on and within your hand are experiencing the same angular velocity and the same centrifugal force. Provided your bunch of keys is not too large, the centre of mass of the 'hand-and-key' system lies within your hand.

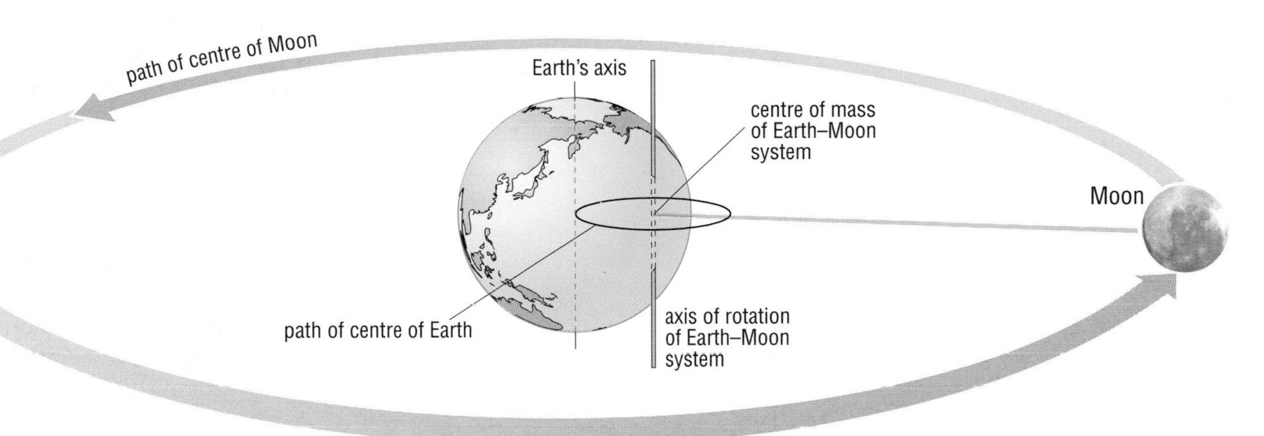

Figure 2.2 Rotation of the Earth–Moon system (not to scale). The Moon orbits the Earth about their common centre of mass (located within the Earth) once every 27.3 days. The centre of the Earth also rotates about this centre of mass once every 27.3 days, describing a very much smaller orbit (fine black line), as do all other points on and within the Earth. Note that the orbits are shown as circular for simplicity, whereas in fact they are elliptical (see later text); note also that the Earth's own central rotation axis is shown here as perpendicular to the plane of the Moon's orbit, which happens twice every 27.3 days – see Figure 2.8.

In the text which follows, you do not need to understand the details of the explanation related to Figures 2.3 and 2.4. However, you do need to be aware of the relationship embodied in Equation 2.2 on p. 55, i.e. that tide-producing forces are inversely proportional to the *cube* of the Earth–Moon distance, and that the tide-producing forces are greatest along the small circles shown in Figure 2.4(a).

The total centrifugal force acting on the Earth–Moon system exactly balances the forces of gravitational attraction between the two bodies, so the system is in equilibrium, i.e. we should neither lose the Moon, nor collide with it, in the near future. The centrifugal forces are directed parallel to a line joining the centres of the Earth and the Moon (see red arrows on Figure 2.3, overleaf). Now consider the gravitational force exerted by the Moon on the Earth. Its magnitude will not be the same at all points on the Earth's surface, because they are not at the same distance from the Moon. Points nearest the Moon will experience a greater gravitational pull from the Moon than those on the opposite side of the Earth. Moreover, the direction of the Moon's gravitational pull at all points will be directed towards its centre (see blue arrows on Figure 2.3), so it will not be exactly parallel to the direction of the centrifugal forces, except along the line joining the centres of the Earth and Moon.

The resultant (i.e. the composite effect) of the two forces is known as the **tide-producing force**. Depending upon its position on the Earth's surface with respect to the Moon, this force is directed into, parallel to, or away from, the Earth's surface. Its direction and relative strength (not strictly to scale) is shown by thick purple arrows on Figure 2.3.

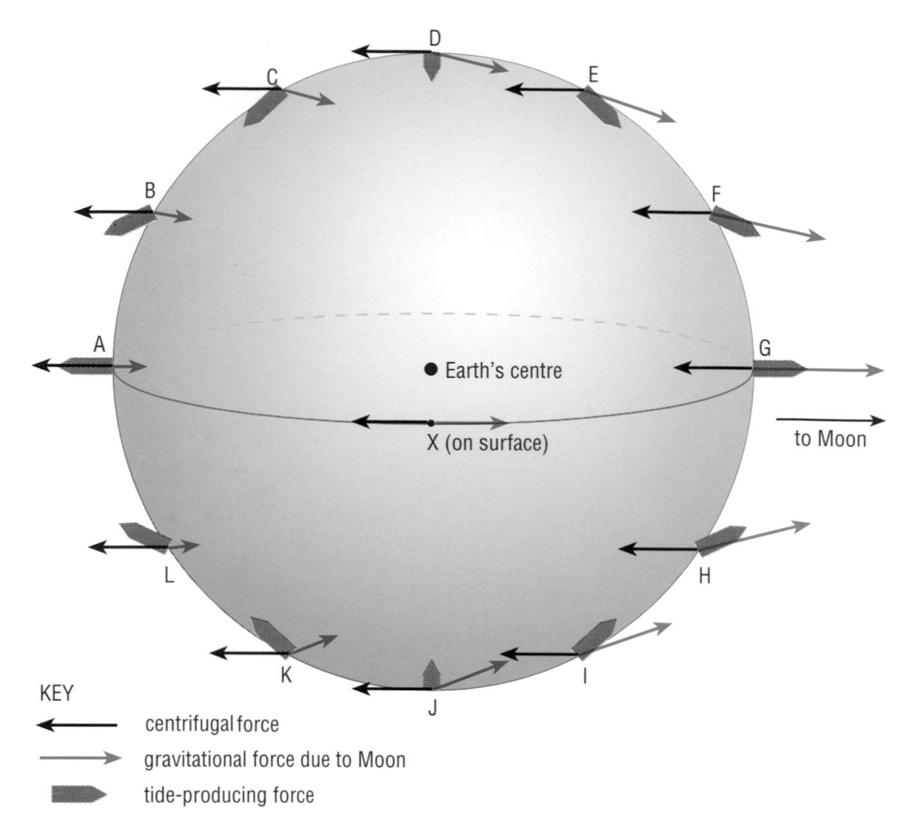

Figure 2.3 The derivation of the tide-producing forces (not to scale), for a hypothetical water-covered Earth. The centrifugal force has exactly the same magnitude and direction at all points, whereas the gravitational force exerted by the Moon on the Earth varies in both magnitude (inversely with the square of the distance from the Moon) and direction (directed towards the Moon's centre, but shown with the angles exaggerated for clarity). The tide-producing force at any point (thick purple arrows) is the *resultant* of the gravitational and centrifugal force at that point, and varies inversely with the *cube* of the distance from the Moon (see text).

QUESTION 2.1 What would be the direction and approximate magnitude (within the context of Figure 2.3) of the tide-producing forces at:

(a) a point on the Earth's surface represented by point X on Figure 2.3?

(b) the Earth's centre?

The gravitational force (F_g) between two bodies is given by:

$$F_g = \frac{GM_1M_2}{R^2} \tag{2.1}$$

where M_1 and M_2 are the masses of the two bodies, R is the distance between their centres, and G is the universal gravitational constant (whose value is $6.672 \times 10^{-11}\ \mathrm{N\,m^2\,kg^{-2}}$).

However, we need to reconcile Equation 2.1 with the statement in the caption to Figure 2.3 that the magnitude of the tide-producing force exerted by the Moon on the Earth varies inversely with the *cube* of the distance. Consider the point marked G on Figure 2.3. The gravitational attraction of the Moon at G (F_{gG}) is greater there than that at the Earth's centre, because G is nearer to the Moon by the distance of the Earth's radius (a). The gravitational force exerted by the Moon at the Earth's centre is exactly equal and opposite to the centrifugal force there, so the tide-producing force at the centre of the Earth is zero. Now as the centrifugal force is equal at all points on Earth, and at the Earth's centre is equal to the gravitational force exerted there by the Moon, it follows that we can substitute the expression on the right-hand side of Equation 2.1 (i.e. GM_1M_2/R^2) for the centrifugal force.

The tide-producing force at point G (TPF_G) is given by the force due to gravitational attraction of the Moon at G (F_{gG}) minus the centrifugal force at G, i.e.

$$TPF_G = \frac{GM_1M_2}{(R-a)^2} - \frac{GM_1M_2}{R^2}$$

which simplifies to:

$$TPF_G = \frac{GM_1M_2a\,(2R-a)}{R^2(R-a)^2}$$

Now a is very small compared to R, so $(2R - a)$ can be approximated to $2R$, and $(R - a)^2$ to R^2, giving the relationship:

$$TPF_G \approx \frac{GM_1M_2 2a}{R^3} \qquad (2.2)$$

In other words, the tide-producing force is proportional to $1/R^3$.

Before reading on, have another look at Figure 2.3, and consider at which of the lettered points on that Figure the local tide-producing force would have most effect in generating tides.

You may have considered point G as your answer. Certainly, G is nearest to the Moon, and hence is one of the two points where the difference between the centrifugal force and the gravitational force exerted by the Moon is greatest. However, at point G all the resultant tide-producing force is acting vertically against the pull of the Earth's own gravity, which happens to be about 9×10^6 greater than the tide-producing force. Hence the local effect of the tide-producing forces at point G is negligible. Similar arguments apply at point A, except that the gravitational attraction of the Moon at point A (F_{gA}) is *less* than the centrifugal force, and consequently the tide-producing force at A is equal in magnitude to that at G, but directed *away* from the Moon (Figure 2.3).

The points we need to identify are those where the horizontal component of the tide-producing force, i.e. the **tractive force**, is at a maximum. Such points do not lie directly on a line joining the centres of the Earth and Moon, and so Equation 2.2 becomes slightly more complex. For example, at point P on Figure 2.4(a) the gravitational attraction (F_{gP}) would be, to a first approximation:

$$F_{gP} = \frac{GM_1M_2}{(R - a\cos\psi)^2} \qquad (2.3)$$

The length $a\cos\psi$ is marked on Figure 2.4(a) (ψ is the Greek letter 'psi').

Equations such as 2.3 can be used to show that the tractive force is greatest at points along the small circles defined in Figure 2.4(a), *which have nothing to do with latitude or longitude.*

It is the tractive force that causes the water to move, because this horizontal component (by definition parallel to – i.e. tangential to – the Earth's surface at the location concerned) is unopposed by any other lateral force (apart from friction at the sea-bed, which is negligible in this context). The gravitational force due to the Earth is much greater than the tractive force but acts at right angles to it and so has no effect. The longest arrows on Figure 2.4(b) show where on the Earth the tractive forces are at a maximum when the Moon is over the Equator.

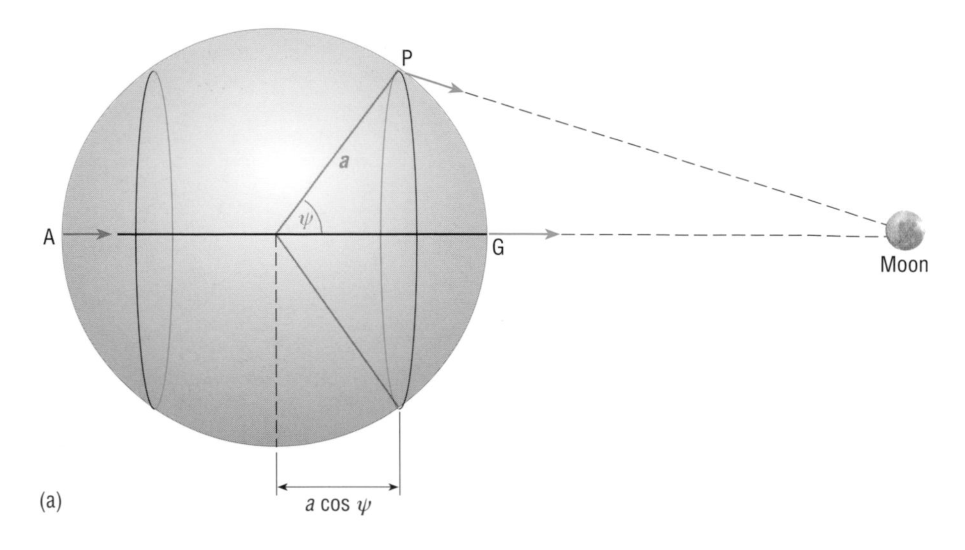

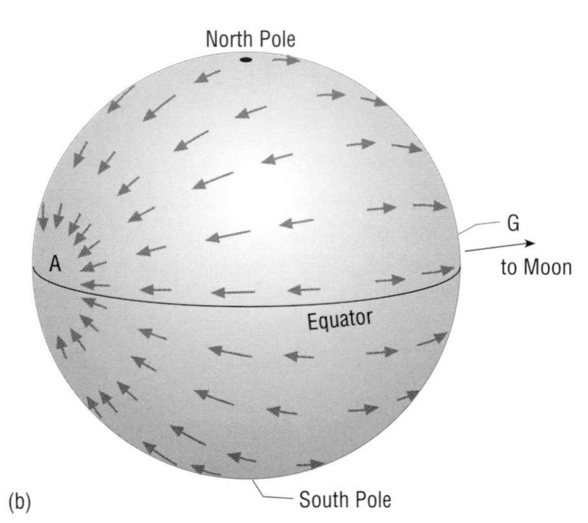

Figure 2.4 (a) The effect of the gravitational force of the Moon at three positions on the Earth. The gravitational force is greatest at G (nearest the Moon) and least at A (furthest from the Moon). At P the gravitational force is less than at G, and can be calculated from Equation 2.3. The tide-producing forces are smallest at A and G, but greatest at P, and all other points on the two small circles. The value for the angle ψ for these circles is 54° 41′. *The circles have nothing to do with latitude and longitude.* For explanation, see text.
(b) The relative magnitudes of the tractive forces (i.e. of the horizontal components of the tide-producing forces, shown as purple arrows on Figure 2.3) at various points on the Earth's surface. The Moon is assumed to be directly over the Equator (i.e. at zero declination, see Section 2.1.1). Points A and G correspond to those on (a) and in Figure 2.3.

In this simplified case, the tractive forces would result in movement of water towards points A and G on Figure 2.4(b). In other words, an equilibrium state would be reached (called the **equilibrium tide**), producing an ellipsoid with its two bulges directed towards and away from the Moon. So, paradoxically, although the tide-producing forces are minimal at A and G, those are the points towards which the water would tend to go. Figure 2.5 shows how such an equilibrium tidal ellipsoid would look in the simplified case we have been considering, i.e. a completely water-covered Earth with the Moon directly above the Equator and the distribution of tractive forces as in Figure 2.4(b).

If you found Figures 2.3 and 2.4 and related text and equations difficult to follow, here is a shorter explanation of why there are two equilibrium tidal bulges (Figure 2.5). The centrifugal force acts in the same direction all over the Earth, i.e. *away from* the Moon (Figure 2.3). Moreover, on the side of the Earth away from the Moon, the gravitational attraction due to the Moon is less than it is on the side of the Earth facing the Moon. The resultant tide-producing force thus acts *away from* the Moon at points such as A on Figure 2.3. That is why there is a tidal bulge away from the Moon as well as a bulge towards it (Figure 2.5). The mathematics of the relationship is such that theoretically the corresponding tide-producing forces on either side of the Earth are equal and opposite.

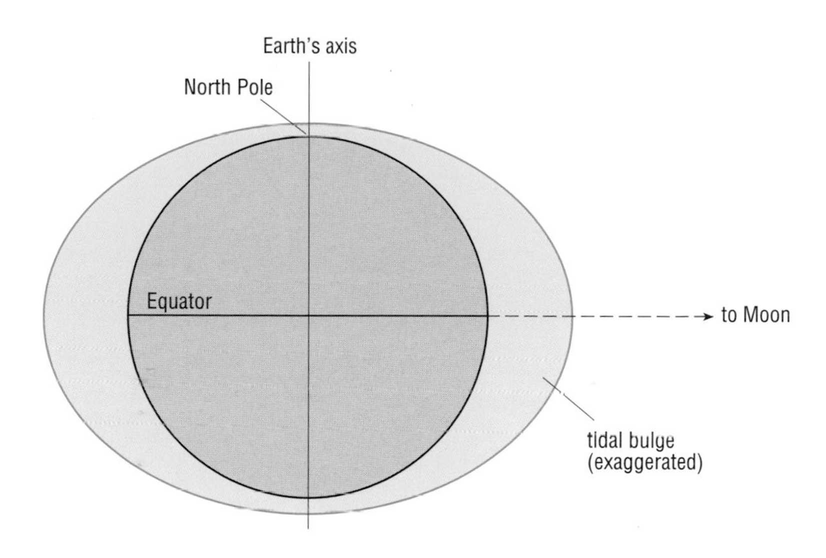

Figure 2.5 The equilibrium tidal ellipsoid (not to scale) as it would appear on a water-covered Earth with the Moon directly above the Equator.

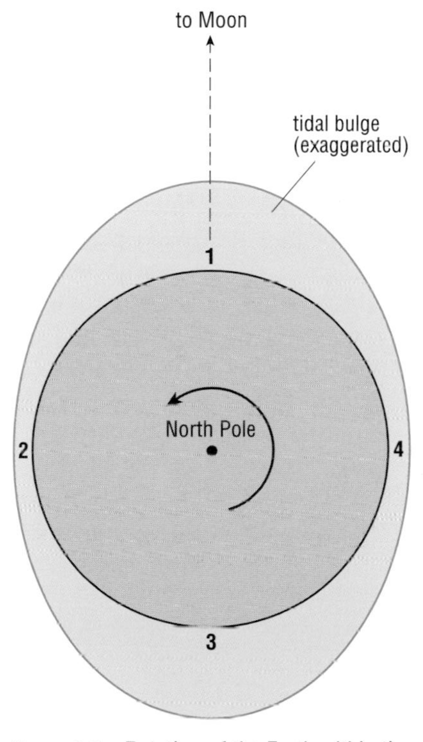

Figure 2.6 Rotation of the Earth within the equilibrium tidal bulge (seen from above the North Pole and not to scale), showing how a point on the Earth's surface would experience two high tides (1 and 3) and two low tides (2 and 4) during each complete rotation of the Earth about its axis.

In practice, the equilibrium ellipsoid does not develop, partly because the Earth is not of course entirely water-covered, but chiefly because the Earth rotates about its own axis. If the two bulges were to maintain their positions relative to the Moon, they would have to travel around the world at the same rate (but in the opposite direction) as the Earth rotates about its axis. Any point on the Earth's surface would thus encounter two high and two low tides during each complete rotation of the Earth (i.e. each day), as illustrated in Figure 2.6.

In fact, Figure 2.6 is an oversimplification. Can you see why (apart from the idealized tidal bulges)?

Figure 2.6 shows both Moon and tidal bulges remaining stationary during a complete rotation of the Earth. That cannot be the case, for the Moon continues to travel in its orbit as the Earth rotates. Because the Moon revolves about the Earth Moon centre of mass once every 27.3 days, in the same direction as the Earth rotates upon its own axis (which is once every 24 hours), the period of the Earth's rotation with respect to the Moon is 24 hours and 50 minutes. This is the **lunar day**.

What effect would this have upon the interval between successive high tides and successive low tides, in Figure 2.6?

The interval between successive high (and low) tides would be about 12 hours 25 minutes – and the interval between high and low tide would be close to 6 hours 12½ minutes. This is the reason why the times of high tides at many locations are almost an hour later each successive day (Figure 2.7, overleaf).

The equilibrium tidal concept also brings out another very important aspect of tidal wave motions.

Looking at Figures 2.5 and 2.6, would you say that tidal waves are more likely to travel as deep- or as shallow-water waves?

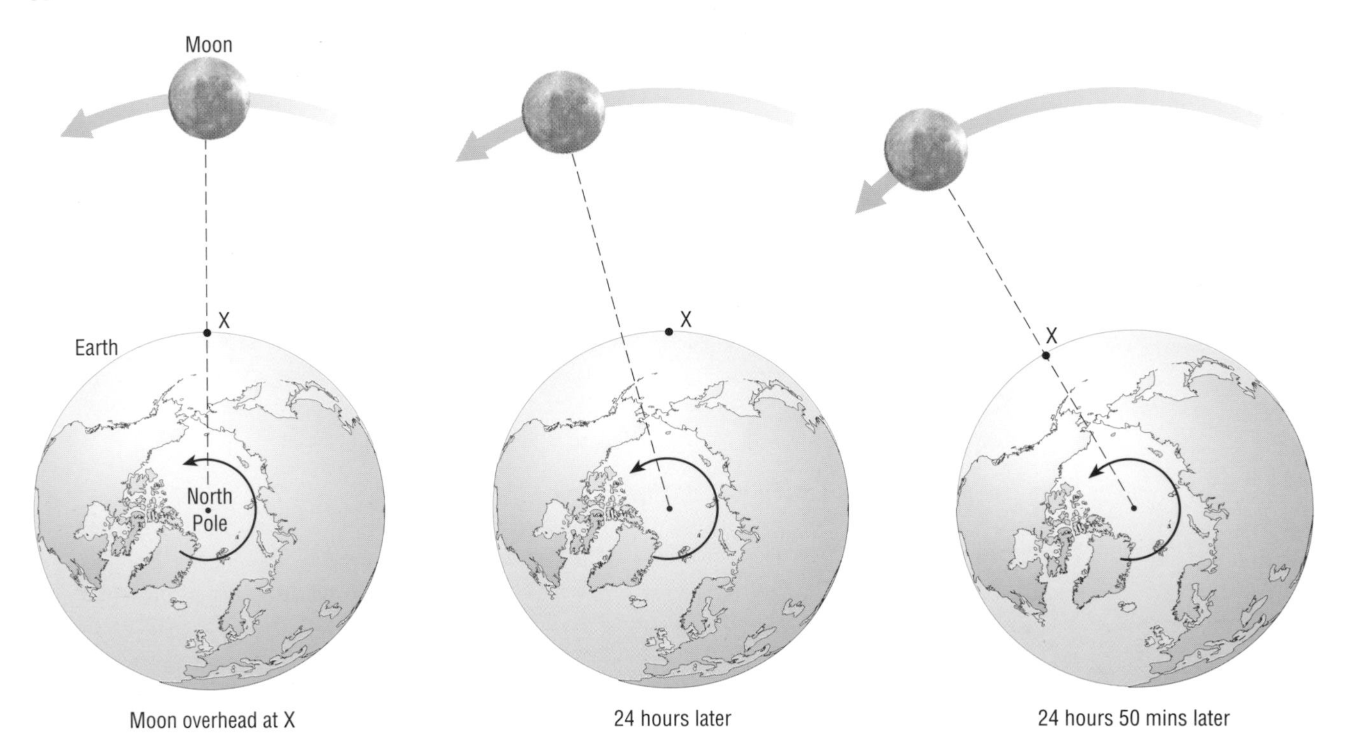

Figure 2.7 The relationship between a solar day of 24 hours and a lunar day of 24 hours and 50 minutes as seen from above the Earth's North Pole. Point X on the Earth's surface when the Moon is directly overhead comes back to its starting position 24 hours later. Meanwhile, the Moon has moved on in its orbit, so that point X has to rotate further (another 50 minutes' worth) before it is once more directly beneath the Moon. (Diagram not to scale.)

There are two 'peaks' (high tide, 1 and 3 in Figure 2.6) and two 'troughs' (low tide, 2 and 4 in Figure 2.6) for one Earth circumference, which is about 40 000 km. So the wavelength of the bulges in Figures 2.5 and 2.6 is of the order of half the Earth's circumference (~20 000 km). Even in the real oceans, tidal wavelengths are many thousands of km, and the average depth of the ocean basins is less than 4 km, i.e. much less than 1/20 of the wavelength (Section 1.2.3). So tidal waves must travel as shallow-water waves, and their speed is governed by Equation 1.4, i.e. the shallower the water the slower they travel. Moreover, just as the height of wind-generated waves increases as they are slowed down on 'feeling' the sea-bed (Section 1.5), so also does tidal range increase as the tidal waves are slowed down over the continental shelf. Tidal ranges are greater and tidal currents are therefore faster in shallow seas and along coasts than in the open oceans (cf. Figures 2.14 and 2.15).

QUESTION 2.2

(a) Using a value of 40 000 km for the Earth's circumference and a period of 24 hr 50 min. (the lunar day), calculate the speed at which the tidal bulges would have to move relative to the Earth's surface along the Equator, in order to 'keep up' with the Moon and so maintain an equilibrium tide. (Assume for simplicity that the Moon is directly overhead at the Equator.)

(b) According to Equation 1.4, how deep would the oceans have to be to allow the tidal bulges to travel as shallow-water waves at the speed you calculated in part (a)?

Your answer to Question 2.2 shows that in practice an equilibrium tide cannot occur at low latitudes on Earth – though it could in principle do so at high latitudes, where distances round the Earth are much less.

The concept of the equilibrium tide was developed by Newton in the seventeenth century, and we have seen that it demonstrates the fundamental periodicity of the tides on a semi-diurnal basis of 12 hours and 25 minutes (Figures 2.6 and 2.7), also that tidal waves must travel as shallow-water waves in the oceans. We can use this concept to explore other aspects of tidal phenomena too, even though the actual tides cannot behave like the equilibrium tide (see Section 2.3) because of the existence of continents.

2.1.1 VARIATIONS IN THE LUNAR-INDUCED TIDES

The relative positions and orientations of the Earth and Moon are not constant, but vary according to a number of interacting cycles. As far as a simple understanding of the tide-generating mechanism is concerned, only two cycles have a significant effect on the lunar tides.

1 The Moon's declination

The Moon's orbit is not in the plane of the Earth's Equator, but is inclined to it (Figure 2.8). This means that a line joining the centre of the Earth to that of the Moon makes an angle ranging from zero up to 28.5° on either side of the equatorial plane (see later text). This angle is the **declination** of the Moon. The result is that, to an observer on Earth, successive paths of the Moon across the sky appear to rise and fall over the 27.3-day period of rotation of the Moon about the Earth (strictly, about the centre of mass of the Earth–Moon system, Figure 2.2), in a similar way to the seasonal variation of the Sun's apparent daily path across the sky over the course of a year (i.e. lower in the sky in winter, higher in the summer, see Figure 2.11).

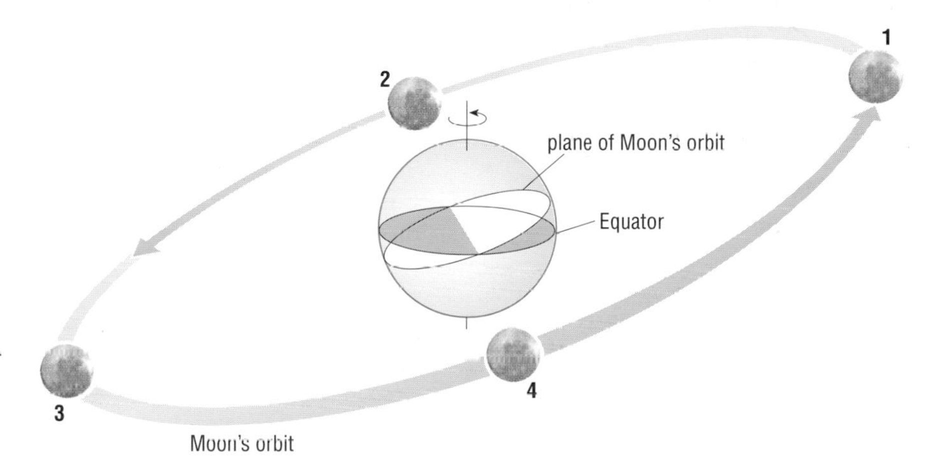

Figure 2.8 Declination of the Moon results from the plane of the Moon's orbit being at an angle to that of the Earth's Equator (shaded). For numbers, see text. Diagram is not to scale, and the Moon's orbit is shown as circular for simplicity (see item 2 overleaf).

At which of the numbered positions of the Moon in its orbit in Figure 2.8 is the declination at a maximum and at which is it zero? What is the time interval between the successive numbered positions on Figure 2.8?

Declination is maximum at positions 1 and 3, and zero at positions 2 and 4 when the Moon is overhead at the Equator. The interval between successive numbered positions in Figure 2.8 is close to seven days (27.3/4). Since the maximum lunar declination is 28.5°, the Moon can never be seen directly overhead poleward of latitude 28.5° N or 28.5° S. So, for example, in southern Britain at about 50° N, the Moon (like the Sun) is always seen in the southern sky. Conversely, in Tasmania for example, at about 40° S, the Moon (like the Sun) is always seen in the northern sky.

When the Moon is at any angle of declination other than zero, the plane of the two tidal bulges will be offset with respect to the Equator, and their effects at a given latitude will be unequal, particularly at mid-latitudes. Hence the heights reached by the semi-diurnal (i.e. twice daily) high tides will show diurnal (i.e. daily) inequalities (Figure 2.9); these will be greatest when the Moon is at maximum declination.

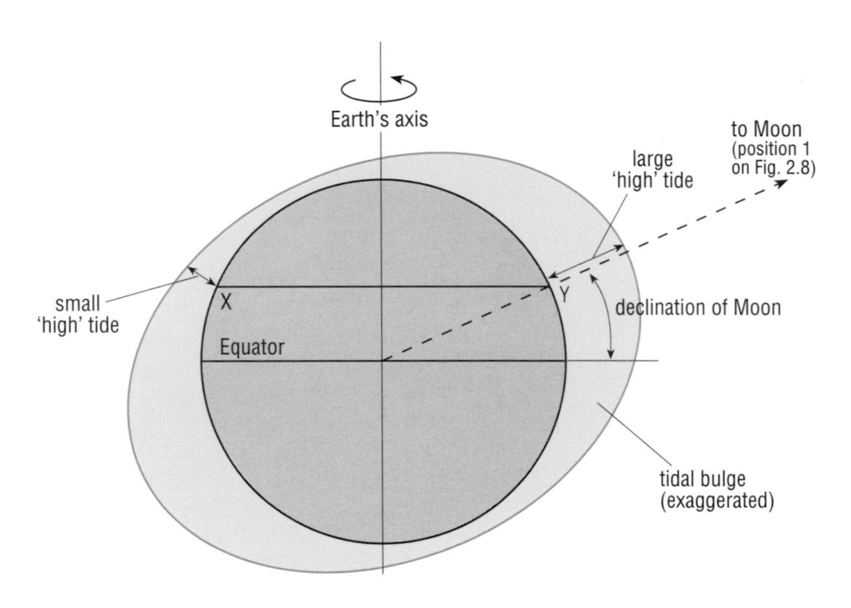

Figure 2.9 The production of unequal tides (tropic tides – see text) at mid-latitudes consequent upon the Moon's declination. An observer at Y will experience a higher high tide than an observer at X; 12 hours and 25 minutes later, their observations will be reversed.

QUESTION 2.3 Assuming that Figure 2.9 shows maximum declination, what will be the extent of the diurnal tidal inequality due to the Moon as seen by a coastal observer at about 28° 30´ south latitude: (a) roughly seven days and (b) roughly 14 days after the situation shown?

a. Nil declination will be zero.
b. The same as in fig 2.9 but on the opposite side of the Earth.

Your answer to Question 2.3 emphasizes the cyclical nature of this diurnal tidal inequality. At maximum declination, the Moon is approximately above one of the Tropics (latitude 23.4° N or S), the diurnal inequality is greatest all over the world, and the tides are known as **tropic tides** (Moon at positions 1 and 3 on Figure 2.8); whereas at minimum (zero) declination (when the Moon is above the Equator), there is no diurnal inequality anywhere in the world and the tides are called **equatorial tides** (Moon at positions 2 and 4 on Figure 2.8).

2 The Moon's elliptical orbit

The orbit of the Moon around the Earth–Moon centre of mass is not circular but elliptical, and the Earth is not at the centre of the ellipse, but at one of the foci (Figure 2.10). The consequent variation in distance from Earth to Moon results in corresponding variations in the tide-producing forces. When the Moon is closest to Earth, it is said to be in **perigee**, and the Moon's tide-producing force is increased by up to 20% above the average value. When the Moon is furthest from Earth, it is said to be in **apogee**, and the tide-producing force is reduced to about 20% below the average value. The difference in the Earth–Moon distance between apogee and perigee is about 13%, and tidal ranges are greater when the Moon is at perigee.

The Moon's elliptical orbit itself *precesses*, i.e. it rotates, as illustrated in Figure 2.10, and it takes 18.6 years to complete a full precessional cycle. In addition, the plane of the lunar orbit makes an angle of 5° with the plane of the Earth's orbit around the Sun (known as *the plane of the ecliptic*). The plane of the Earth's Equator is at an angle of 23.4° to the plane of the ecliptic (see Figure 2.11), so the maximum declination of the Moon ranges from 18.4° (23.4° – 5°) to 28.4° (23.4° + 5°), during the course of the 18.6-year precession cycle. We shall not go into the details of how these additional complications affect the tides other than to observe briefly that:

1 The 18.6-year period of the precession cycle can be identified in long-term tidal records.

2 The combined effects of the relationships depicted in Figures 2.2 and 2.10 cause small variations in the Earth–Moon period of 27.3 days (Figure 2.2), as well as in the declination cycle (27.2 days) and the perigee–apogee–perigee cycle (27.5 days).

3 The elliptical form of the lunar orbit causes the Moon to travel faster at perigee than at apogee, leading to variations in tidal cycles round the average of 12 hours 25 minutes (or 24 hours 50 minutes).

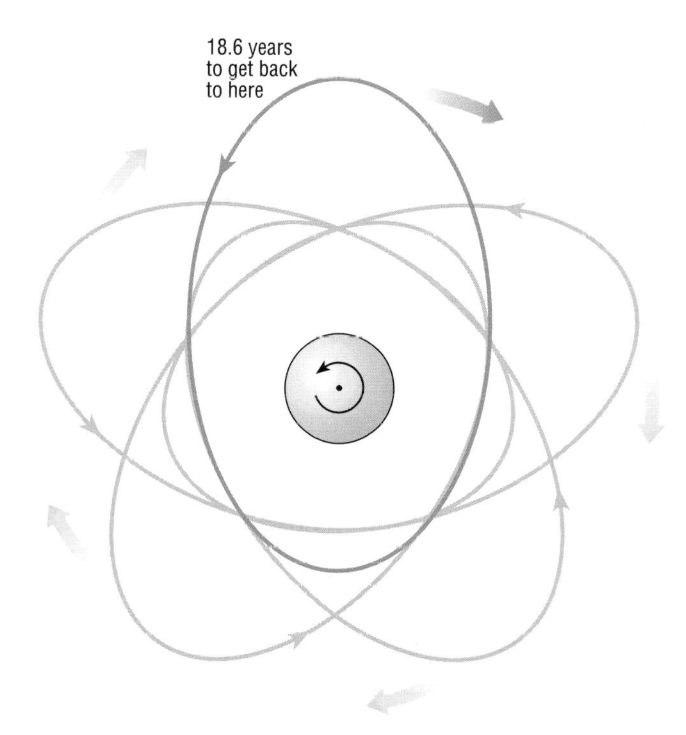

Figure 2.10 Successive positions of the lunar orbit in the 18.6-year precession cycle seen from above the North Pole (not to scale). Note that the sense of rotation of the orbit is *opposite* to that of both the Earth about its axis and of the Moon about the Earth.

We now turn to consideration of the way in which the Sun influences the tides.

2.2 TIDE-PRODUCING FORCES – THE EARTH–SUN SYSTEM

Like the Moon, the Sun also produces tractive forces and two equilibrium tidal bulges. Although enormously greater in mass than the Moon, the Sun is some 360 times further from the Earth, so the magnitude of its tide-producing force is about 0.46 that of the Moon. As we saw in Section 2.1, tide-producing forces vary directly with the mass of the attracting body, but are inversely proportional to the cube of its distance from Earth. The two solar equilibrium tides produced by the Sun sweep westwards around the globe as the Earth spins towards the east. The solar tide thus has a semi-diurnal period of twelve hours.

Just as the relative heights of the two semi-diurnal lunar tides are influenced by the Moon's declination, so there are diurnal inequalities in the solar-induced components of the tides because of the Sun's declination.

The Sun's declination varies over the seasonal yearly cycle, and ranges up to 23.4° either side of the equatorial plane. This angle of 23.4° is the angle between the plane of the Earth's Equator and the plane of the ecliptic (Section 2.1.1) and is therefore also the tilt of the Earth's axis (Figure 2.11).

As in the case of the Moon's orbit round the Earth, the orbit of the Earth around the Sun is elliptical. When the distance between Earth and Sun is at a minimum the Earth is said to be at **perihelion**; when it is at a maximum, the Earth is said to be at **aphelion**. However, the difference in Earth–Sun distance between perihelion and aphelion is only about 4%, compared with an approximate 13% difference in Earth–Moon distance between lunar perigee and apogee. Characteristics of the Earth's orbit round the Sun change cyclically over periods of tens of thousands of years, and these will of course affect the tides, but not on time-scales which concern us for the purposes of this Volume.

QUESTION 2.4 According to Figure 2.11, at what time(s) of the year will the solar-induced component of the tide be at its strongest?

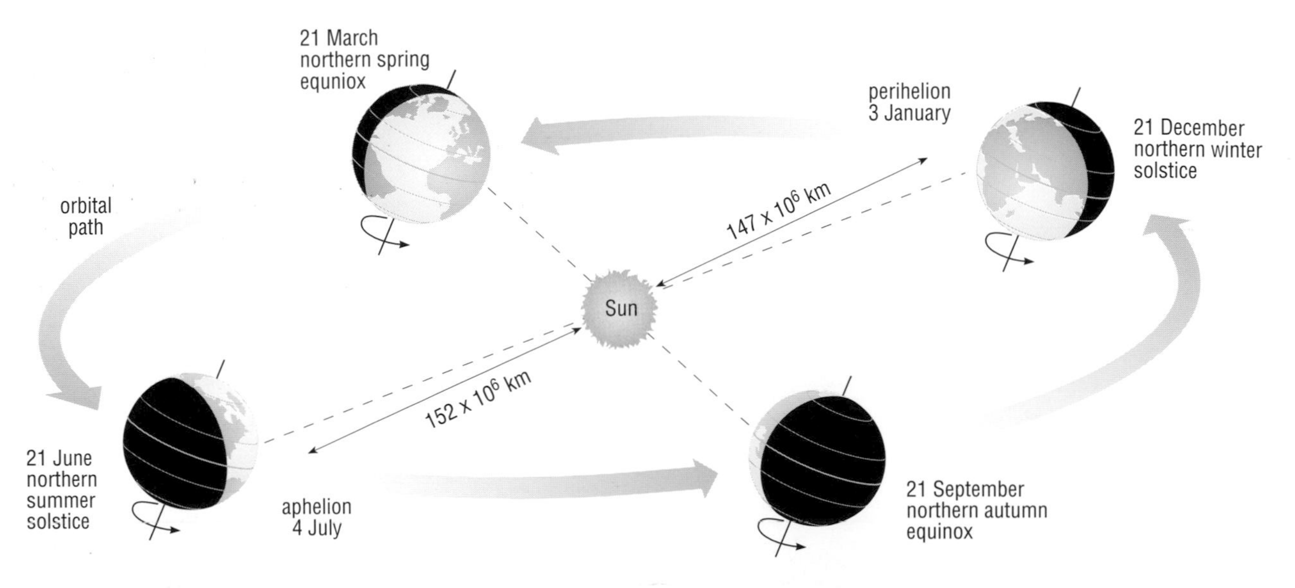

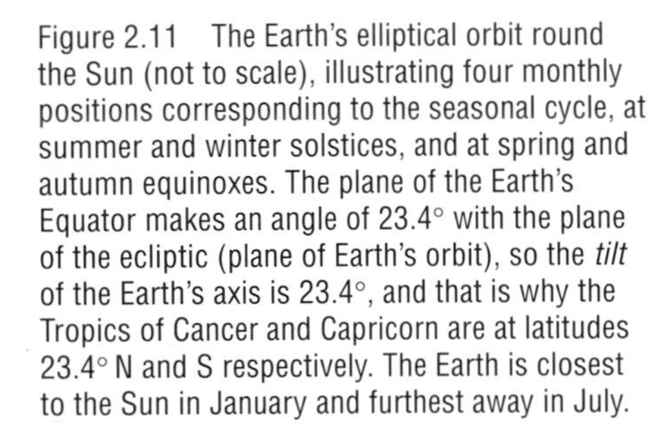

Figure 2.11 The Earth's elliptical orbit round the Sun (not to scale), illustrating four monthly positions corresponding to the seasonal cycle, at summer and winter solstices, and at spring and autumn equinoxes. The plane of the Earth's Equator makes an angle of 23.4° with the plane of the ecliptic (plane of Earth's orbit), so the *tilt* of the Earth's axis is 23.4°, and that is why the Tropics of Cancer and Capricorn are at latitudes 23.4° N and S respectively. The Earth is closest to the Sun in January and furthest away in July.

2.2.1 INTERACTION OF SOLAR AND LUNAR TIDES

In order to understand the interaction between solar and lunar tides, it is helpful to consider the simplest case, where the declinations of the Sun and Moon are both zero. Figure 2.12 (overleaf) shows these conditions, looking down on the Earth from above the North Pole. In Figure 2.12(a) and (c), the tide-generating forces of the Sun and Moon are acting in the same directions, and the solar and lunar equilibrium tides coincide, i.e. they are in phase, so that they reinforce each other. The tidal range produced is larger than the average, i.e. the high tide is higher and the low tide is lower. Such tides are known as **spring tides**. When spring tides occur, the Sun and Moon are said to be either in *conjunction* (at new Moon – Figure 2.12(a)) or in *opposition* (at full Moon – Figure 2.12(c)). There is a collective term for both situations: the Moon is said to be in **syzygy** (pronounced 'sizzijee').

In Figure 2.12(b) and (d), the Sun and Moon act at right angles to each other, the solar and lunar tides are out of phase, and do not reinforce each other. The tidal range is correspondingly smaller than average. These tides are known as **neap tides**, and the Moon is said to be in **quadrature** when neap tides occur. Inshore fishermen sometimes refer to spring and neap tides by the descriptive names of 'long' and 'short' tides respectively.

The complete cycle of events in Figure 2.12 takes 29.5 days and the reason why this cycle is different from the Earth–Moon rotation period of 27.3 days (Figure 2.2) can be seen by reference to Figure 2.13(a) (which is analogous to Figure 2.7). It is simply that in the 27.3 days taken by the Moon to make a complete orbit of the Earth, the Earth–Moon system has also been orbiting the Sun. For the Moon to return to the same position relative to *both* Earth and Sun, it must move further round in its orbit, and that takes an extra 2.2 days or so.

Figure 2.13(b) is a summary diagram of the combined motions of Earth and Moon about the Sun. It shows how both the Moon and the centre of the Earth trace out undulating paths as they themselves rotate about their common centre (the centre of mass of the Earth–Moon system, Figure 2.2). The diagram also illustrates the 29.5-day spring–neap cycle of Figure 2.12, a period sometimes called the *synodic month* but more commonly known as the **lunar month** (i.e. the period between successive new Moons). The 27.3-day period of rotation of the Moon about the Earth–Moon centre of mass is known as the *sidereal month*.

QUESTION 2.5

(a) What is the time interval between two successive neap tides?

(b) What is the state of the tide 22 days after the Moon is in syzygy?

(c) How soon after the new Moon might a tide of 'average' range be expected?

(d) Figure 2.12 illustrates the simplest case of zero declination for both Sun and Moon. Bearing this in mind, what astronomical phenomena would be observed on the Earth's Equator if the Sun, Moon and Earth were in the positions shown in Figure 2.12(a) and (c) respectively?

It is crucial to realize from Figure 2.12 that spring and neap tides must each occur at about the same time *all over the world*, because the Earth rotates *within* the tidal bulges (cf. Figure 2.6), which themselves move only in response to orbital motions of Moon and Earth. For the same reason, the tropic and equatorial tides (Section 2.1.1) must also occur at the same times.

New Moon

Earth

solar tide

lunar tide

Moon

Sun

(a)

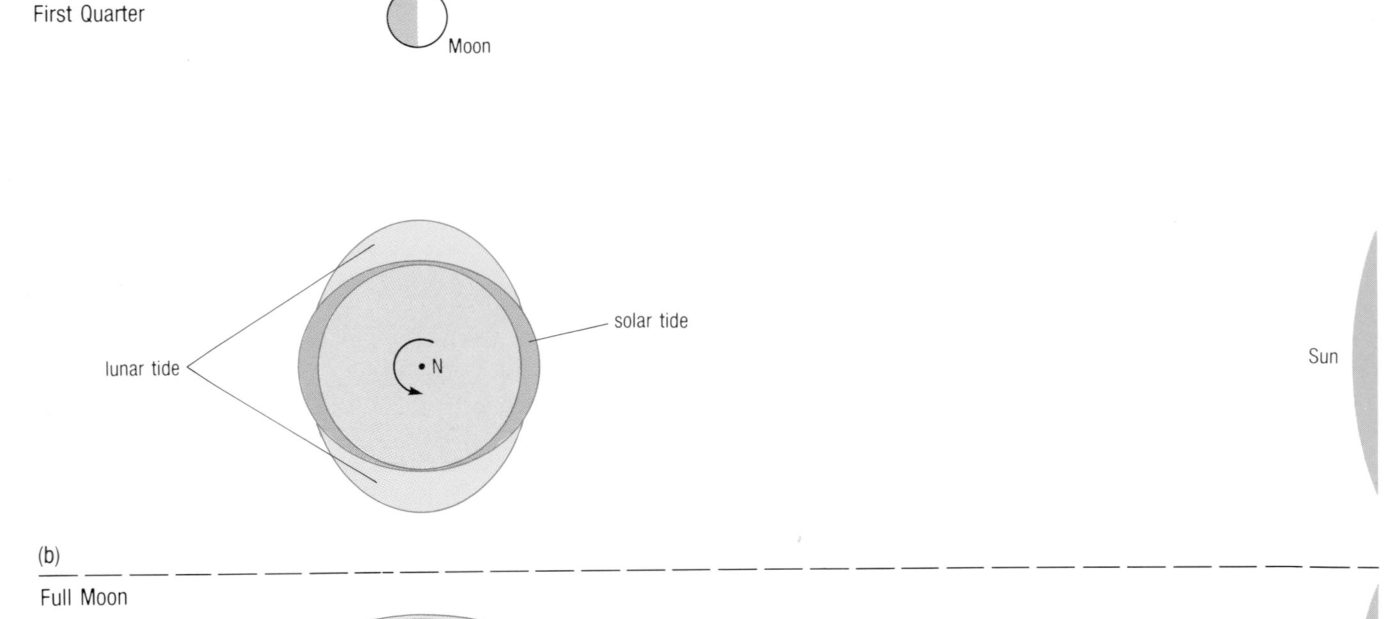

Full Moon

Moon

N

Sun

(c)

Third (or last)
Quarter

N

Sun

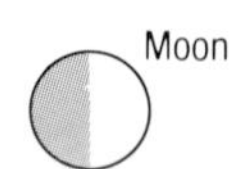

(d)

Figure 2.12 Diagrammatic representation (not to scale) of the interaction of the solar and lunar tides, as seen from above the Earth's North Pole, showing direction of rotation of the Earth (arrowed) and the tidal bulges caused by the Moon and the Sun.
(a) New Moon. Moon in syzygy (Sun and Moon in *conjunction,* i.e. positioned above the same line of Earth's longitude). Spring tide.
(b) First quarter. Moon in quadrature (overhead positions of Sun and Moon separated by 90° of Earth's longitude). Neap tide.
(c) Full Moon. Moon in syzygy (Sun and Moon in *opposition*, i.e. overhead positions separated by 180° of Earth's longitude). Spring tide.
(d) Third (or last) quarter. Moon again in quadrature (see (b)). Neap tide.

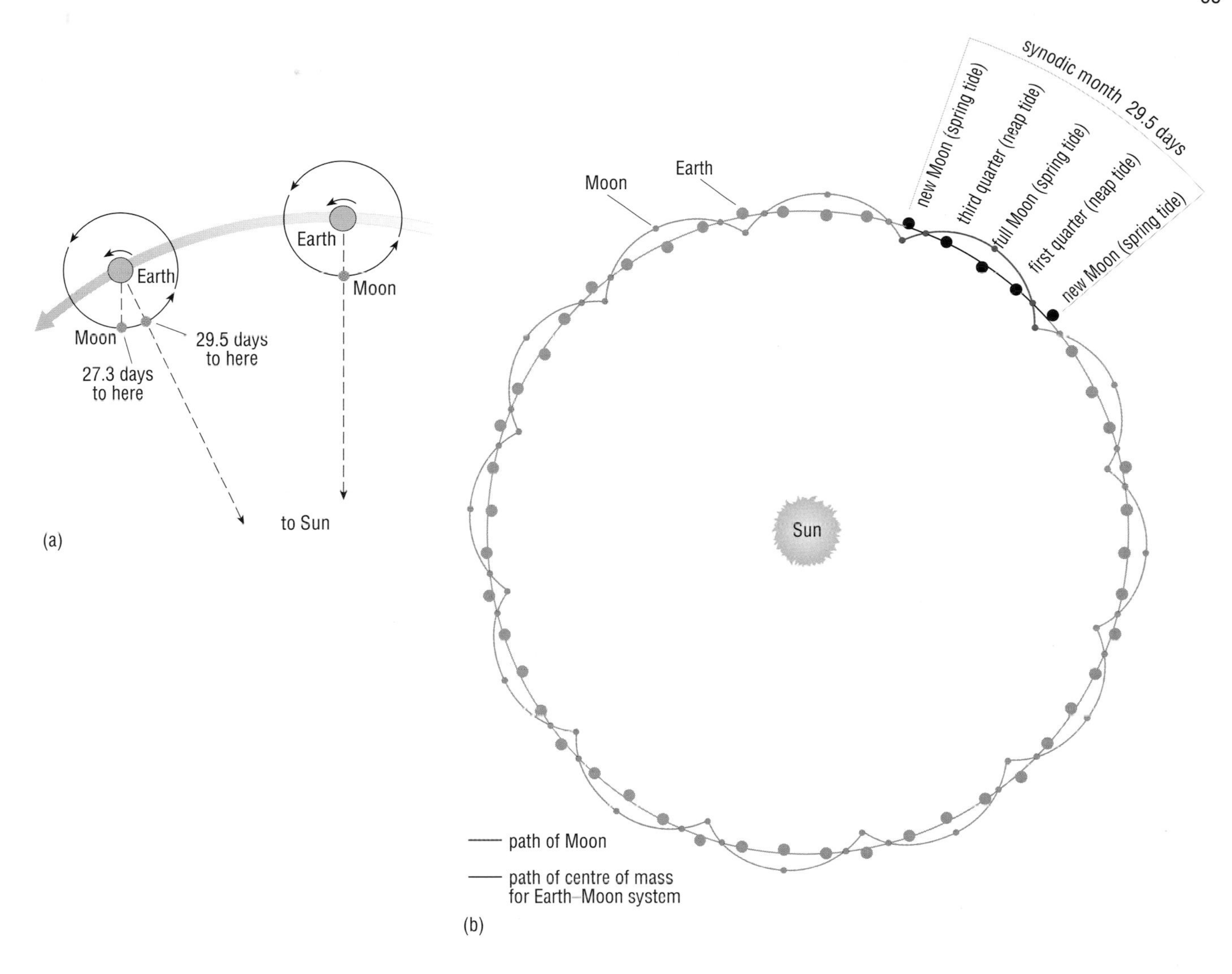

Figure 2.13 (a) Diagram (not to scale), illustrating how the Moon (shown here in conjunction) must travel further round its orbit to return to the same position relative to both Earth and Sun, because the Earth has also moved in its orbit relative to the Sun. For simplicity both orbits are portrayed as circular.
(b) Summary diagram (not to scale) of combined motions of Earth and Moon about the Sun and about the centre of mass of the Earth–Moon system. For simplicity, orbits are assumed to be circular. *Note:* synodic month = lunar month (see p. 63).

The regular changes in the declinations of the Sun and Moon, and their cyclical variations in position with respect to the Earth, produce very many harmonic constituents, each of which contributes to the tide at any particular time and place. One interesting situation is the 'highest astronomical tide', i.e. that which would create the greatest possible tide-producing force, with the Earth at perihelion, the Moon in perigee, the Sun and Moon in conjunction and both Sun and Moon at zero declination. Such a rare combination would produce tidal ranges greater than normal, all over the world. For example, at Newlyn, Cornwall, the normal tidal range is about 3.5 m, the mean spring tidal range about 5 m, and the highest astronomical tidal range about 6 m. However, there is no immediate need to sell any seaside property which you may own – the next such event is not due until about AD 6580.

2.3 THE DYNAMIC THEORY OF TIDES

When Newton formulated the equilibrium theory of tides in the seventeenth century, he was conscious that it was only a static treatment of the problem and thus only a rough approximation. He was well aware of discrepancies between the predicted equilibrium tides and the observed tides, but did not pursue the matter any further. The equilibrium theory is of limited practical value, even though certain of its predictions are correct, notably that spring and neap tides will occur at new and full Moon (Figure 2.12), that the range of spring tides will typically be two or three times that of neap tides (cf. Figure 2.1), and that tidal inequality is related to declination (Figure 2.9).

There are a number of reasons why actual tides do not behave as equilibrium tides.

1 As discussed earlier, the wavelength of tidal waves is long relative to depth in the oceans, so *they travel as shallow-water waves* (Section 2.1) and as we have seen (Question 2.2) their speed is governed by $c = \sqrt{gd}$ (Equation 1.4). The speed of any wave longer than a few km is therefore limited to about 230 m s^{-1} in the open ocean, less in shallower seas. This is much slower than the linear velocity of the surface of the rotating Earth with respect to the Moon: 448 m s^{-1} at the Equator (Question 2.2(a)). (In fact, this linear velocity decreases with distance from the Equator: to *c*. 230 m s^{-1} at about latitudes 60° N and 60° S, to 78 m s^{-1} at 80° latitude, and zero at the poles themselves.)

2 In any case, the Earth rotates on its axis far too rapidly for either the inertia of the water masses or the frictional forces at the sea-bed to be overcome fast enough for an equilibrium tide to occur. A time-lag in the oceans' response to the tractive forces is thus inevitable, i.e. there is a tidal *lag*, such that high tide commonly arrives some hours after the passage of the Moon overhead. Because the linear velocity of the surface of the Earth with respect to the Moon decreases polewards (cf. (1) above), the tidal lag is greatest at low latitudes (*c*. 6 hours), decreasing to zero at about latitude 65° – but the precise lag is always constant for a particular location. In addition, at most localities, spring tides occur a day or two after both full and new Moon (cf. Figure 2.12), and the time difference (in days) between the meridian (overhead) passage of full or new Moon and the occurrence of the highest spring high tide is sometimes called the *age of the tide*.

3 The presence of land masses prevents the tidal bulges from directly circumnavigating the globe, and the shape of the ocean basins constrains the direction of tidal flows. In fact, the only region of the oceans where a westward-moving tidal bulge could travel unimpeded around the world is the Southern Ocean surrounding Antarctica.

4 Except at the Equator, all lateral (horizontal) water movements (including tidal currents) are subject to the **Coriolis force**, which deflects winds and currents *cum sole* (literally 'with the Sun'), i.e. to the right, or clockwise, in the Northern Hemisphere, and to the left, or anticlockwise, in the Southern Hemisphere.

The **dynamic theory of tides** was developed during the eighteenth century by scientists and mathematicians such as Bernoulli, Euler and Laplace. They attempted to understand tides by considering ways in which the depths and configurations of the ocean basins, the Coriolis force, inertia, and frictional forces might influence the behaviour of fluids subjected to rhythmic forces resulting from the orbital relationships of Earth, Moon and Sun.

As a consequence of the many and varied factors involved, the dynamic theory of tides is intricate, and solutions of the equations are complex. Nevertheless, the dynamic theory has been steadily refined, and computed theoretical tides are very close approximations to the observed tides.

The combined constraint of ocean basin geometry and the influence of the Coriolis force (items 2 and 4 on p. 66) results in the development of **amphidromic systems**, in each of which the crest of the tidal wave at high water circulates around an **amphidromic point** once during each tidal period (Figures 2.14 and 2.15 overleaf). The tidal range is zero at each amphidromic point, and increases outwards away from it.

In each amphidromic system, **co-tidal lines** can be defined, which link all the points where the tide is at the same stage (or phase) of its cycle. The successive co-tidal lines radiating outwards from the amphidromic point thus indicate the passage of the tidal wave crest around it.

Cutting across co-tidal lines, approximately at right angles to them, are **co-range lines**, which join places having the same tidal range. Co-range lines form more-or-less concentric circles about the amphidromic point, representing larger and larger tidal ranges the further away they are from it. Figure 2.14 shows the amphidromic systems for the North Sea, and Figure 2.15 shows the computed world-wide amphidromic systems for the dominant tidal component resulting from the diurnal influence of the Moon (see also Section 2.3.1).

QUESTION 2.6

(a) Assume that a high tide coincides with the co-tidal lines marked zero (i.e. '0') on Figure 2.14. At what stage of the tidal cycle is:

1 The Wash?

2 The Firth of Forth?

(b) Which of (1) and (2) has the greater tidal range?

Inspection of Figures 2.14 and 2.15 shows that, with a few exceptions, the tidal waves of amphidromic systems tend to rotate anticlockwise in the Northern Hemisphere and clockwise in the Southern Hemisphere. At first sight, this pattern of rotation appears to conflict with the principle that the Coriolis force deflects moving fluid masses *cum sole*, but we need to bear in mind that the direction of motion of tidal *waves* is not synonymous with the movement of individual parcels of water.

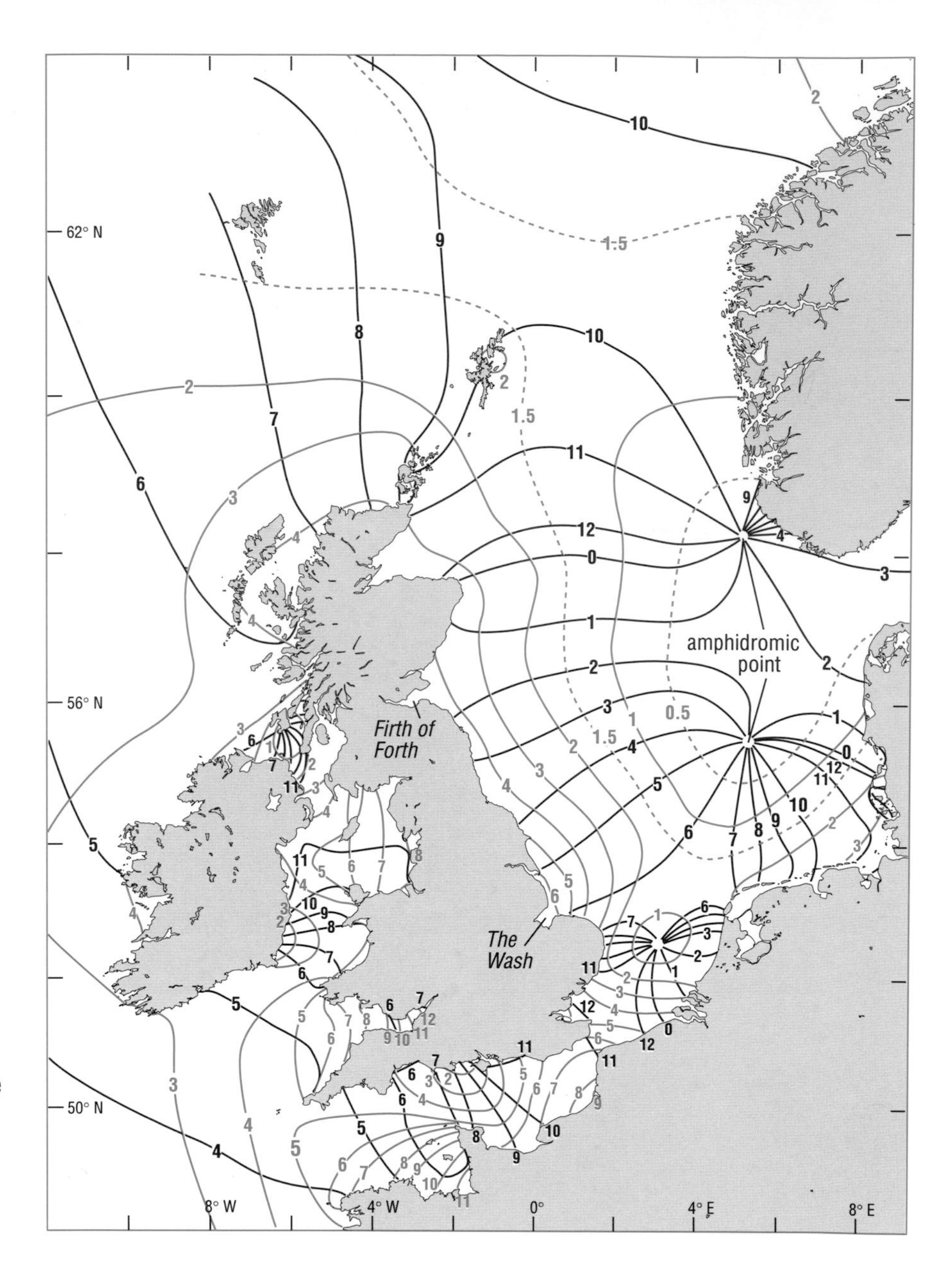

Figure 2.14 Amphidromic systems around the British Isles. The figures on the co-tidal lines (red) indicate the time of high water (in hours) after the Moon has passed the Greenwich meridian. Blue lines are co-range lines, with tidal range in metres.

Consider the enclosed basin shown in Figure 2.16 (on p. 70). The 'bent' arrows in Figure 2.16(a) show how water moving in response to the flooding tide, i.e. in the tidal currents, is deflected to the right by the Coriolis force (the basin is in the Northern Hemisphere), and the water is piled up on the eastern side. Conversely, when the tide ebbs, the water becomes piled up on the western side (Figure 2.16(b)). Hence, because the tidal wave is constrained by land masses, an *anticlockwise* amphidromic system is set up (Figure 2.16(c) and (d)).

It is also very important to remember that tidal waves behave as shallow-water waves, so their orbital motions are flattened like those in Figure 1.8(d). Tidal currents are the horizontal water movements that accompany the rise and fall of the tides as the tidal wave *form* rotates about the amphidromic point, and of course tidal currents change direction during the tidal cycle (see Section 2.4.1).

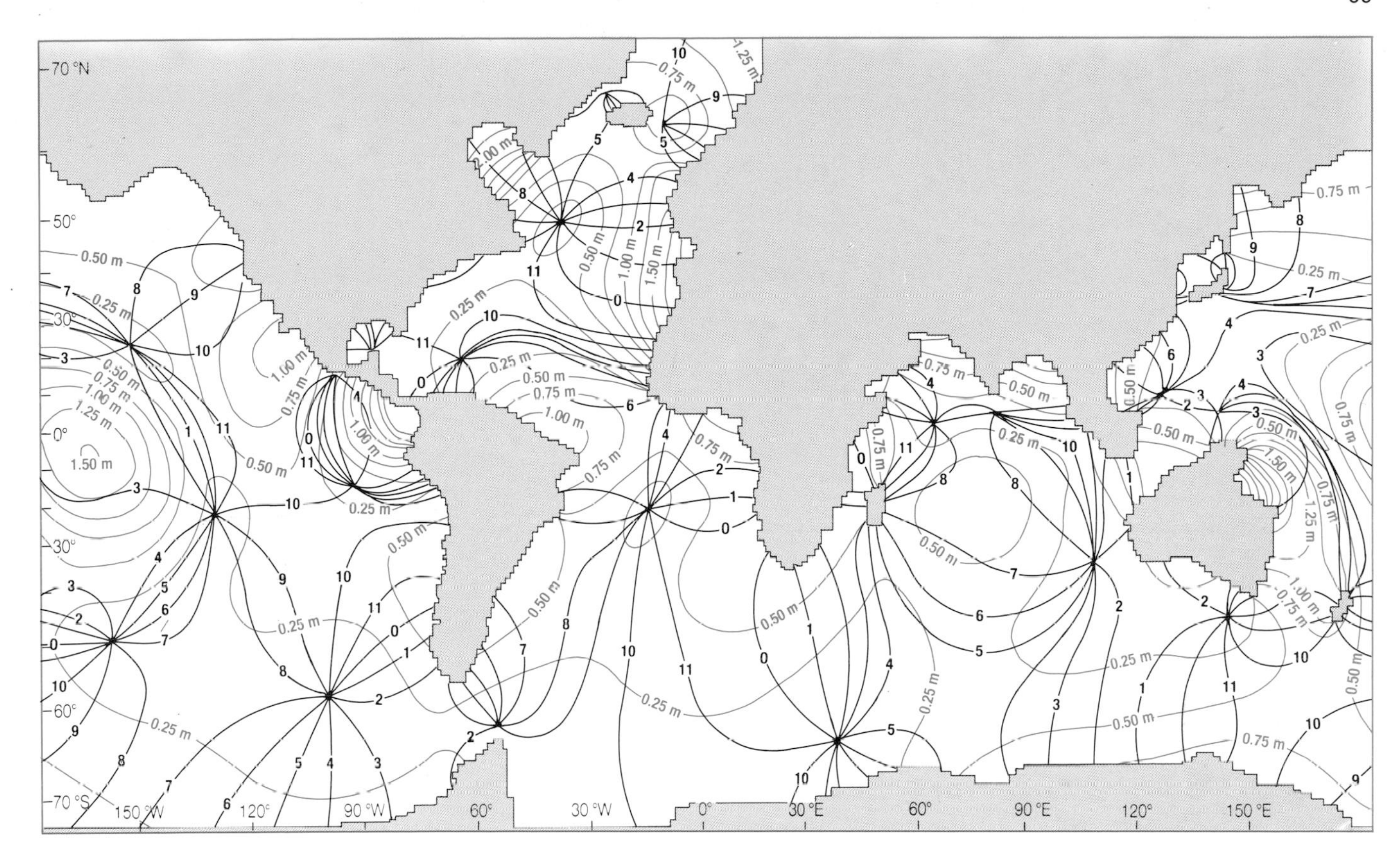

Figure 2.15 Computer-generated diagram of world-wide amphidromic systems for the dominant semi-diurnal lunar tidal component M_2 (see Table 2.1 on p. 71). Blue lines are co-range lines and red lines are co-tidal lines.

The main exceptions to the general pattern of rotation of tidal waves round amphidromic points shown on Figure 2.15 are amphidromic systems less obviously constrained by land masses, e.g. in the South Atlantic (centred on 20° S, 15° W), mid-Pacific (centred on 20° S, 130° W), and North Pacific (centred on 25° N, 155° W); or in certain cases where the amphidromic system rotates about an island, e.g. Madagascar.

QUESTION 2.7

(a) Locate on Figure 2.15 the amphidromic systems identified above, and state how they are exceptions to the general pattern.

(b) Locate the amphidromic system centred near 65° E, 5° N in the north-west Indian Ocean. In what way is this also anomalous?

a. They go the opposite way to the theory.

b, it's clockwise in the Northern Hemisphere.

Tidal waves in amphidromic systems are a type of **Kelvin wave**, in which the amplitude is greatest near coasts (Figure 2.16). Kelvin waves occur where the deflection caused by the Coriolis force is either constrained (as at coasts) or is zero (as at the Equator).

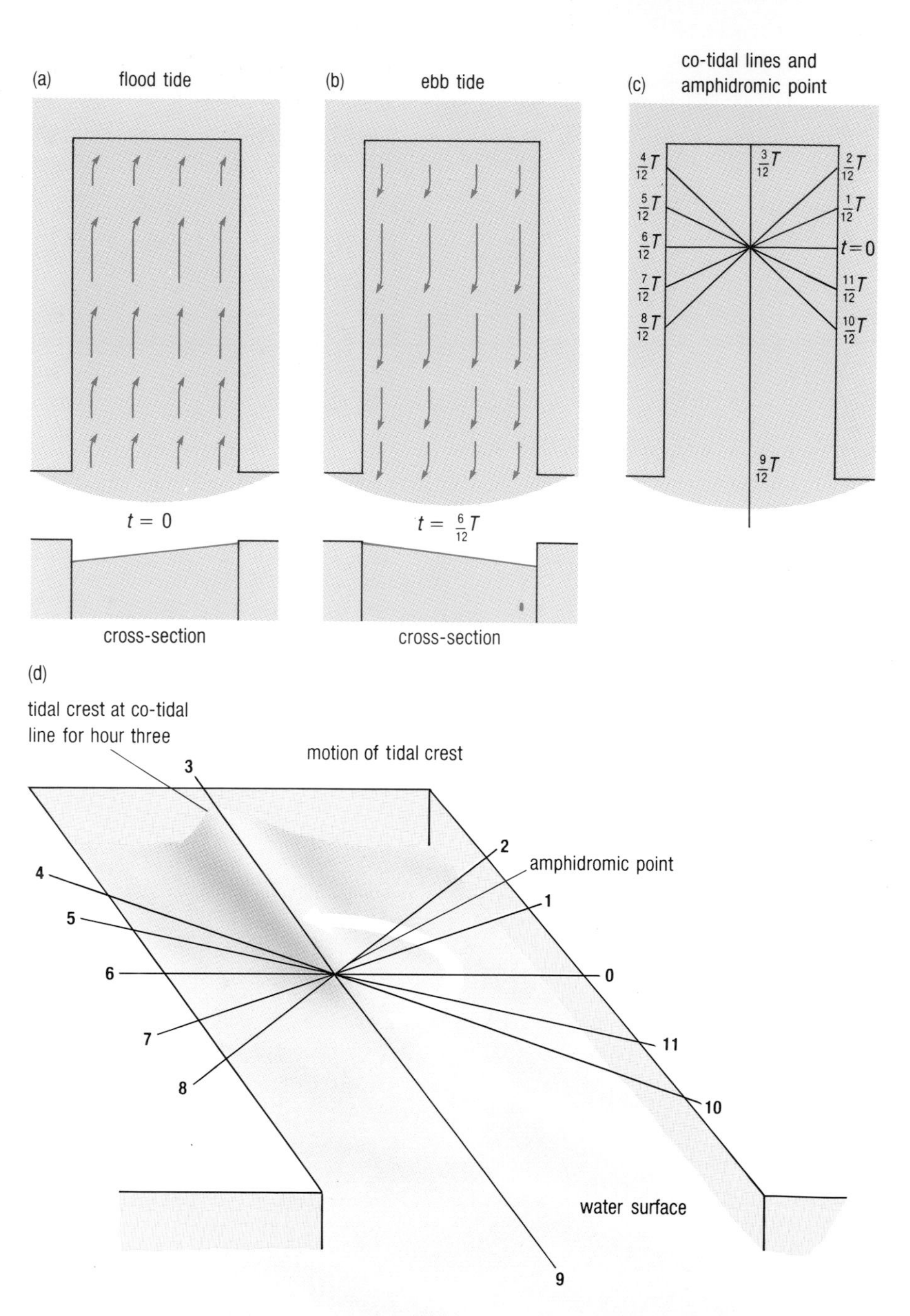

Figure 2.16 The development of an amphidromic system. The hypothetical basin shown is in the Northern Hemisphere.
Note: In (a) and (b), the 'bent' arrows show how the Coriolis force deflects the moving water, 'piling it up' against the sides of the basin.
(a) Flood tide. Water is deflected by the Coriolis force to the right, i.e. towards the east.
(b) Ebb tide. Returning water is deflected by the Coriolis force to the right, i.e. towards the west.
(c) An anticlockwise amphidromic system is established. Times, t, are in twelfth parts of the tidal period, T (= 12 hr 25 min.).
(d) The tidal wave travels anticlockwise. Numbers on co-tidal lines correspond to values of t in (c).

2.3.1 PREDICTION OF TIDES BY THE HARMONIC METHOD

The harmonic method is the practical application of the dynamic theory of tides and is the most usual and satisfactory method for the prediction of tidal heights. It makes use of the knowledge that the observed tide is the sum of a number of harmonic constituents or **partial tides**, each of whose periods precisely corresponds with the period of some component of the relative astronomical motions between Earth, Sun and Moon. For any coastal location, each partial tide has a particular amplitude and phase. In this context, phase means the fraction of the partial tidal cycle that has been completed at a given reference time. It depends upon the period of the tide-producing force concerned, and upon the lag (Section 2.3) of the partial tide for that particular location.

The basic concept is analogous to that illustrated in Figure 1.9, though with a great many more component wave motions (partial tides). The wave form that represents the *actual* tide at a particular place (e.g. Figure 2.1) is the resultant or sum of all of the *partial* tides at that place. An example using just two partial tides is illustrated in Figure 2.17: the combination of a diurnal and a semi-diurnal component produces two unequal high tides (H and h) and two unequal low tides (L and l) each day, and the time interval between the higher low tide (l) and the lower high tide (h) is significantly shorter than that between H and l or L and h. Tides like these, characterized by high and low tides of unequal height, are known as *mixed tides*, and are common, for example, along the Pacific coast of North America (see also Figure 2.18).

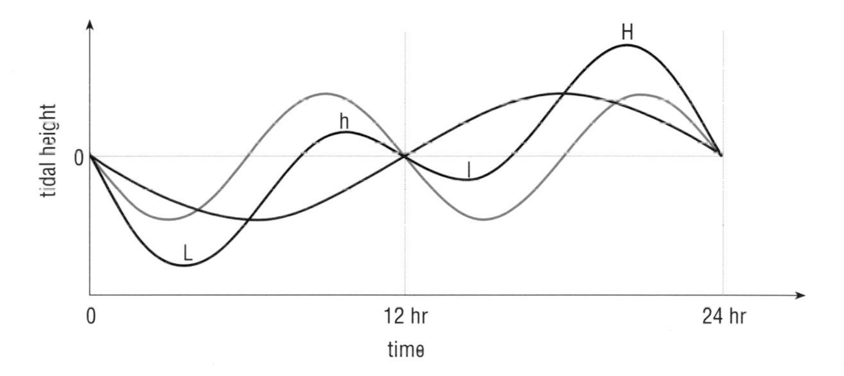

Figure 2.17 Mixed tide (purple) produced by the combination of a diurnal (red) and a semi-diurnal (light blue) partial tidal constituent. H and h = high tides; L and l = low tides. For simplicity, the semi-diurnal period is shown as 12 hr, whereas the M_2 period is 12 hr 25 min.

In order to make accurate tidal predictions for a location such as a seaport, the amplitude and phase for each partial tide that contributes to the actual tide must first be determined from analysis of the observed tides. This requires a record of measured tidal heights obtained over a time that is long compared with the periods of the partial tides concerned. As many as 390 harmonic constituents have been identified. Table 2.1 shows the nine most important of these: four semi-diurnal, three diurnal and two longer-period constituents.

Table 2.1 Some principal tidal constituents. The coefficient ratio (column 4) is the ratio of the amplitude of the tidal component to that of M_2.

Name of tidal component	Symbol	Period in solar hours	Coefficient ratio ($M_2 = 100$)
Semi-diurnal:			
Principal lunar	M_2	12.42	100
Principal solar	S_2	12.00	46.6
Larger lunar elliptic	N_2	12.66	19.2
Luni-solar	K_2	11.97	12.7
Diurnal:			
Luni-solar	K_1	23.93	58.4
Principal lunar	O_1	25.82	41.5
Principal solar	P_1	24.07	19.4
Longer period:			
Lunar fortnightly	M_f	327.86	17.2
Lunar monthly	M_m	661.30	9.1

Semi-diurnal partial tides result from tide-producing forces that are symmetrically distributed over the Earth's surface with respect to the Sun and Moon, as illustrated in Figures 2.3 and 2.4. M_2 and S_2 are the most important ones, because they control the spring–neap cycle (Figure 2.12). The last column of Table 2.1 shows that S_2 has only 46.6 per cent of the amplitude of M_2, because although the Sun is much more massive than the Moon it is also much further away (Section 2.2).

Diurnal tides are principally a consequence of lunar and solar declination (Figures 2.8 and 2.11), and relate to the diurnal inequalities described on p.60. Thus, on Figure 2.9 there is a high diurnal tide at point Y and a low diurnal tide at point X. Half a day later, the diurnal tide will be high at X and low at Y, as the Earth rotates within the equilibrium tidal bulge. The tidal range at both X and Y will be lowest when the Moon is at zero declination. However, at locations where the semi-diurnal influence is minimal (p.73), only diurnal tides occur (see Figure 2.18(a), p.74) and tidal ranges are smallest when lunar declination is zero.*

Before moving on, it is worth noting some regularities among other constituents in Table 2.1. The luni-solar diurnal partial tide, K_1, has twice the period of its semi-diurnal counterpart K_2, but has much greater amplitude, while the average of K_1 and P_1 is exactly 24 hours. Small departures of the periods of some semi-diurnal and diurnal constituents (e.g. N_2, P_1) from a simple relationship with those of M_2 and S_2 result mainly from complications related to the orbits of Moon and Earth (Figures 2.10 and 2.11). With regard to the longer cycles listed in Table 2.1, the lunar fortnightly period (M_f) works out to 13.66 days, almost exactly half the 27.3-day period of the Moon's rotation about the Earth–Moon centre of mass; while the lunar monthly period (M_m) is very close to the perigee–apogee cycle of 27.5 days mentioned in relation to Figure 2.10. There are of course still longer cycles, an obvious example being the 18.6-year period related to precession of the lunar orbit (Figure 2.10); and there are shorter-period constituents as well (see Section 2.4).

Even using the few major constituents in Table 2.1, analysis of tidal records and production of tide-tables for a port for an entire year used to be a very time-consuming activity. In the early years of harmonic analysis, they were computed by hand. The first machine to do the job was invented by Lord Kelvin in 1872. Electronic computers are admirably suited to this repetitive procedure, and tide-tables for individual ports all over the world now take little time to prepare.

The precision achieved by radar altimeters (Section 1.7.1) is such that tidal ranges in the deep oceans can be determined using information on tidal amplitude and phase extracted from the satellite data. Results are in good agreement with predicted values, and are nowadays supplemented by tidal data from the deep-sea pressure gauges mentioned in Section 1.7, placed at strategic locations in the oceans, far from land.

* The period between high and low 'lunar' diurnal tides is much shorter than that between high and low 'solar' diurnal tides. That is because the Moon's declination changes from zero to maximum and back every two weeks or so (Figure 2.8), whereas the same change in solar declination takes about six months (Figure 2.11).

2.4 REAL TIDES

Having examined the theory, let us see how the actual tides behave in different places. Every partial tide has its own set of amphidromic systems, and their amphidromic points do not necessarily coincide.

Suppose you were at a coastal location close to the amphidromic points of both S_2 and M_2, but far from those of O_1 and K_1. Would you expect the tidal period to be predominantly diurnal or semi-diurnal?

It would be predominantly diurnal. The tidal range increases with distance from the amphidromic point (Figure 2.16), so in this case, the tidal range due to the semi-diurnal constituents would be small relative to that due to the diurnal constituents.

Tides can in fact be classified according to the ratio (F) of the sum of the amplitudes of the two main diurnal constituents (K_1 and O_1) to the sum of the amplitudes of the two main semi-diurnal constituents (M_2 and S_2). Some examples are illustrated in Figure 2.18.

QUESTION 2.8

(a) From Figure 2.18, what are the main differences between tidal cycles characterized by high and low values of the ratio F?

(b) Would you expect the interval between spring tides to be 14.75 days (i.e. half of 29.5 days) at all times, and at all locations, irrespective of the other types of tidal fluctuation?

Figure 2.18 shows only a selection of the many possible types of tides that can occur. The actual tides at any particular location result principally from the combination of amplitude and phase of the diurnal and semi-diurnal constituents (Table 2.1) at that location. A high value of F (say above 3.0) implies a diurnal tidal cycle, i.e. only one high tide occurs daily, and fluctuations in tidal range are largely due to changes in the Moon's declination (Figure 2.9). Low values of F (say less than 0.25) imply a semi-diurnal tide, and the fluctuations in tidal range are mainly due to the relative positions of Sun and Moon, giving the spring–neap variation (Figure 2.12), and variations in lunar declination have only a relatively small effect.*

Between these two extremes are the mixed tidal types, where daily inequalities are important, and there can be considerable variations in the amplitudes of, and time intervals between, successive high tides. The middle two tidal records in Figure 2.18(a) show diurnal inequalities where typical 'large tides' alternate with 'half-tides' (cf. Figure 2.17), and there is an additional contribution to the diurnal inequality resulting from the changing declination of the Moon, i.e. the change from tropic to equatorial tides and back again (Section 2.1.1). For example, the transition between tropic and equatorial tides can be seen at around days 6–9 and 19–22 in the record for San Francisco (Figure 2.18(a)), as lunar declination passes through zero. However, changes in the Moon's declination have less effect at higher latitudes, and diurnal inequalities are therefore not an obvious feature of tides around Britain, for example.

The configuration of an ocean basin determines its natural resonant period (Section 1.6.4), and along open ocean coasts the type of tide (Figure 2.18(a)) depends upon whether the adjacent ocean responds more readily to diurnal or semi-diurnal constituents of the tide-producing forces. In the Atlantic Ocean and most of the Indian Ocean, the response is mainly semi-diurnal, though the natural period of the Gulf of Mexico appears to be about 24 hours, and diurnal tides predominate there. In the Pacific Ocean, the diurnal response is more significant and tides are usually of the mixed type, though they are predominantly diurnal in northern and some western parts of the Pacific (cf. Figure 2.18(a)).

*With increasing values of F (signifying greater influence of diurnal constituents) the time interval between maximum (or minimum) tidal ranges decreases from about 15 days (approximately half of 29.5 days) to about 13.5 days (approximately half of 27.3 days).

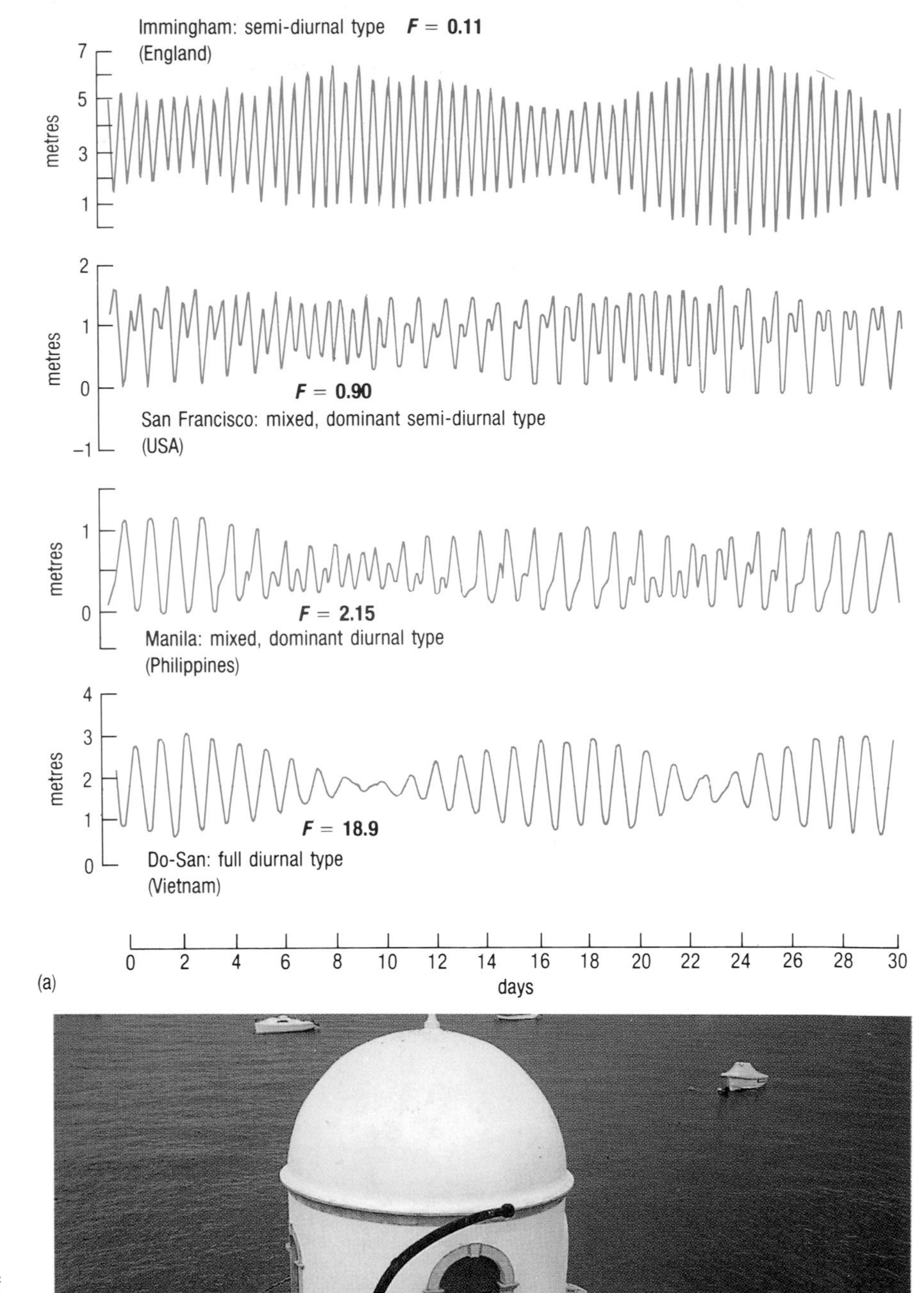

Figure 2.18 (a) Examples of different types of tidal curves, in England, the USA, the Philippines, and Vietnam. Note that the vertical scales are not the same in each record. For full explanation, see text.
(b) The tidal measuring station at Cascais harbour, Portugal, in use since the late 19th century, is one of the longest-serving stations in the European tidal gauge network. The recorder is at the top of a 'stilling' well which dampens the oscillations of swell and other wind waves entering the harbour.

It is worth mentioning here that, at any particular location, the highest and lowest spring tides will occur at the same times of day (~6 hr 25 min. apart). That is because the alternation of spring and neap tides is determined by the Sun (Figure 2.12) and the period of the S_2 constituent is 24 hr (Table 2.1). As you might expect, a similar relationship applies to neap tides. The feeding and reproductive behaviour of many marine animals, especially those living in nearshore and shallow shelf waters, is 'tuned' to tidal cycles, notably the 29.5-day lunar or synodic month (the spring–neap period, Section 2.2.1) – see also Section 2.4.1.

In shallow water, local effects can modify tidal constituents such as M_2, particularly by producing harmonics whose frequencies are simple multiples of the frequency of the constituent concerned. These harmonics result from frictional interactions between the sea-bed and the ebb and flow of the tide – especially in shallow waters. For example, the quarter-diurnal constituent M_4 (twice the frequency of M_2) and the one-sixth-diurnal constituent M_6 (three times the frequency of M_2) are generated in addition to the semi-diurnal constituents. In most locations, the effect of these two harmonics is insignificant compared with the principal constituents, but along the Dorset and Hampshire coasts of the English Channel each has a larger amplitude than usual. Moreover, the two harmonics are in phase, and their combined amplitude is significant when compared to that of M_2. (Just west of the Isle of Wight, M_2 is about 0.5 m, M_4 about 0.15 m, and M_6 about 0.2 m.) The additive effect of all three constituents causes the double high waters at Southampton and the double low waters at Portland. However, there is no truth in the popular myth that double high water at Southampton is caused by the tide flooding at different times around either end of the Isle of Wight.

The Mediterranean and other enclosed seas (e.g. Black Sea, Baltic Sea) have small tidal ranges of about 0.5 m or less, because they are connected to the ocean basins only by narrow straits. The tidal waves of the major amphidromic systems (Figure 2.15) cannot themselves freely propagate through these restricted openings. However, interaction between Atlantic tides and the shallow-water shelf region near Gibraltar for example, results in the generation of internal waves, which *do* propagate into the Mediterranean (Figure 1.23(b)) – and the internal waves seen in Figure 1.23(a) in the South China Sea may have a similar cause. By contrast, it is unlikely that similar packets of internal waves would occur where the Bosphorus connects to the Black Sea because the tidal range in the adjacent Mediterranean is negligible.

2.4.1 TIDES AND TIDAL CURRENTS IN SHALLOW SEAS

We saw at the end of Section 2.1 that tidal range and tidal currents increase as tidal waves are slowed in shallow water (Equation 1.4), and on continental shelves tidal currents typically reach speeds of 1 to 2 knots (0.5 to 1 m s^{-1}). Tidal currents of necessity have the same periodicities as the vertical tidal oscillations, and in restricted channels (e.g. estuaries) they flow in one direction for one half of the tidal cycle and in the opposite direction for the other half of the cycle. Such back and forth motions are the principal reason why tides are not very good at flushing away pollutants: what you discard or discharge on the falling tide may come back to you when the tide rises again (see also Section 2.4.3). In wide bays and estuaries and in the open sea, however, the Coriolis force causes the currents to constantly change direction so that the water particles tend to follow a more-or-less elliptical path rather than having a simple to-and-fro motion.

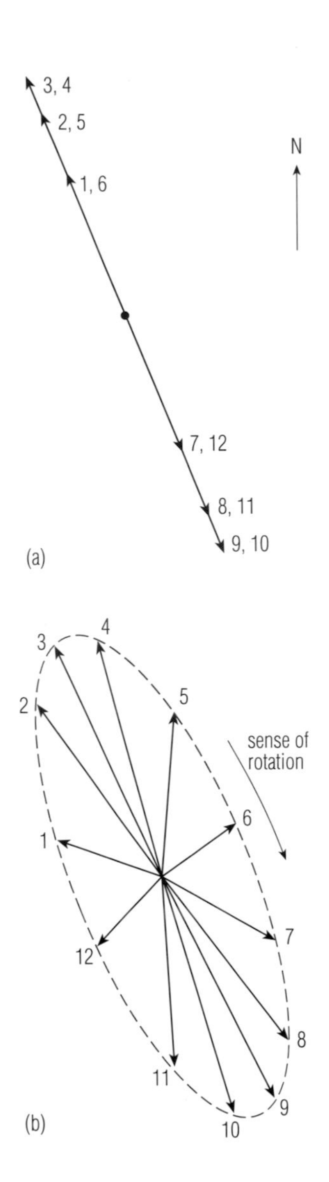

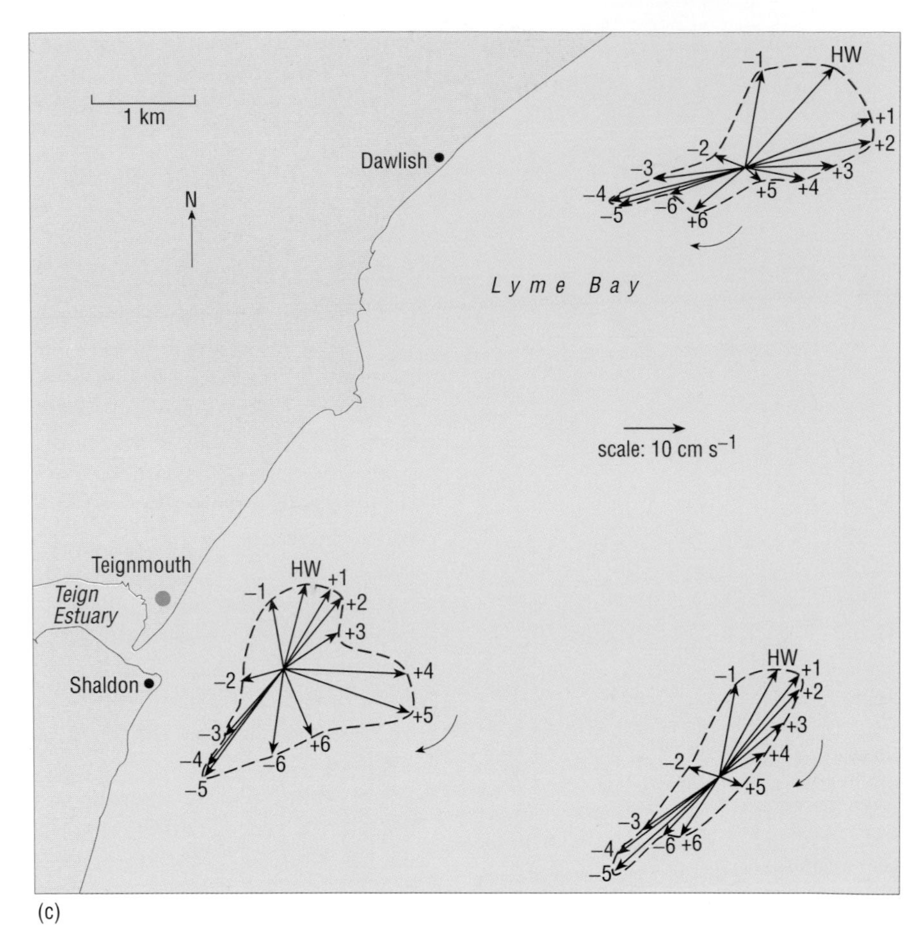

Figure 2.19 A variety of tidal ellipses. Each shows the change in direction and speed of the tidal current *at one location* (i.e. *not* the path of a water particle).
(a) Representation of a linear ebb–flow–ebb tidal current system.
(b) A more typical tidal ellipse, showing changing directions of tidal current during a complete tidal cycle. In both (a) and (b), arrows represent current speed and direction, and length of arrows is proportional to current velocity at the relevant time. Numbers refer to lunar hours (62 minutes) measured after an arbitrary starting time in the cycle.
(c) Three irregular and asymmetrical tidal ellipses in Lyme Bay, from current meter measurements averaged over a lunar month in summer. Arrows represent current speed and direction, and numbers show time in hours either side (+/–) of predicted high water (HW) at Devonport, Plymouth.

Tidal current patterns can be conveniently represented by diagrams in which the direction and speed of current flows, measured *at specific locations* at intervals throughout the tidal cycle, are recorded by arrows of appropriate length plotted from a common origin, i.e. vector arrows. Figure 2.19(a) shows a simple to-and-fro motion, with tidal currents flowing NNW throughout half of the tidal cycle, and SSE during the other half. The arrows at each interval are of different lengths because the currents wax and wane with the tidal ebb and flow. In Figure 2.19(b), the currents display the more usual elliptical pattern, increasing in speed as they swing from WNW to NNW, decreasing again as they swing back to ESE, and then speeding up again in southerly directions, before completing the cycle. The sense of rotation of tidal ellipses may be either clockwise or anticlockwise, but rotations *cum sole* tend to be favoured if there are no constraining land masses.

Figure 2.19(c) shows three typically irregular asymmetrical tidal ellipses, drawn from current meter measurements made in Lyme Bay, off south-west

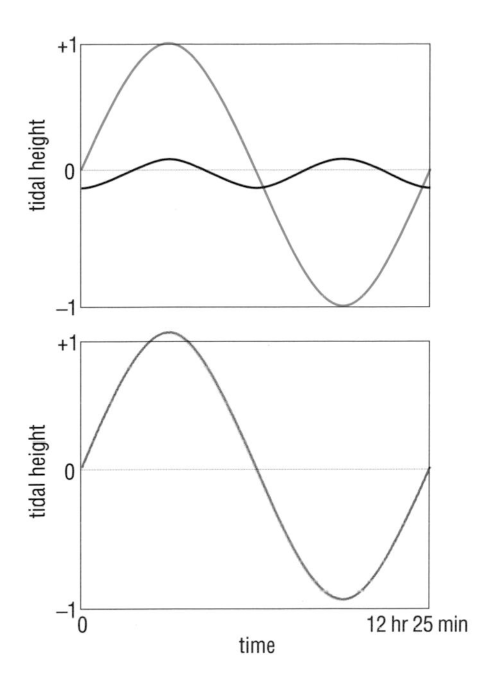

Figure 2.20 The combination of the semi-diurnal M_2 and quarter-diurnal M_4 tidal constituents. When the semi-diurnal M_2 (blue) and the quarter-diurnal M_4 (red) tidal constituents are in phase, the flood tide is strengthened and the ebb tide is weakened (purple curve, lower picture).

England. The dominant tidal flows are generally between north-east and south-west, modified off the mouth of the Teign estuary by an easterly component resulting from the flow of the river (see also Chapter 6).

On continental shelves and in shallow seas generally, it is usual for the tidal ellipses (Figure 2.19) to be asymmetrical, because the peak ebb and flow tidal currents tend to be unequal, i.e. complete reversals of tidal current flow are rare. Part of the reason for this is the interaction between tidal constituents of different periods. Figure 2.20 shows such an interaction, between the larger semi-diurnal M_2 constituent and the smaller quarter-diurnal M_4 constituent (Section 2.4). In this example, the two constituents are in phase such that the flood tidal current is strengthened because the two 'crests' coincide, while the ebb tidal current is weakened because the 'trough' of M_2 coincides with the 'crest' of M_4. Asymmetry and/or distortion of the tidal ellipse also occurs if a persistent current is superimposed on the tidal flow, for example the ellipse off the Teign estuary in Figure 2.19(c).

The patterns of tidal current flow in shelf seas are additionally modified by factors such as the shape of coastlines, bottom topography, and local weather conditions, as well as fronts (see Figure 2.22). All of these can reinforce the effect illustrated in Figure 2.20 and further contribute to distortion and asymmetry of the idealized tidal ellipses (Figure 2.19). The result is that there are **residual currents**, long-term net movements of water in fairly well-defined directions. Residual currents can be of considerable significance for the movement of sediments (see Chapter 4), though their speeds are typically only a few cm s^{-1}.

The effect of the sea-bed upon tidal current velocity in shallow water is illustrated in Figure 2.21, which shows a series of current velocity profiles during a tidal cycle. Retardation of the flow towards the bottom of the profile is a consequence of friction with the sea-bed which produces vertical **current shear**, i.e. change of current velocity with height above the bed. Tidal ellipses at the surface and near the bed are often out of phase, so that the surface and near-bed currents turn (from flood to ebb or *vice versa*) at different times. The result of such a 'phase difference' is particularly clear in the velocity profile for hour 3 in Figure 2.21.

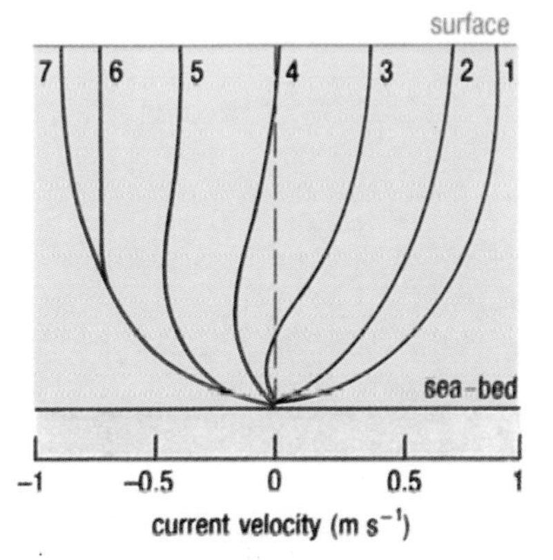

Figure 2.21 A series of tidal current velocity profiles, showing vertical current shear due to retardation of the flow close to the sea-bed. The numbers refer to time in lunar hours after an arbitrary starting time, and only half a tidal cycle is shown. Note that at hour 3 water at the surface is moving in one direction while water near the bed is moving in the other.

The turbulence resulting from friction with the sea-bed causes vertical mixing of the water column, which can extend to the surface in areas where the water is shallow and/or where the tidal current is strong enough. In other areas, where tidal currents are weaker and/or the water is deeper, less mixing occurs, and **stratification**, with layers of different densities, can develop when surface waters are warmed in summer. The inclined boundaries or **fronts** between contrasting areas of mixed and of stratified waters typically have gradients of between 1 in 100 and 1 in 1000 and are often sharply defined, with marked differences in water density on either side of the front (Figure 2.22). Density must increase downwards in both stratified and unstratified waters, but the average density on the stratified side of the front is less than that of the mixed water column on the other side. Turbulent mixing of surface waters by winds will break down upper layers of the stratification and will reinforce mixing by tidal currents on the other side. During winter in mid- and high latitudes, cooling and mixing by strong winds breaks down the stratification completely, causing the fronts to disappear. Fronts are generally zones of convergence of surface water and are often visible as lines of froth and/or floating debris. They are also generally regions of elevated nutrient concentrations and hence of high biological production.

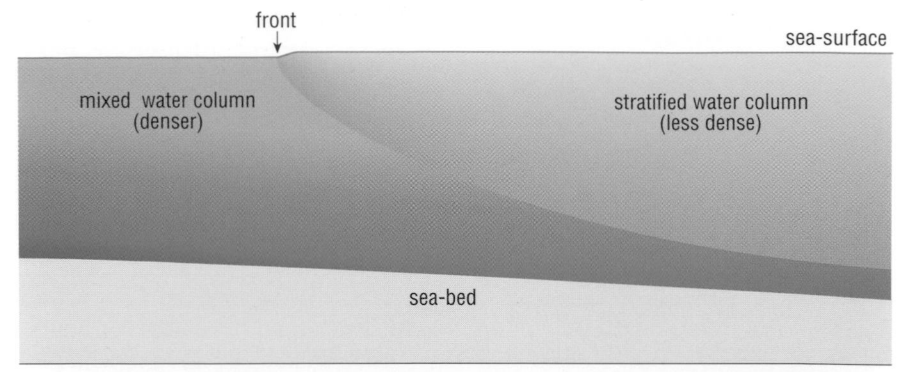

(a)

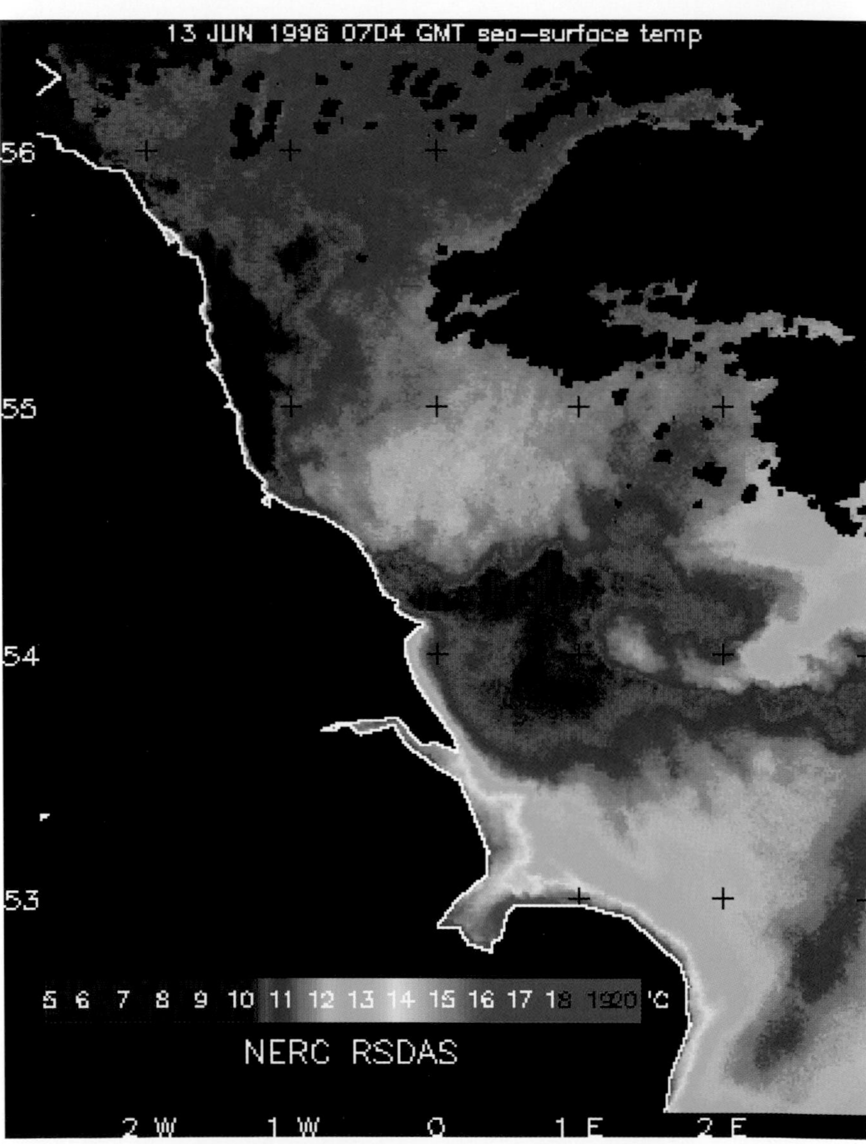

(b)

Figure 2.22 (a) Diagrammatic section (with greatly exaggerated vertical scale) through a tidal front between stratified and tidally mixed waters in a shallow sea (such as the North Sea). Note that fronts are typically zones of transition rather than sharp boundaries.
(b) Satellite image of sea-surface temperature in the North Sea in June 1996, showing the Flamborough Head Front at about 54° 30′ N as a wavy boundary zone between well-mixed cooler water (<10 °C) to the south, and warmed stratified water (surface >11 °C) to the north. (The colour scale has been chosen to show the position of the front as clearly as possible.) The front tends to disappear in winter, when winds are strong.

Tidal waves are progressive waves (Section 1.1.1), so we would expect tidal currents to be strongest at high and low tides, i.e. as the crest and trough of the tidal wave pass through. This is the case in the open ocean and along straight coastlines where cliffs enter relatively deep offshore waters (though there is some frictional retardation, leading to lateral current shear, and tidal currents close to shore are in general slower). Where coastlines are irregular, with bays and estuaries and/or there is a shelving sea-bed and relatively shallow offshore waters, tidal waves entering the bays and estuaries can be envisaged as being somewhat analogous to the long low swell waves that slide gently up beaches, often without breaking (cf. Figure 1.18(d)) – though of course on a time-scale of hours rather than seconds or minutes. In these circumstances, as anyone who has observed a tidal estuary can testify, tidal currents are minimal at both high and low water; i.e. there is *slack water* at those times (the 'turn of the tide'), and tidal currents tend to be strongest at around mid-tide during both ebb and flow phases.

Tidal currents in shallow seas are utilized by some bottom-dwelling (**demersal**) fish populations to save energy while migrating between their feeding and spawning grounds – a good example is afforded by plaice in the North Sea. When the tide is running in the required direction, the fishes swim with the current a few metres above the bottom. At slack water, they descend to the sea-bed and remain there during the other half of the tidal cycle (while the tidal currents flow in the opposite direction), ascending into the water column when the tide runs favourably once more. This *selective tidal stream transport*, as it is known, has been well documented for several decades, both by electronic tagging of fishes and by trawl catches at different tidal states.

In some larger embayments, reflection of the progressive tidal wave entering the basin will result in a standing wave being established if the basin is of appropriate length (Section 1.6.4). Under these circumstances, the tidal wave is reflected back to the entrance of the basin to coincide with the arrival of the next tidal wave. The result is to increase the amplitude of the tidal wave, and tidal ranges in such embayments can be very large. The length (270 km) and average depth (60 m) of the elongate Bay of Fundy, Nova Scotia, give it a natural resonant period almost exactly that of the semi-diurnal tide. As a result, there is a strong resonant oscillation, a tidal range of some 15 m at the head of the bay, and strong tidal currents, especially during mid-tide (cf. Figure 1.20(c) and (d)).

QUESTION 2.9

(a) Use Equation 1.18 to show that the dimensions of the Bay of Fundy are such that its resonant period is close to that of the semi-diurnal tide, and explain why this equation was used in preference to Equation 1.17.

(b) What is the wavelength of the standing wave in the Bay of Fundy?

(c) Why should we expect the tidal range near the mouth of the Bay of Fundy to be much less than 15 m?

In the larger North Sea, the tidal oscillations are partly determined by the dimensions of the North Sea basin (which has a natural resonant period of about 40 hours, cf. Equation 1.18), and partly by the progressive semi-diurnal tides entering from the Atlantic (Figure 2.14). As a result, a standing wave with three nodes tends to develop in the North Sea. However, as the basin is large enough for the water to be deflected by the Coriolis force, the nodes of the three standing waves have become the amphidromic points of Figure 2.14. As a result, the progress of the tidal waves around the amphidromic points in the North Sea resembles that in Figure 2.16(d).

Some British bays and estuaries have relatively large tidal ranges (see Figure 2.24 on p. 83), often because of resonance. In the Wash, for example, the range is nearly 7 m. In the Bristol Channel it is about 12 m, which is very large. Here, resonance is reinforced by the funnelling effect as the tidal wave travels up the narrowing Severn estuary – its crest length shortens and its height increases, cf. Equation 1.15, Section 1.5.1.

Resonance is also possible on continental shelves open to the ocean (i.e. not enclosed, like the North Sea). The **continental shelf** bordering most continental regions is overlain by water rarely more than 200 m deep, and it extends to the **shelf break**, the edge of the shelf, which is effectively the top of the **continental slope**, where water depths increase relatively rapidly (see Chapter 3). Resonance is theoretically possible where the shelf width (the distance from the coast to the shelf break) is about one-quarter of the tidal wavelength (or simple multiples thereof, e.g. 3/4, 5/4). The relationship is identical to that shown in Figure 1.20(c), where 'basin length' = continental shelf width, and there is a node at the shelf break. In water depths of about 100 m, the tidal wavelength for M_2 (the principal lunar semi-diurnal component, Table 2.1) is about 1400 km. A shelf width of some 350 km is thus required for resonance to occur, and most continental shelves are narrower than this. Nonetheless, the wider the shelf, the more closely the conditions approach those required for resonance, and there is a rough correlation between shelf width and nearshore tidal range. Increased tidal range means increased tidal current speeds. For example, mean near-surface spring tidal current speeds around the British Isles exceed 1.5 m s^{-1} in places.

Strong tidal currents can be produced where the flow is constrained by the presence of islands, narrow straits or headlands. This is because of the requirement for **continuity**, i.e. volumes of water flowing into and out of a given space per unit time must be equal. If a current is forced to become narrower, it will speed up (a shoaling sea-floor can have a similar effect). Where the Cherbourg peninsula of north-west France reaches out towards the Channel Island of Alderney, spring tides can routinely generate currents of 10 knots (*c*. 5 m s^{-1}) and interaction between tidal currents and other currents can result in confused seas – even white-capping – on an otherwise calm day. The sea off the tip of Portland Bill, Dorset, can present similar problems. Such areas are marked on navigational charts as 'overfalls', and are particularly to be avoided when waves are steepened by opposition to such tidal currents.

Currents associated with tidal flows include so-called 'hydraulic currents'. Water tends to 'pile up' at the entrances to narrow straits, leading to a downward slope of the sea-surface in the direction of flow. This slope causes a horizontal pressure gradient along the strait, generating a 'hydraulic' component of the current. The tidal currents causing the legendary Lofoten *Maelstrom* off the northern coast of Norway, for example, are probably enhanced by hydraulic pressure gradients along channels between the Lofoten Islands. Renowned for centuries in Scandinavian folk lore, the *Maelstrom* gained world-wide notoriety for dangerous currents and whirlpools through the stories of Edgar Allan Poe (*Descent into the Maelstrom*, 1841) and Jules Verne (*20 000 Leagues under the Sea*, 1869). Sadly, as is often the case, reality does not quite live up to the legend. The tidal currents have been said to run at speeds of 5 or 6 m s^{-1}. Although there are no current meter records against which to check these estimates, speeds nearer to 3 m s^{-1} are considered more likely by modern observers. Eddies and zones of lateral current shear appear on satellite images of the area, and the *Maelstrom* of legend could be a fictional amalgam of such eddies, especially as the tidal currents have long been known to rotate during the tidal cycle.

2.4.2 STORM SURGES

An additional complication in the prediction of tidal heights is that meteorological conditions can considerably change the height of a particular tide, and the time at which it occurs. The wind can hold back the tide, or push it along, and changes in atmospheric pressure can also affect the water level.

QUESTION 2.10 What is the effect on the local sea-level of a fall in atmospheric pressure of 50 millibars, as might occur when a severe storm passes, given that a head of water 10 m in height exerts a pressure equivalent to 1 atmosphere (1 bar)?

Thus, not only wind changes but changes in atmospheric pressure can cause the actual water level to be very different from the predicted value, especially during storms. The combined effects of wind and low atmospheric pressure can lead to exceptionally high tides, termed **positive storm surges**, which threaten low-lying coastal regions with the prospect of flooding. On the other hand, abnormally low tides, termed **negative storm surges**, may occur during periods of high atmospheric pressure, especially if there are strong offshore winds. Although less common, these surges can cause problems in shallow seas for large ships such as supertankers which have a relatively deep draught.

The most catastrophic positive surges are those caused by tropical cyclones (typhoons and hurricanes) or by severe depressions in temperate latitudes. One of the worst in recent history struck the north coast of the Bay of Bengal in 1970, killing 250 000 people; a subsequent surge in 1985 caused the loss of 20 000 lives. The well-documented North Sea storm surge of 1953 led to sea-levels locally up to 3 m above normal and caused 1800 deaths in Holland and 300 in England. In this case (as with most positive surges), high spring tides, strong onshore winds and very low barometric pressure all combined to produce an abnormal rise in local sea-level. In 1986, more than 30 years after this disaster, a barrier 8 km long was built across the eastern Scheldt, completing the final stage of the Delta Project which is intended to protect the Netherlands from another such flood catastrophe. The Thames Barrage provides similar protection for the low-lying areas in and around London. Early warning of storm surges is now routine in many parts of the world (including eastern England and vulnerable parts of the Indian sub-continent), because accurate meteorological and tidal data have become more readily available, and forecasting is aided by satellite tracking of storms as well as by computer-modelling of past surges. Britain's storm-surge warning service is based at the Proudman Oceanographic Laboratory at Bidston on Merseyside, where the nation's tide-tables are also compiled.

Storm surges in the North Sea can, in theory, add as much as 4 m to the normal tidal height, but fortunately most storm surges (of which there are, on average, about five per year) increase high tide levels by only about 0.5 to 1 m. They are usually associated with eastward-moving depressions, and follow a three-phase pattern:

1 The first signs are evident as a relatively small positive storm surge in the North Atlantic, with water being displaced by south-westerly winds to the north-east Atlantic.

2 At the same time as the events in (1), a negative surge is experienced on the east coast of Britain as the south-westerly winds displace water to the north-east corner of the North Sea. This negative surge travels southwards down the east coast and swings eastward across the southern part of the North Sea, following the amphidromic system shown in Figure 2.14.

3 As the depression moves across Britain and out over the North Sea, the wind veers (i.e. swings in a clockwise direction) to blow from the north-west. The next high tide, by now travelling southwards down the North Sea, is thus reinforced not only by the wind but by the Atlantic surge referred to in (1) above, which by this time is displacing water into the northern part of the North Sea. This large positive surge travels down the east coast of Britain, and reaches a maximum in the south-western corner of the North Sea. The problem is compounded partly by the funnelling effect imposed by the basin shape (cf. Equation 1.15), and partly by the 'piling up' of water onshore because of the Coriolis force (Section 2.3): water being driven southwards by the strong winds is deflected to the right, towards the east coast of England. In addition, the arrival of the surge may coincide with the arrival of the low pressure area in the centre of the depression, thus increasing local sea-level still further.

Storms occur every winter and you might wonder why floods caused by storm surges are not more common. The answer is that severe flooding will only happen when low pressure and strong onshore winds coincide with a high spring tide – a storm surge at low tide can be considered a non-event.

2.4.3 TIDES IN RIVERS AND ESTUARIES

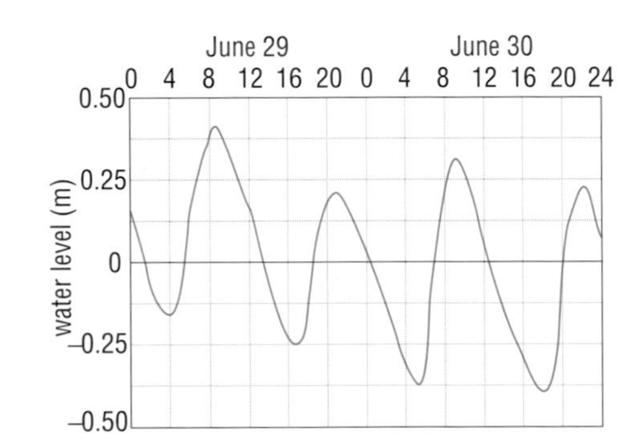

Figure 2.23 A tidal curve for the Hudson River estuary near Albany, New York, showing a typical estuarine tide with peaks tending to catch up with the preceding trough. Numbers on the horizontal axis are time in hours.

Sea-levels have risen since the last glacial period, and many river valleys have become inundated by the sea, forming tidal estuaries or *rias*. Tidal waves propagate up the estuaries, the speed of travel depending upon the water depth (Equation 1.4). Hence the wave crest (high water) will travel faster than the wave trough (low water). As a result, there is an asymmetry in the tidal cycle, with a relatively long time interval between high water and the succeeding low water, and a shorter interval between low water and the next high tide (Figure 2.23).

The maximum speeds of the tidal currents in estuaries may not always be in phase with the tidal crests and troughs. Thus, at the estuary mouth, the tide may behave as a progressive wave and the maximum speed of the flooding tide will coincide with high water; whereas further up-river, the tide may behave as a standing wave and high tide will be a time of slack water (i.e. zero current). However, the ebb current will invariably persist for longer than the flood, partly as a result of the asymmetry of the estuarine tidal cycle referred to above (Figure 2.23), and partly because the freshwater discharge into the river results in a net seaward discharge of water – so ebb tidal currents in estuaries can actually be stronger than flood tidal currents, despite the tidal asymmetry. Many towns and cities sited near such estuaries rely upon this net seaward flow to carry away sewage, a strategy which has sometimes only mixed success (cf. beginning of Section 2.4.1).

In some tidal rivers, where either the river channel narrows markedly, or the gradient of the river bed steepens, the tidal range can be increased and a **tidal bore** may develop. The formation of tidal bores has features in common with the propagation of waves against a counter-current (Section

1.6.1). The rising tide may force the tidal wave-front to move faster than a shallow-water wave can freely propagate into water of that depth, according to Equation 1.4 (cf. Section 2.3, item 1). When this happens, a shock wave is formed, analogous to the 'sonic boom' that occurs when a pressure disturbance is forced to travel faster than the speed of sound. The tidal bore propagates as a *solitary* wave with a steep leading edge, moving upstream as a rolling wall of water. Most tidal bores are relatively small, of the order of 0.5 m high, but some can be up to ten times that height. The Severn River bore in England reaches heights of 1–2 m on spring tides, whereas the Amazon bore (the *pororoca*) reaches about 5 m, and moves upstream at about 6 m s^{-1} (12 knots). Other rivers where bores develop include the Colorado, Trent, Elbe, Yangtze and the Petitcodiac, which flows into the Bay of Fundy, notable for its large tidal range (Section 2.4.1).

2.4.4 TIDAL POWER

Power can be generated by holding incoming and outgoing tides behind a dam, using the head of water so produced to drive turbines for electricity generation. The tidal range determines the potential energy available at any locality, and on average must exceed 5 m for electricity generation to be economic. Suitable locations are limited to those where such tidal ranges exist and where dams can feasibly be built (Figure 2.24).

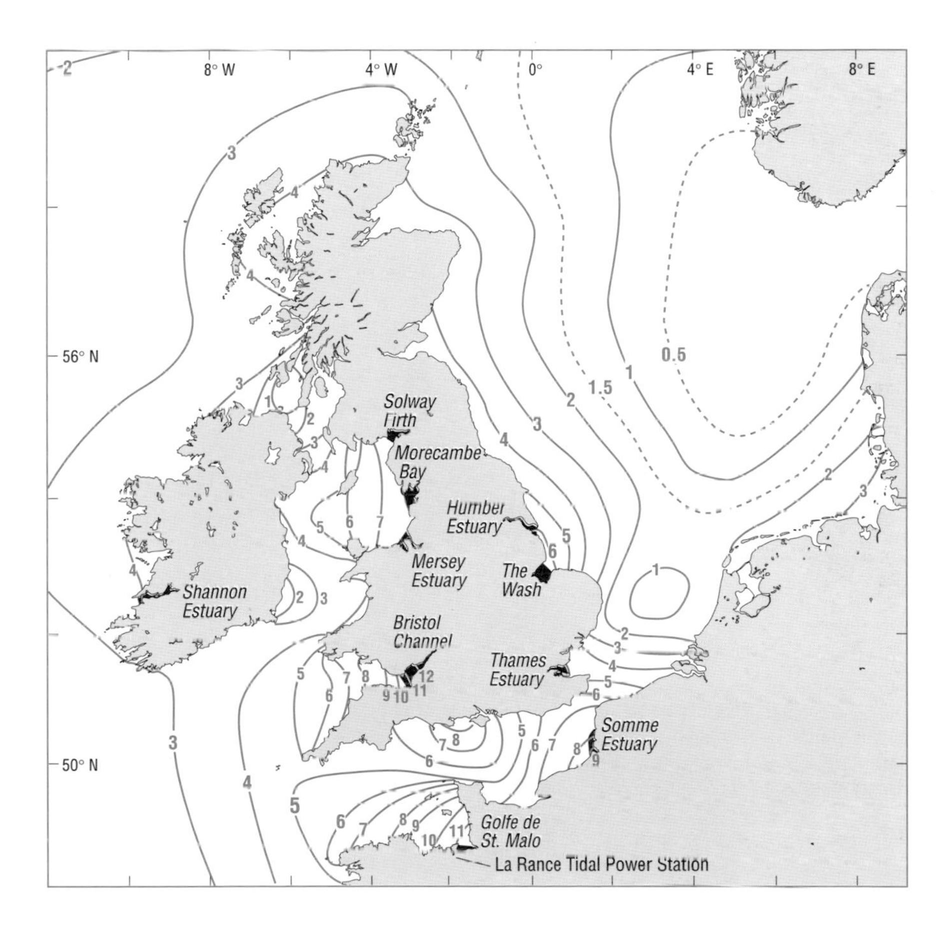

Figure 2.24 Tidal range (in metres), and the sites of actual (La Rance) and potential tidal-power barrages.

One such site is the Rance estuary in Brittany (Figure 2.25), operating since 1966. A much larger scheme for Britain's Severn estuary has been proposed and discussed many times. Although such a scheme would produce an appreciable proportion (in the order of 6%) of Britain's electrical power requirements, dam construction would dramatically affect the patterns of currents and sediment movements, and ecological disturbance would be inevitable – all factors to be considered whenever schemes of this kind are planned.

It has also been proposed that the reversing flows of strong tidal currents in narrow straits and channels could be harnessed simply by installing turbines in the water. One location where this has been investigated is among the Orkney and Shetland Islands off northern Scotland, where tidal currents commonly exceed $2\,\mathrm{m\,s^{-1}}$. The idea is simple enough in principle, but the practical problems are considerable.

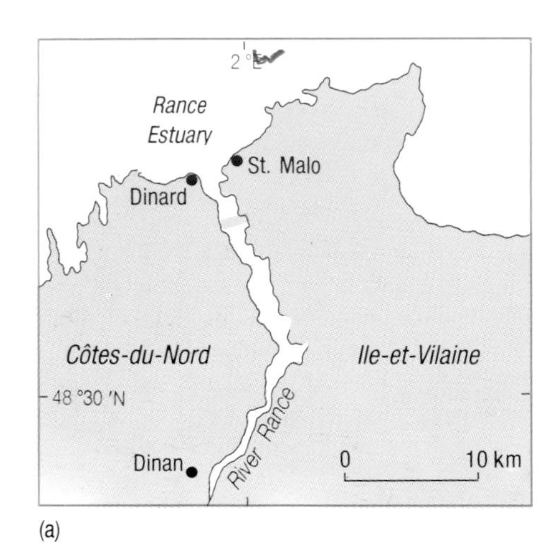

(a)

(b)

Figure 2.25 (a) The location of La Rance tidal power station. It has been producing about 550×10^6 kW h annually since 1966.
(b) An aerial view of La Rance.

2.5 SUMMARY OF CHAPTER 2

1 Tides are long-period waves, generated by gravitational forces exerted by the Moon and Sun upon the oceans. They behave as shallow-water waves because of their very long wavelengths. Tidal currents are the horizontal water movements corresponding to the rise and fall (flood and ebb) of the tide.

2 A centrifugal force, directed away from the Moon, results from the Earth's (eccentric) rotation (period 27.3 days) around the Earth–Moon centre of mass, which is within the Earth. This centrifugal force is exactly balanced *in total* by the gravitational force exerted on the Earth by the Moon. However, gravitational force exceeds centrifugal force on the 'Moon-side' of Earth, resulting in tide-producing forces directed towards the Moon, whereas on the other side of the Earth centrifugal force exceeds gravitational force, resulting in tide-producing forces directed away from the Moon.

3 Tractive forces (horizontal components of tide-producing forces) are maximal on two small circles either side of the Earth, and produce two (theoretical) equilibrium tidal bulges – one directed towards the Moon, and the other directed away from it. As the Earth rotates with respect to the Moon (with a period of 24 hours 50 minutes), the equilibrium tidal bulges would need to travel in the opposite direction (relative to the surface of the rotating Earth) in order to maintain their positions relative to the Moon. The elliptical orbit of the Moon about the Earth causes variation in the tide-producing forces of up to 20% from the mean value.

4 With the Moon overhead at the Equator, the equilibrium tidal bulges would be in the same plane as the Equator, and at all points the two bulges would theoretically cause two equal high tides daily (equatorial tides). The Moon has a declination of up to 28.5° either side of the Equator, and when the plane of the tidal bulges is offset with respect to the Equator, there are two unequal, or tropic, tides daily. The declination varies over a 27.2-day cycle.

5 The Sun also produces tides which show inequalities related to the Sun's declination (up to 23.4° either side of the Equator), and vary in magnitude due to the elliptical orbit of the Earth around the Sun. The Sun's tide-producing force has about 46% of the strength of the Moon's. Solar tides combine with and interact with lunar tides. When Sun and Moon are in syzygy, the effect is additive, giving large-ranging spring tides; but when Sun and Moon are in quadrature, tidal ranges are small (neap tides). The full cycle (a lunar month), includes two neaps and two springs, and takes 29.5 days.

6 Tidal speed is limited to about 230 m s^{-1} in the open oceans (less in shallower seas), and land masses constrain tidal flow. Water masses have inertia and experience friction with coasts and the sea-bed, so they do not respond instantaneously to tractive forces. The Coriolis force, and constraining effects of land masses, combine to impose amphidromic systems upon tides. High tidal crests circulate (as Kelvin waves) around amphidromic points which show no change in tidal level, i.e. tidal range increases with distance from an amphidromic point. Amphidromic systems tend to rotate in the opposite direction to the deflection caused by the Coriolis force.

7 The actual tide is made up of many constituents (partial tides), each corresponding to the period of a particular astronomical motion involving Earth, Sun or Moon. Partial tides can be determined from tidal measurements made over a long time at individual locations, and the results used to compute future tides. Actual tides are classified by the ratio (F) of the summed amplitudes of the two main diurnal constituents to the summed amplitudes of the two main semi-diurnal constituents.

8 Tidal rise and fall are produced by lateral water movements called tidal currents. Tidal current vectors typically display 'tidal ellipses' rather than simple to-and-fro motions.

9 Areas of low atmospheric pressure cause elevated sea-levels, whereas high pressure depresses sea-level. A strong wind can hold back a high tide or reinforce it. Storm surges are caused by large changes in atmospheric pressure and the associated strong winds. Positive storm surges may result in catastrophic flooding.

10 In estuaries, the tidal crest travels faster than the tidal trough because speed of propagation depends upon water depth; hence the low water to high water interval is shorter than that from high water to low water. Tidal bores develop where tides are constrained by narrowing estuaries and the wave-front is forced by the rising tide to travel faster than the depth-determined speed of a shallow-water wave. Where tidal ranges are large and the water can be trapped by dams, the resultant heads of water can be used for hydro-electric power generation.

Now try the following questions to consolidate your understanding of this Chapter.

QUESTION 2.11 Write an expression for the tide-producing force at point P on Figure 2.4(a), using the terms as defined for Equations 2.1, 2.2 and 2.3. It is not essential to try to simplify or approximate the expression.

QUESTION 2.12 Which of the following statements are true?

(a) 'In syzygy' has the same meaning as 'in opposition'.

(b) Neap tides would be experienced during an eclipse of the Sun.

(c) Spring tides do not occur in the autumn.

(d) The lowest sea-levels of the spring–neap cycle occur at low tide while the Moon is in quadrature.

QUESTION 2.13 Briefly summarize the factors accounting for differences between the equilibrium tides and the observed tides.

QUESTION 2.14 How will each of the following influence the tidal range at Immingham (Figure 2.18(a)):

(a) The Earth's progress from perihelion to aphelion?

(b) The occurrence of a tropic tide?

(c) A 30 millibar rise in atmospheric pressure?

CHAPTER 3 INTRODUCTION TO SHALLOW-WATER ENVIRONMENTS AND THEIR SEDIMENTS

'Down I come with the mud in my hands,
And plaster it over the Maplin Sands.'

From *The River's Tale* by Rudyard Kipling.

Shallow-water environments are the coastal and shallow marine regions which form the interface between the deep ocean basins and exposed land surfaces. Some of them, such as beaches, tidal flats and estuaries, are very familiar, not least because their natural beauty and associated leisure opportunities make them popular holiday locations. Less familiar are the offshore (sub-tidal) waters of the continental shelf (Section 2.4.1). Figure 3.1(a) shows areas of continental shelf around the world at the present day, occupying some 15 per cent of the total area covered by seawater.

Continental shelves are areas of submerged continental crust that has been stretched and thinned by the processes of **sea-floor spreading** and **plate tectonics**, and has subsided **isostatically**. They are often called **shelf seas**, especially when enclosed, and these have their own names, e.g. Gulf of Carpentaria, Hudson Bay, the North Sea, the English Channel; but most are bounded on one side by land, on the other by open ocean. Continental shelves vary enormously in width, from tens to hundreds of kilometres, which is in part related to their different tectonic settings. Along **passive (aseismic) margins**, shelves tend to be broad, whereas along **active (seismic) margins** they are much narrower, although in the western Pacific some volcanic **island arcs** enclose quite extensive areas of shelf sea, e.g. the Yellow Sea and East China Sea.

Most continental shelves are formed of accumulations of sediments eroded from adjacent land areas and deposited upon the thinned (stretched) continental crust bordering them, often reaching thicknesses in excess of 1 km. They are generally almost flat, sloping gently seawards from the coast to water depths of between about 100 and 250 m (occasionally as much as 500 m) at the shelf break. The average gradient is only about 0.1° (Figure 3.1(b)), though the sea-bed is not always smooth. There may, for example, be small-scale sand ripples or larger-scale gentle undulations in the form of sand-waves or sand-banks, and in places gravels or even bare rock may be exposed. All these features result from the combined action of waves and currents.

3.1 SEDIMENTS OF SHALLOW-WATER ENVIRONMENTS

A large proportion of all the sediments in the oceans (along with most of the dissolved constituents in seawater) ultimately come from the weathering and erosion of continental rocks. In other words, they are *terrigenous* sediments. We can recognize two types of weathering:

1 *Physical (or mechanical) weathering*: the fragmentation of rocks, for example by frost-shattering (when water freezes and expands in cracks), or by the cumulative effects of alternate heating and cooling in arid climates at low latitudes, especially in deserts.

2 *Chemical weathering*: the decomposition of rocks through the reaction of surface waters and groundwater with minerals in rocks, resulting in the formation of new minerals and the removal of material in solution.

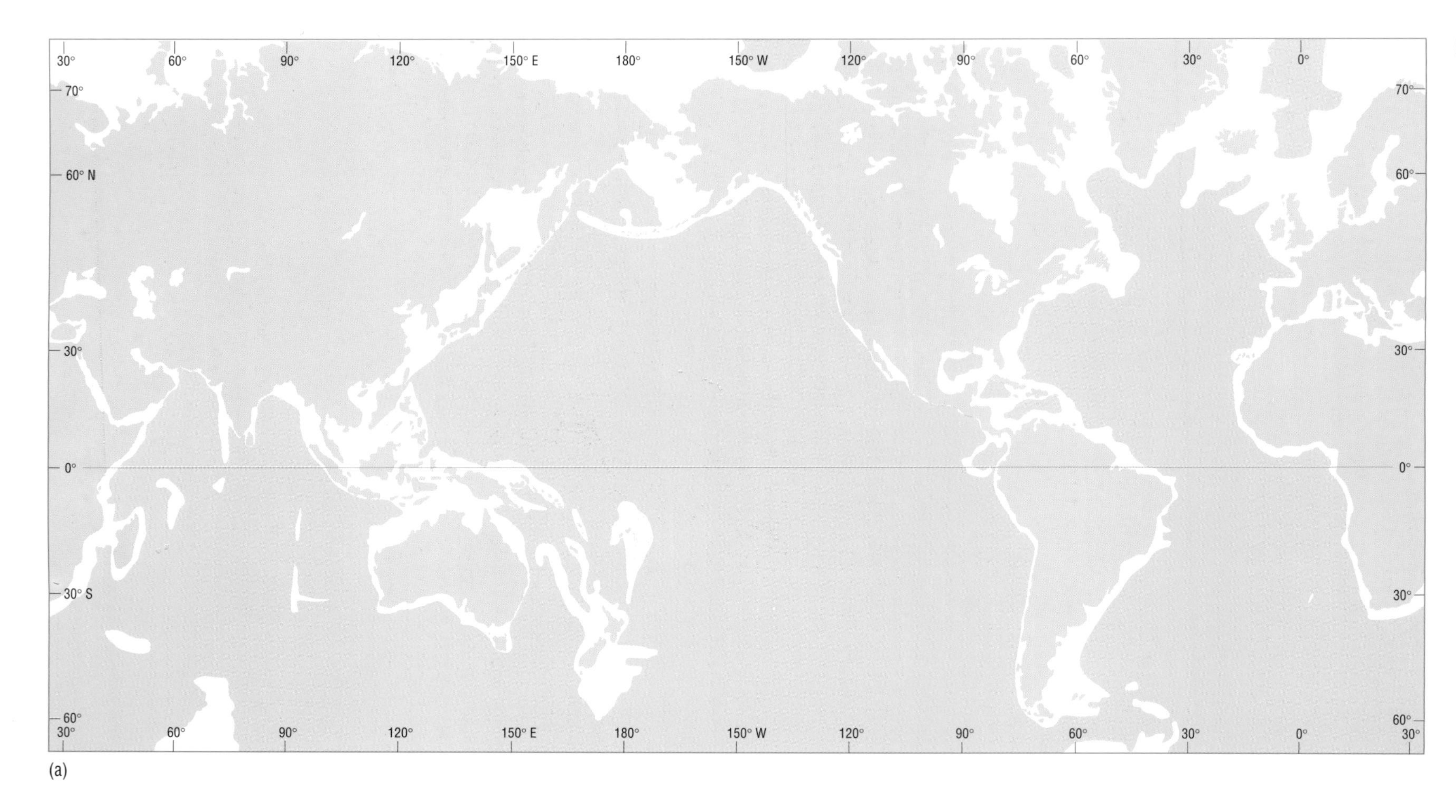

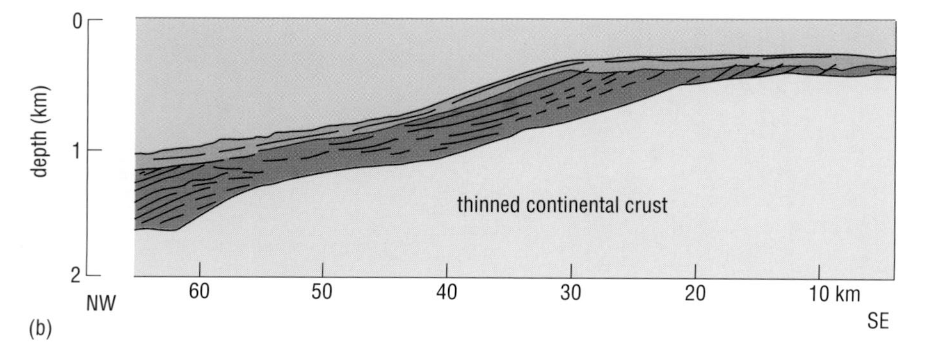

Figure 3.1 (a) Generalized map of continental shelf areas (and some other shallow-water areas) of the world. The edge of the white area is the shelf break (Section 2.4.1), where the gradient steepens at the top of the continental slope. (Remember that this map projection exaggerates areas at high latitudes.)
(b) Seismically determined cross-section through the continental shelf edge north-west of Scotland. Vertical exaggeration is × 8. The shelf break is quite clearly visible and the continental slope has a relatively gentle gradient (1 in 25). Successive layers of sediment can be seen, as can the fact that at some time of low sea-level in the past the uppermost sediments were eroded away; the sediments shown in light brown were deposited later. Continental slopes elsewhere are often steeper than this, descending into water 2–3 km deep or more, and shelf sediment thicknesses are commonly greater than shown here.

Weathering by biological agencies is mainly chemical, as many organisms secrete compounds that attack rocks; but some is physical – for example, the growth of plant roots in cracks can also bring about physical fragmentation. The rate at which rocks become weathered depends on many factors, including climatic conditions, the strength, porosity and permeability of the rock, and the composition, size and shape of the constituent mineral grains. The products of weathering ultimately reach the sea, transported (at the present time) mainly by water, but also by wind and ice. The transported particles and fragments abrade each other and erode the surfaces over which they pass, thus contributing further to the breakdown of continental rocks. In this context, time and distance are important factors: the longer the journey to the sea, the more chance there is for mineral grains to be rounded and reduced in size by abrasion, and sorted according to density and size and shape, and for chemically less stable minerals to break down.

The commonest solid products of weathering are rock fragments, quartz and clay minerals. Quartz (SiO_2) is the only common mineral of igneous and metamorphic rocks that is both hard (resistant to abrasion) and chemically stable at the Earth's surface. It is the dominant mineral in the sand deposits of most beaches. Clay minerals, formed mainly by chemical weathering of feldspars and iron- and magnesium-bearing minerals in igneous and metamorphic rocks, are the most important constituents of muds. Clay mineral particles typically take the form of very small flakes (less than 2 μm in size), which stick together easily and give mud its glutinous properties. Sands, silts and muds are the predominant sediments of continental shelves (silts are mostly mixtures of clay minerals and very small quartz particles).

Where there is a plentiful supply of terrigenous sediments, their distribution depends on how they are transported, sorted and distributed by water. Where sediment supply is negligible, however, biogenic sediments (i.e. of biological origin) will accumulate, or older terrigenous sediments (deposited in the past) will predominate.

Among the more important factors controlling the supply of sediments are climate (latitude) and topography. Climate determines whether weathering is mainly physical or chemical, and it also determines the terrestrial vegetation. Although it can contribute to weathering (as mentioned above), vegetation generally slows down erosion by binding or protecting weathering products, and by limiting run-off.

In general, would you expect physical or chemical weathering to be dominant in low (i.e. tropical) latitudes?

These are warm, humid regions where chemical weathering would be expected to predominate. Physical weathering is more characteristic of high latitudes (and high altitudes).

In mountainous areas, rapid erosion will generally provide a plentiful supply of mainly coarse-grained sediment. Lowland areas, on the other hand, are more slowly eroded because the generally slow-moving rivers that cross them can transport only fine-grained and dissolved material.

Where the supply of terrigenous sediments is negligible, for example along arid coastlines, shallow-water carbonates (deposits of calcium carbonate, $CaCO_3$) may accumulate. These are mainly of biogenic origin, and in many tropical and sub-tropical regions – especially where the water is warm and clear enough for reef-building corals to grow – the sediments are dominated by remains of corals, molluscs and other shelly organisms, including carbonate-precipitating ('coralline') algae (cf. Figure 3.2(a)). Carbonate sands and muds of inorganic origin can also form in these regions, the result of direct precipitation of calcium carbonate from the warm seawater (Figure 3.2(b)).

Inorganic carbonates do not form at higher (colder) latitudes, but biogenic carbonates can still be abundant (Figure 3.2(a)), and shell sands – generally the remains of molluscs and barnacles, etc., rather than of corals – may accumulate where terrigenous sediment supplies are low. For example, beach and bay sediments at John O'Groats, Scotland, and Sitka Sound, Alaska, contain more than 90 per cent and 66 per cent of calcium carbonate, respectively. It is worth noting here that although reef-building corals require the warm clear shallow waters of tropical regions, non-reef-building corals are widely distributed in deeper ocean waters beyond the shelf break. One example is *Lophelia*, which became a focus of environmental controversy when oil exploration began off north-western Britain during the late 1990s. *Lophelia* is a colonial coral that grows in water depths ranging from 100 m down to 600 m or more, and can form reef-like structures or mounds tens of metres high. It is found over an area extending from the Porcupine Seabight (south-west of the British Isles) to the Norwegian fjords.

(a)

(b)

Figure 3.2 (a) A rock pool on the west coast of Scotland. The encrustations are formed by the coralline alga *Lithothamnion*.
(b) Inorganic grains of $CaCO_3$ (ooliths) from the Great Bahama Bank (magnification × 10). Such grains are generally 0.2–0.6 mm in diameter, formed by the precipitation of $CaCO_3$ from seawater around a mineral grain or shell fragment which acts as a nucleus.

3.1.1 THE SUPPLY OF SEDIMENTS TO SHELF SEAS AND OCEANS

Figure 3.3 provides an inventory of the amounts of materials supplied to the oceans from different sources. The combined total of particulate and dissolved material entering the oceans annually from continental sources amounts to some 26×10^9 tonnes. Individual fluxes are only approximate, because the problems of sampling and measuring such highly variable systems make inventories of this kind difficult to compile. Nonetheless, Figure 3.3 indicates the relative importance of different sediment sources, and shows that rivers supply about 85 per cent of the total solid material entering the oceans.

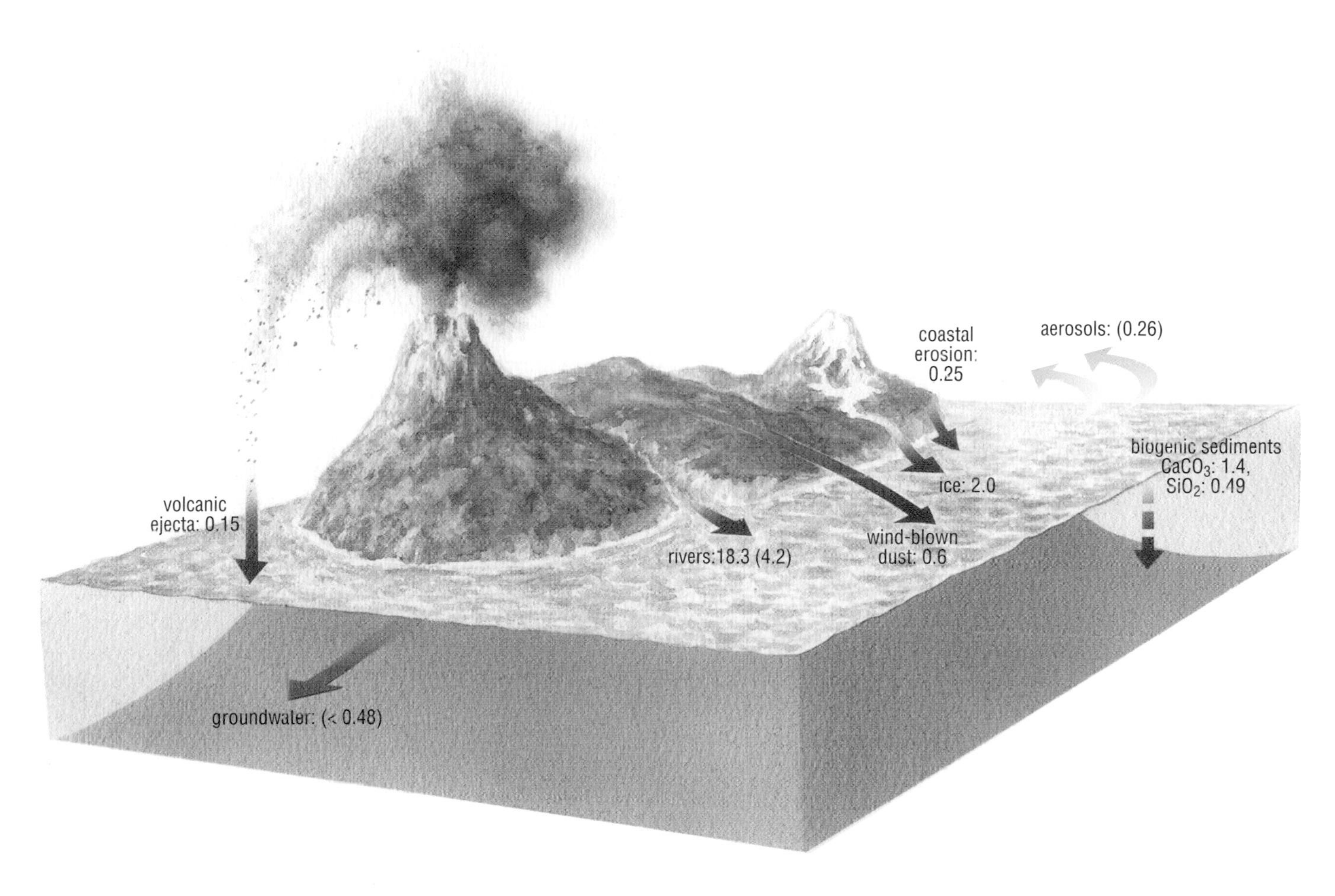

Figure 3.3 The annual transfer of sedimentary materials to the oceans in 10^9 tonnes per year. Numbers in brackets refer to material in solution. Biogenic sediments are the skeletal remains of organisms precipitating calcium carbonate or silica from solution. They occur on continental shelves as well as in the deep sea. **Aerosols** form from breaking waves and recycle dissolved constituents which then re-enter the sea via rain, rivers and groundwater.

Where does most of this material end up?

The dissolved material will be dispersed ocean-wide, but the bulk of the solid products of erosion are deposited in coastal regions and on the continental shelf. At the present day, sediments are supplied to the sea by ice only at high latitudes where glaciers and ice-sheets enter the sea. 95 per cent of such material comes from Antarctica and is deposited within 20 km of the Antarctic coast. By contrast, wind-blown dust is distributed virtually world-wide, though most originates in subtropical latitudes, where Trade Winds blow off continental desert areas such as the Sahara. Much volcanic material is deposited near its source, but large eruptions can eject fine ash and dust high into the atmosphere, to be dispersed globally by winds.

Some of the finer-grained (clay) material that enters the sea in rivers remains in suspension and is carried to the open ocean where, together with dust from deserts and volcanoes, it contributes to the **pelagic sediments** of the deep ocean floor. However, the major contributors to deep ocean pelagic sediments are the **planktonic** organisms that live in surface waters, especially those planktonic algae that secrete shells or tests of calcium carbonate or silica (SiO_2). When the organisms die, their skeletal remains sink to the sea-bed, large areas of which are covered with predominantly biogenic sediments, consisting of calcareous and/or siliceous remains. Biogenic sediments of planktonic origin can of course also accumulate on continental shelves, in favourable circumstances. The famous white Chalk (limestone) of south-east England, for example, consists almost entirely of the remains of microscopic planktonic algae that were deposited some 90 million years ago, in water no more than a couple of hundred metres deep.

Sediments accumulating near the shelf break can be destabilized by earthquakes or large storms, which cause them to collapse and slide or slump down the continental slope into the deep ocean. These 'submarine landslides' often give rise to **turbidity currents**, dense mixtures of sediment and water which can travel at speeds of $20\,m\,s^{-1}$ or more down the continental slope and out across the ocean floor. Slumps, slides and turbidity currents transfer large volumes of sediment from the continental shelf to the deep ocean, and are responsible for eroding **submarine canyons** in the continental slope, as well as for contributing layers of **turbidite** deposits that are interlayered with pelagic sediments in the thick sedimentary sequences forming the **continental rise** at the base of the continental slope, and the flat **abyssal plains** that cover large areas of the deep sea-bed.

Potentially catastrophic tsunamis (Section 1.6.3) can be generated by large slumps. These can be either of rock, from the failure of unstable slopes of volcanic islands (which are especially common in the Pacific) or of destabilized sediment from continental margins, discussed above. For example, enigmatic bouldery deposits far above present-day sea-level in eastern Scotland may have been carried and dumped there by tsunamis generated when a series of giant slumps occurred off the coast of Norway several thousands of years ago. Similar deposits found along the coast of New South Wales (Australia) are attributed to a huge tsunami that struck in the late 18th century (i.e. little more than 200 years ago), and is believed by some authorities to have resulted from slumping, by others to a meteorite impact in the sea nearby.

3.1.2 VARIATIONS IN SUPPLY AND DISTRIBUTION OF SEDIMENTS OVER TIME

Rivers are by far the dominant source of sediments to the oceans, but the supply is by no means evenly distributed around the ocean margins, as Figure 3.4 demonstrates.

QUESTION 3.1 Examine Figure 3.4.

(a) Why is more sediment discharged into the Atlantic from Brazil than is discharged into the Pacific from Ecuador and Peru?

(b) Why is sediment discharge so high in the western Pacific?

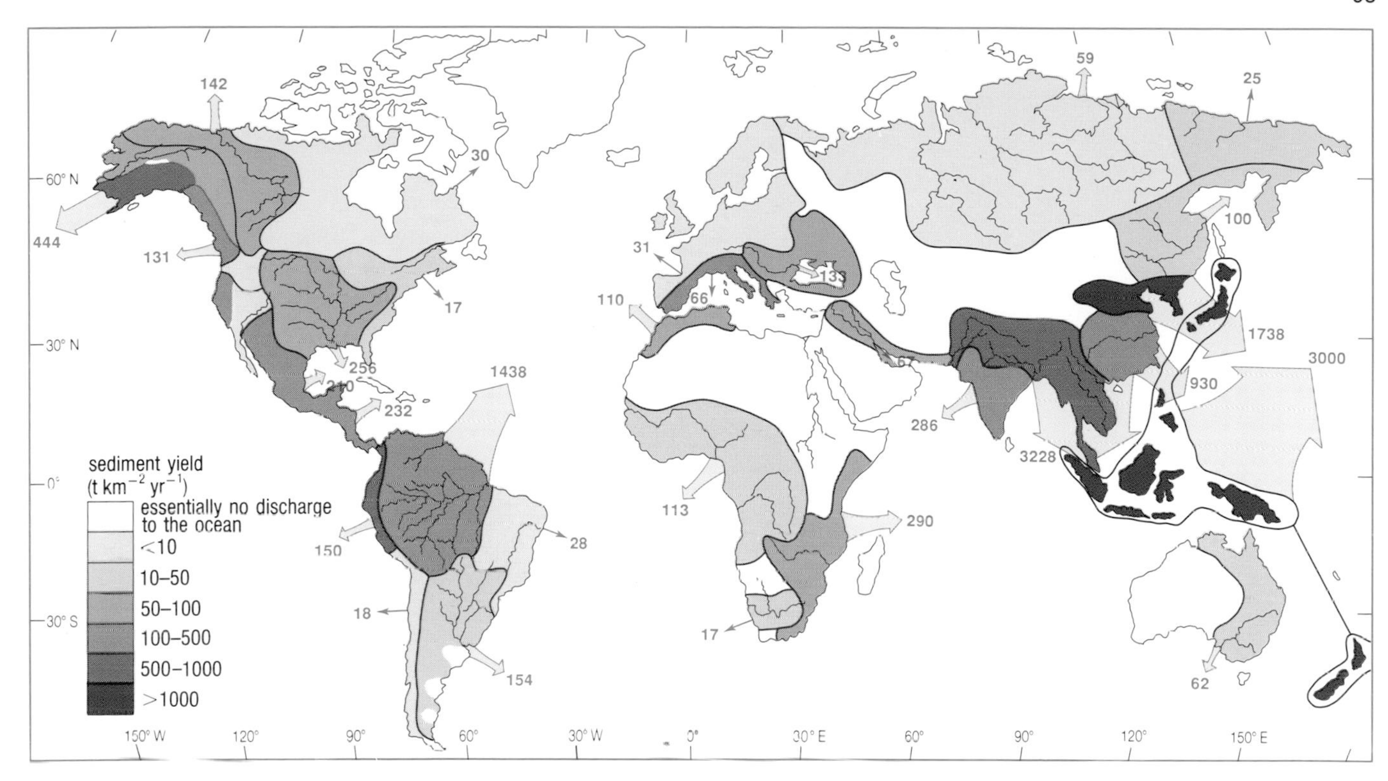

Figure 3.4 Blue numbers and arrows show average annual discharge (in 10^6 tonnes yr^{-1}) of suspended sediment from the major drainage basins of the world. Sediment discharge is proportional to the widths of the arrows. The sediment yields of the various basins are shown in the key, and red lines show boundaries of the main drainage basins. Note that the sediment loads of the Nile and Colorado rivers are largely trapped in reservoirs behind dams, and so do not reach the sea.

Locally, the patterns of sediment supply and distribution can change very rapidly – on a seasonal or even a diurnal time-scale. Storms, floods or droughts can greatly affect the amount of sediments transported to the sea and distributed in the coastal zone, as anyone living near a beach or estuary will know. Rapid and often profound changes can be caused by human activity (e.g. deforestation, urbanization, dam-building) in both the supply and distribution of sediments. In comparison with such activity, natural seasonal fluctuations may be insignificant – dams across major rivers (e.g. the Nile and Colorado) can do in a few years what millennia of natural change would not have achieved.

Over periods of the order of 10^4–10^5 years, global climatic changes provide the dominant controls on rates of weathering and erosion, because this time-scale is relatively short in geological terms, and so topography remains a fairly constant factor. Thus, the most important changes to have affected the sediments found on the sea-floor today were those resulting from the climatic fluctuations that occurred over the past couple of million years (Pleistocene or Quaternary) and which were responsible for the repeated waxing and waning of the ice-sheets. These changes obviously caused variations in the supply of sediments, but the related changes in sea-level were more significant. Alternating **transgressions** and **regressions** of the sea over the continental shelves exposed them to alternating periods of submergence and of subaerial and glacial erosion, and caused coastal environments to migrate back and forth across them. During glacial maxima, sea-level was 100 m or more below its present level, so coastal sediments were deposited near the shelf-edge. This greatly increased the frequency of turbidity currents and hence the rate of sedimentation on continental rises and in the deep sea.

During these periods of low sea-level, both rivers (at low latitudes) and glaciers (at high latitudes) extended much further towards the open ocean, and river and glacial sediments were deposited on coastal plains that are now submarine continental shelf. When the ice melted and the sea inundated the shelves once more, these terrestrial deposits remained on the shelves, below sea-level, as **relict sediments**. Many relict sediments were originally mixtures of gravels, sands and muds. Like all shelf sediments, they are subject to constant reworking by waves and currents, which winnow away the finer-grained sands and muds, to be deposited elsewhere. The waves and currents are not usually strong enough to shift the gravels, which are left behind (as *residual deposits*), and in many places (including the North Sea and English Channel) are nowadays extracted for use by the construction industry (see Chapter 8).

Shallow-water environments are transient, existing only for relatively brief periods of geological time. Thus, shelf seas are not permanent features; they advance and retreat on the same time-scales as global changes in sea-level. Marine erosion is cutting back the sea cliffs of parts of eastern England at rates measured in metres per year, while marine deposition is extending the shoreline of some other parts at a similar rate. Most present-day estuaries originated when sea-levels rose close to their present levels, following the end of the last major **glaciation**, about 12 000 years ago. Sea-level began to stabilize about 5000 to 6000 years ago and estuaries then began to silt up quite rapidly. Today's familiar coastal features around Britain are not the same ones that the Romans encountered when they invaded Britain some 2000 years ago, nor are they the same as those that will be appreciated by the generation that celebrates the year 4000.

Even during periods when sea-level is changing rapidly on the geological time-scale, changes are intermittent rather than continuous. The existence of raised beaches round many coasts, for example (Figure 3.5), is evidence that the sea-level remained essentially constant over periods of perhaps several hundred to a few thousand years – because it takes that long for waves and currents to erode a more or less flat surface at the coast. Similarly, submerged cliffs and river valleys formed during periods of low sea-level may be important features of the continental shelves, especially those that receive little sediment, such as the central part of the shelf off the eastern United States. The present heights and depths of such features are not related merely to past rises and falls of sea-level in response to changes in the volume of ice-caps and glaciers; they result also from isostatic adjustments caused by differential loading of the crust by ice or sediments.

For the remainder of this Volume, we shall concentrate on the physical and chemical principles that underlie the transport and deposition of sediments, with particular reference to the near-shore environment. Understanding these processes can help to safeguard coastal environments against the worst excesses of human intervention in sedimentary cycles. There is a natural tendency towards equilibrium between the rate at which sediment is supplied to coastal regions, and the redistribution of this sediment by water movement. This equilibrium needs to be understood by marine engineers when they build jetties and breakwaters to protect harbours, construct groynes to prevent beach erosion, or lay submarine cables and pipelines. Otherwise, the equilibrium may be disrupted with disastrous consequences either for the constructions themselves, or for adjacent stretches of coastline, and sometimes even for both.

Figure 3.5 Raised beach near Westward Ho! in Devon. Pebbles and boulders overlie a raised shore platform cut across inclined layers (slates) forming the bedrock. The present-day shore platform is visible in the distance.

3.2 SUMMARY OF CHAPTER 3

1 The sediments of most continental shelves are predominantly terrigenous and consist of rock fragments, quartz sands and clay-rich muds. Where terrigenous sediments are scarce or absent, carbonate sediments of biogenic or inorganic origin may occur, especially in low latitudes.

2 River transport is the most important means of bringing terrigenous sediments to the ocean margins, but ice transport, wind transport and volcanic eruptions may be important locally.

3 Periodic falls in sea-level during the Quaternary resulted in the deposition of river and glacial sediments on areas of the continental shelf which are today covered by the sea. These relict sediments are now being reworked by waves and tidal currents. Changes of sea-level lead to changes in, and migrations of, shallow-water environments on time-scales of hundreds to thousands of years.

Now try the following question to consolidate your understanding of this Chapter.

QUESTION 3.2 Which of the following types or sources of shallow marine sediments are most likely to be restricted to particular bands of latitude?

(a) Relict sediments.

(b) Glacial sediments.

(c) Wind-blown dust.

(d) Volcanic debris.

(e) Carbonate sediments.

CHAPTER 4 PRINCIPLES AND PROCESSES OF SEDIMENT TRANSPORT

'Turning the shingle, returning the shingle,
Changing the set of the sand ...'

From *The Dyke* by Rudyard Kipling.

In this Chapter, we consider in general terms the physical conditions which lead to the erosion, transport and deposition of sediment. Sediments are moved about more in shallow water than in the deep sea, because surface waves can affect the sea-bed (Section 1.2.3) and tidal currents are typically stronger in shelf seas than in the open ocean, because of increased tidal ranges (Section 2.4.1). Sediment transport and deposition are also more easily studied in shallow water, but the principles governing these processes are as valid in the deep ocean as they are in any estuary or on any beach – or anywhere else where there is moving water.

4.1 FACTORS CONTROLLING THE MOVEMENT OF SEDIMENT

Table 4.1 The classification of sedimentary particles according to size (modified from the Wentworth scale). Note that this is a *geometric* scale, each size division differing from those below and above by a factor of two, except for the smallest and largest sizes.

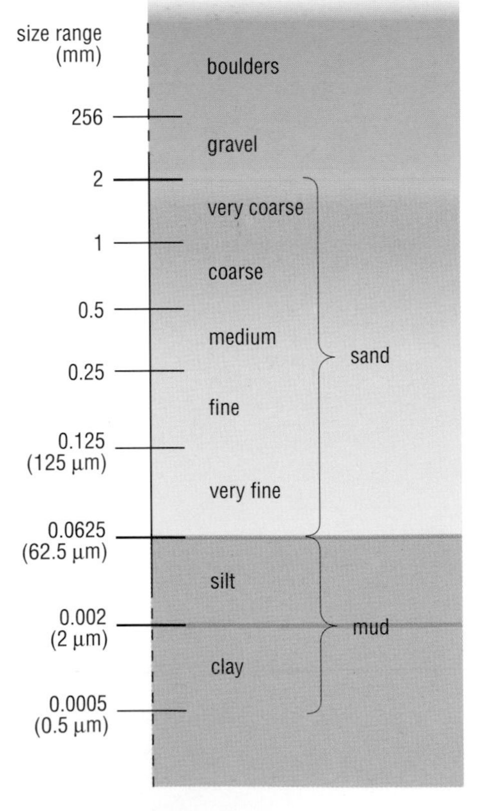

From your own observations, you know that gentle waves breaking on a sandy beach are capable of washing sand grains up and down the beach but do not normally shift pebbles. Plainly, one of the most important parameters controlling the transport and deposition of sediments is grain size, and it is convenient to classify sediment particles on this basis. To those who work with sediments, the terms mud, clay, silt, sand and gravel have specific grain-size limits, and Table 4.1 presents a widely used classification system. You may find it useful to refer back to Table 4.1 as you read though this Section, to keep in mind the particle sizes involved when a particular type of sediment is being discussed.

When water flows over a surface fast enough, sediment particles on the surface are picked up and transported, to be deposited again when the speed of flow diminishes. Particles too large to be lifted into suspension may be rolled or bounced along the surface by the water flow.

In more detail, the four modes of transport in water are *sliding*, *rolling*, *saltation* and *suspension* (Figure 4.1). *Sliding* particles remain in continuous contact with the bed, merely tilting to and fro as they move. *Rolling* grains also remain in continuous contact with the bed, whereas *saltating* grains 'jump' along the bed in a series of low trajectories. Sediment particles in these three categories collectively form the **bedload**. The **suspended load** consists of particles in *suspension*, that is, particles that follow long and irregular paths within the water and seldom come in contact with the bed, until they are deposited when the flow slackens. Sliding and rolling are prevalent in slower flows, saltation and suspension in faster flows.

Notes to Table:

* Some older scales use 4 μm as the dividing line between silt and clay on this scale.

** Mud is a non-specific term used to describe cohesive mixtures of clays, silt and sometimes even very fine sand (see also Section 4.1.2).

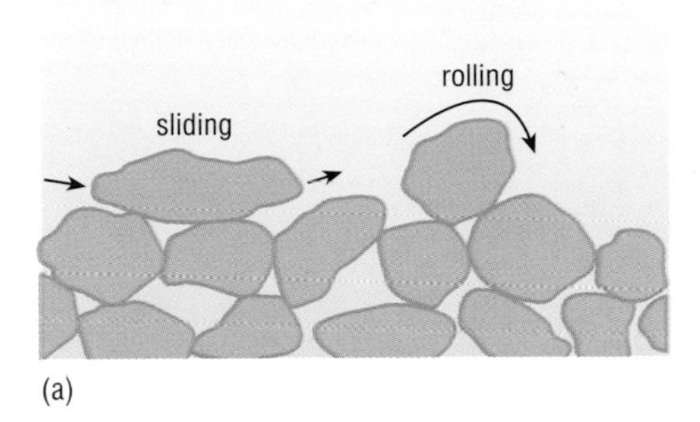

(a)

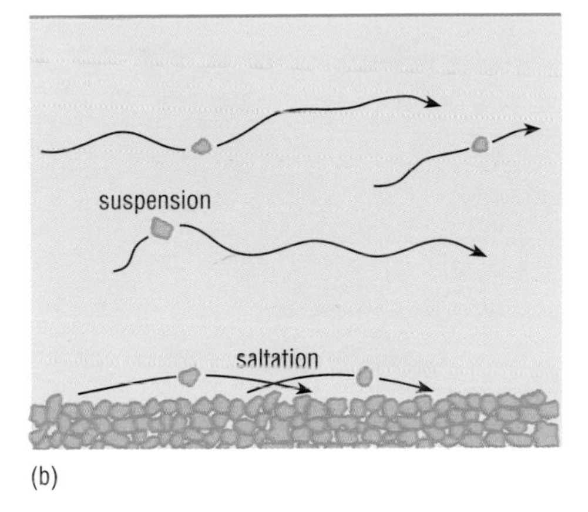

(b)

Figure 4.1 Modes of sediment transport in water.
(a) Sliding and rolling.
(b) Saltation and suspension. Broadly similar patterns of transport occur under wind, though of course flow velocities need to be much greater to move particles of comparable size, because of the much lower density and viscosity of air.

4.1.1 FRICTIONAL FORCES AND THE BOUNDARY LAYER

Water (indeed, any fluid) flowing near a solid surface is slowed down by friction along the boundary – it experiences a current shear (Section 2.4.1, Figure 2.21), and the region of flow influenced by proximity to the surface is called the **boundary layer**. A boundary layer develops wherever a fluid moves over a surface, whether it be water over the sea-bed, winds over the sea-surface, or syrup over a tabletop.

If you could measure the speed of a current at intervals above the bed, you would find that it varies systematically (Figure 4.2). In theory, provided that no sediment on the bed is moving, the imperceptibly thin layer of water in direct contact with the bed is also stationary – its speed should be zero. However, the layer of water immediately above this does move, albeit very slowly, and slides over the lower layer. With increasing distance from the bed, successive layers of water move a little faster as the effects of friction with the bed decrease. There is thus a velocity gradient – a change of velocity with depth, or velocity *shear* – and a graph of flow velocity against height above the bed (velocity profile) would resemble Figure 4.2. However, the *rate* at which speed increases gradually lessens with increasing distance from the bed, as the influence of friction with the bed begins to die out. Eventually, the speed stops increasing and reaches a more or less constant value, at the top of the boundary layer.

The top of each layer of water is acted upon by a **shear stress** (frictional force) due to the layer above (which is moving faster and tending to drag it along) as well as by a shear stress due to the layer below (which is moving more slowly and tending to drag it back). To determine whether sea-bed sediment is likely to be moved by a current, we need to know the value of the shear stress *actually at the bed*. As you will see later, because speed increases systematically with distance from the bed, we can use the rate of increase to determine the shear stress at the bed, rather than measuring it directly (which would be impractical).

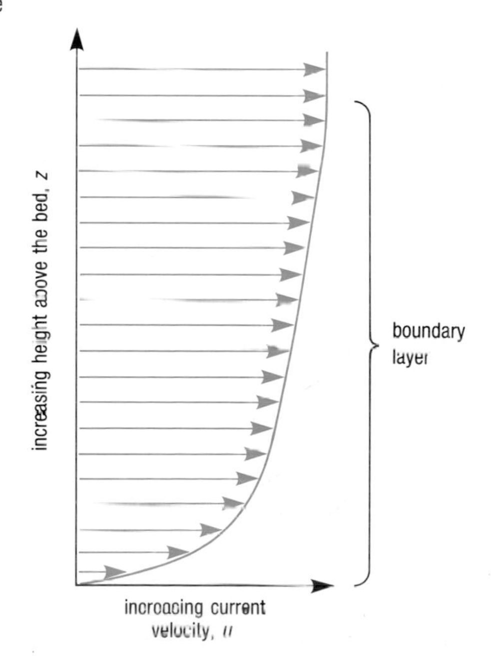

Figure 4.2 The velocity profile for steady current flow over a bed, showing current shear (length of arrows proportional to velocity) in the boundary layer.

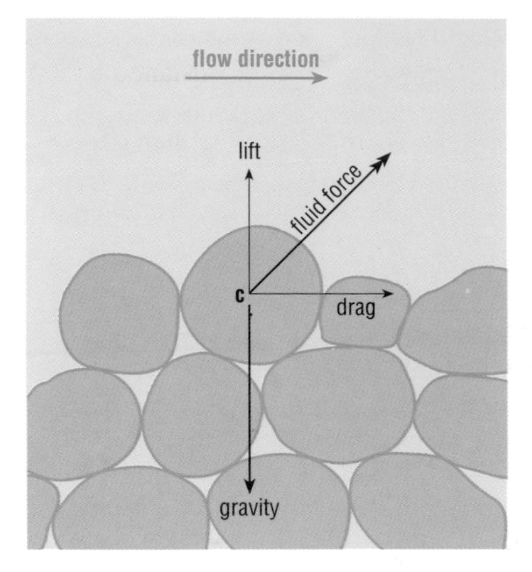

Figure 4.3 The forces acting on a stationary sediment grain resting on a bed of similar grains in a flow. Point c is the centre of gravity of the grain. The drag and lift forces do not necessarily act through the centre of gravity, because the flow around the grain will be influenced by the shape of the surrounding grain surfaces.

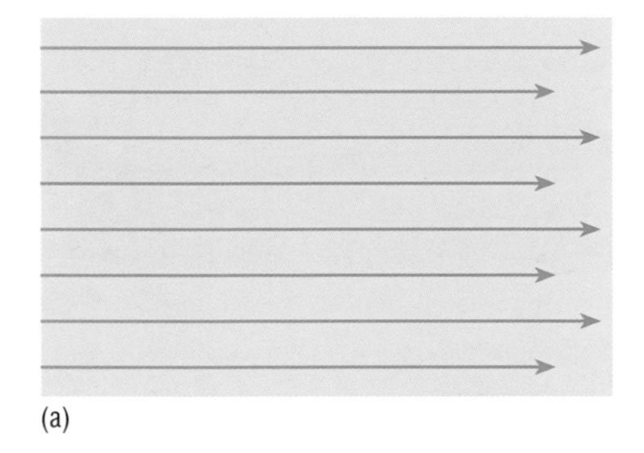

(a)

(b)

Figure 4.4 Diagrammatic illustration of the difference between
(a) laminar flow and
(b) turbulent flow in which complex multi-directional eddies are superimposed on the overall flow direction.

The shear stress exerted by any moving fluid is proportional to the square of the speed of flow (cf. wind stress, Section 1.1.2). In addition to the frictional drag, particles are subjected to a lifting force by the moving current – analogous to the lift on an aircraft wing (cf. Figure 4.3) – but in theoretical treatments this is often either ignored or considered as part of the drag (shear stress).

Flow in boundary layers may be laminar or turbulent (Figure 4.4); in the atmosphere and the oceans, winds and currents are nearly always turbulent, but may be laminar close to a boundary (see Section 4.2.3).

In the oceans, most of the erosion and deposition (and most of the transport) of sediment takes place in the *benthic boundary layer* adjacent to the sea-bed, which can be several tens of metres thick – so in shallow water it can occupy the whole water column. In any reasonably fast-flowing tidal current (or river for that matter), larger eddies can be seen as *boils* that form as they surge ('boil') upwards to interact with the water surface. They are transient features, lasting no more than tens of seconds at most, and are constantly developing and decaying.

The extent to which sediment movement takes place depends upon the degree of turbulence (frictional interaction with the bed is greater when flow is turbulent) and the current shear (i.e. the rate of change of current velocity with depth, Figure 4.2), which determines the shear stress. It also depends upon the nature (composition) and roughness of the bed.

In addition to particle size and current speed, other factors controlling sediment transport and deposition are the densities of the particles and of water (in practice, the density *contrast* between particles and water), the viscosity of the water and whether the flow is laminar or turbulent (Figure 4.4). These factors also control the rate at which particles sink, and are therefore important when considering sediment deposition.

In turbulent flow, erratic and variable fluctuations are superimposed on the mean motion (Figure 4.4). Parcels of water follow random eddying paths about the net direction of flow. To deal with turbulence mathematically, the flow velocity can be broken down into three components at right angles to each other (Figure 4.5), each of which at any one time consists of the mean value plus the randomly varying component. The u-component of velocity is horizontal and parallel to the net flow direction; the v-component of velocity is also horizontal, but at right angles to the net flow direction; and the w-component of velocity is vertical. The time-averaged velocity in the net flow direction is given the symbol $\bar{u}$ (*u-bar*).

These complications notwithstanding, we can make the general statement that shear stress is proportional to the product of the density of the water and the square of the time-averaged current speed:

$$\tau_0 \propto \rho \bar{u}^2 \qquad (4.1)$$

where τ_0 is the shear stress at the bed (τ is the Greek letter '*tau*');

ρ is the density of the water – effectively constant for most practical purposes; and

$\bar{u}$ is the time-averaged current velocity.

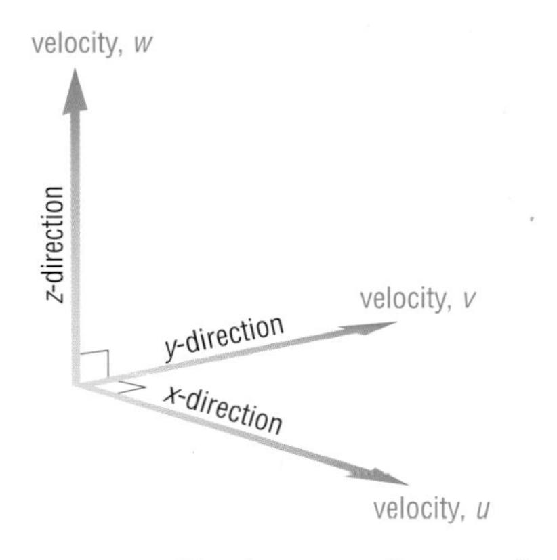

Figure 4.5 The three mutually perpendicular components of velocity into which turbulent flows can be resolved. *u, v* and *w* are the velocity components in, respectively, the *x*-, *y*- and *z*-directions.

Although the density of fluids (e.g. water, air) does vary with temperature, for most practical purposes it can be considered effectively constant and so shear stress depends in practice only on the square of the current (or wind) speed. (Note, however, that the density of air is very much less than that of water, and a wind of *c.* $10\,m\,s^{-1}$ is required to generate a shear stress similar to that of a current of *c.* $0.5\,m\,s^{-1}$.)

If density is in $kg\,m^{-3}$ and velocity is in $m\,s^{-1}$, then τ_0 is in $kg\,m^{-3}\,m^2\,s^{-2}$, which simplifies to $kg\,m^{-1}\,s^{-2}$, which is the same as $N\,m^{-2}$. Shear stress thus has the same units as pressure (since N = newton, the unit of force, $kg\,m\,s^{-2}$), but the shearing force is exerted *parallel* to the bed.

What will happen to the shear stress if the average current velocity is doubled?

From Equation 4.1, if current velocity is doubled, the shear stress will be quadrupled, i.e. it will increase by a factor of four.

The movement of sediment on the sea-bed (Figure 4.1) begins when the shear stress at the bed (τ_0) becomes sufficiently great to overcome the frictional and gravitational forces holding the grains on the bed. This is known as the **critical shear stress**.

4.1.2 COHESIVE AND NON-COHESIVE SEDIMENTS

The relationship between grain size and critical shear stress is not a straightforward linear one. Some sediments are **cohesive** in character, and this has a significant effect on sediment erosion. Cohesiveness results mainly from the presence of clay minerals in the sediment. Clay mineral particles are flaky and typically very small, less than $2\,\mu m$ in size (cf. Table 4.1). In sediments, they tend to form aggregates in which the individual flakes are held together by a combination of electrostatic attraction and the surface tension of the films of water surrounding the flakes. These forces are strong and give muds their glutinous property. Clays increase the overall cohesion of the bed, even when they constitute only a small proportion of the total sediment; cohesion begins to be significant when sediment contains more than about 5–10% of clay by weight. Indeed, the 'stickiness' of many predominantly silty sediments is probably caused by only relatively small amounts of clay. The critical shear stresses required to set cohesive sediments in motion are much greater than might be supposed from their small particle size (see Figure 4.6 on pp. 102–3).

The grains of **non-cohesive sediments** are more equidimensional in shape than those of cohesive sediments. They lack the physico-chemical interactions that exist between clay particles, and so are free to move independently. As you might expect, therefore, critical shear stresses required to set non-cohesive sediments in motion decrease with decreasing particle size. Non-cohesive sands and silts (Table 4.1) nearly everywhere consist largely of quartz of terrigenous origin, though they include the black sands of beaches near volcanoes, and the carbonate sands formed where supplies of terrigenous material are small (Section 3.1). Smaller gravel fragments usually consist mostly of quartz grains, larger ones of rock fragments.

4.2 SEDIMENT EROSION, TRANSPORT AND DEPOSITION

Note that use of the term 'erosion' in this and subsequent Chapters refers only to the process of setting particles on the sea-bed in motion, whether in suspension or as bedload (Figure 4.1). We do not consider here the other aspects of erosion (Section 3.1), namely the abrasive action of sediment particles upon each other and upon the surface over which they are transported – though such abrasion does of course take place.

Figure 4.6(a) (overleaf) illustrates the relationship between the average grain sizes of sediments and the average current speeds required to erode them. The right-hand half of the diagram applies to non-cohesive sediments, so the current speeds (and hence the critical shear stresses) required to set them in motion decrease with decreasing particle size.

QUESTION 4.1 By contrast, the left-hand half of Figure 4.6(a) is at first sight counter-intuitive, for it shows that current speeds (and hence the critical stresses) required for erosion actually *increase* with decreasing average particle size. Why is this?

Figure 4.6(b) illustrates the relationship between average particle sizes of sediments and the current speeds above which they are transported (whether in suspension or in the bedload), and below which they are deposited. The boundary between bedload transport and suspension transport is represented by a broken line, indicating that the transition between these two modes of transport is gradational, on account of the inherent variability of natural sediments and real currents.

Comparing Figure 4.6(a) and (b), you can see that the curves for erosion and deposition lie close together for particles averaging more than about 1 mm in size (and which are non-cohesive), but diverge progressively for particles smaller than this – because as we have seen, the proportion of clay minerals increases as particle size decreases.

Figure 4.6(c) combines the two curves. The most important features of Figure 4.6 are:

1 For particles larger than about 1 mm, the 'deposition' curve lies close to but just below the 'erosion' curve. Current speeds required to set these non-cohesive sediments in motion in the bedload are slightly greater than those at which they are deposited, partly because of the inertia of the particles on the bed, and partly because natural sediments are not of uniform grain size, so that larger particles are liable to interlock with and to 'shelter' smaller ones.

2 Particles smaller than about 0.1 mm are typically not transported in the bedload, but taken directly into suspension. When in suspension, however, they can be transported at current speeds very much smaller than those required to erode them, and the difference increases dramatically with decreasing particle size. Once muds have been eroded, therefore, they can be transported for great distances before being deposited again.

3 Also included for completeness in Figure 4.6(c) is the scale for settling velocities. These are calculated for equidimensional particles of different sizes, and you may have spotted that the scale changes part-way along. This is because for grains finer than about 0.1 mm, flow round the sinking particles

is laminar (Figure 4.4(a)), and according to Stokes' law, the settling velocity is proportional to the *square* of the grain diameter. Above a diameter of about 2 mm, flow round the sinking particles is turbulent (Figure 4.4(b)), Stokes' law no longer applies, and the settling velocity is proportional to the *square root* of the diameter – which means that even quite a large increase in diameter leads to only a rather small increase in settling velocity. For the size range 0.1 to 2 mm, flow round the sinking particles is transitional between laminar and turbulent, and settling velocity is proportional to progressively decreasing powers of the diameter, from d^2 to $d^{\frac{1}{2}}$, where d is the grain diameter. Settling velocity is commonly given the symbol w_S, because settling is *vertical* motion (Figure 4.5).

4.2.1 EROSION OF COHESIVE SEDIMENTS AND YIELD STRENGTH

We have seen that, despite their fine grain size, muds are not easily eroded once they have been deposited. The strong binding forces that hold clay particles together mean that cohesive sediments tend to be lifted as clumps or 'flocs', rather than as individual particles, though they generally disaggregate once in suspension. If the muds have become partially consolidated (e.g. on exposed tidal mud-flats, see Chapter 6), erosion occurs following mass failure of the sediment surface, which is ripped off in large lumps – a process that requires very high shear stresses. The cohesion of very fine-grained sediments is influenced by their water content; by the proportion and particle size of clay minerals in them; and by the salinity of both the overlying water and water trapped between the sediment grains (which in estuaries, for example, is not necessarily equal to that of normal seawater).

The resistance of mud to erosion can be assessed by its **yield strength**, which is the maximum shear stress that the sediment can withstand before failure occurs. One method of measuring yield strength is by using a shear vane. This is a device that has a tip comprising two plates at right angles to one another at the end of the shaft. The tip is inserted into a mud sample and force is applied to rotate the shaft. The yield strength is calculated from the force measured at the moment when the sediment yields and the blades suddenly start to rotate. It requires shear stresses equivalent to those needed to move fine gravel in order to erode cohesive, but relatively uncompacted, muds. As the degree of compaction increases, however, very much higher shear stresses are needed to induce sediment failure. In relatively sheltered waters where wave and current action is not strong, both cohesive and finer non-cohesive sediments may be bound together over large areas by mats formed of **benthic** (i.e. bottom-living) filamentous algae, bacteria and organic material that they produce.

In general, for most muddy sediments in marine environments, the yield strength ranges from about 0.5 to about 5 N m^{-2}, depending partly upon the particle size and partly upon the degree of compaction and binding by organisms. Values in excess of 5 N m^{-2} may be encountered in the older sediments of estuaries or tidal flats, which tend to be both drier and more compacted than 'fresh' muds.

Theories about sediment movement can become very intricate and the detailed measurements required to test them are difficult to obtain. Nonetheless, a reasonable approximation of reality can be arrived at by making some simplifying assumptions. In the next Section, we consider other aspects of the relationship between current speed, shear stress and the movement of sediment.

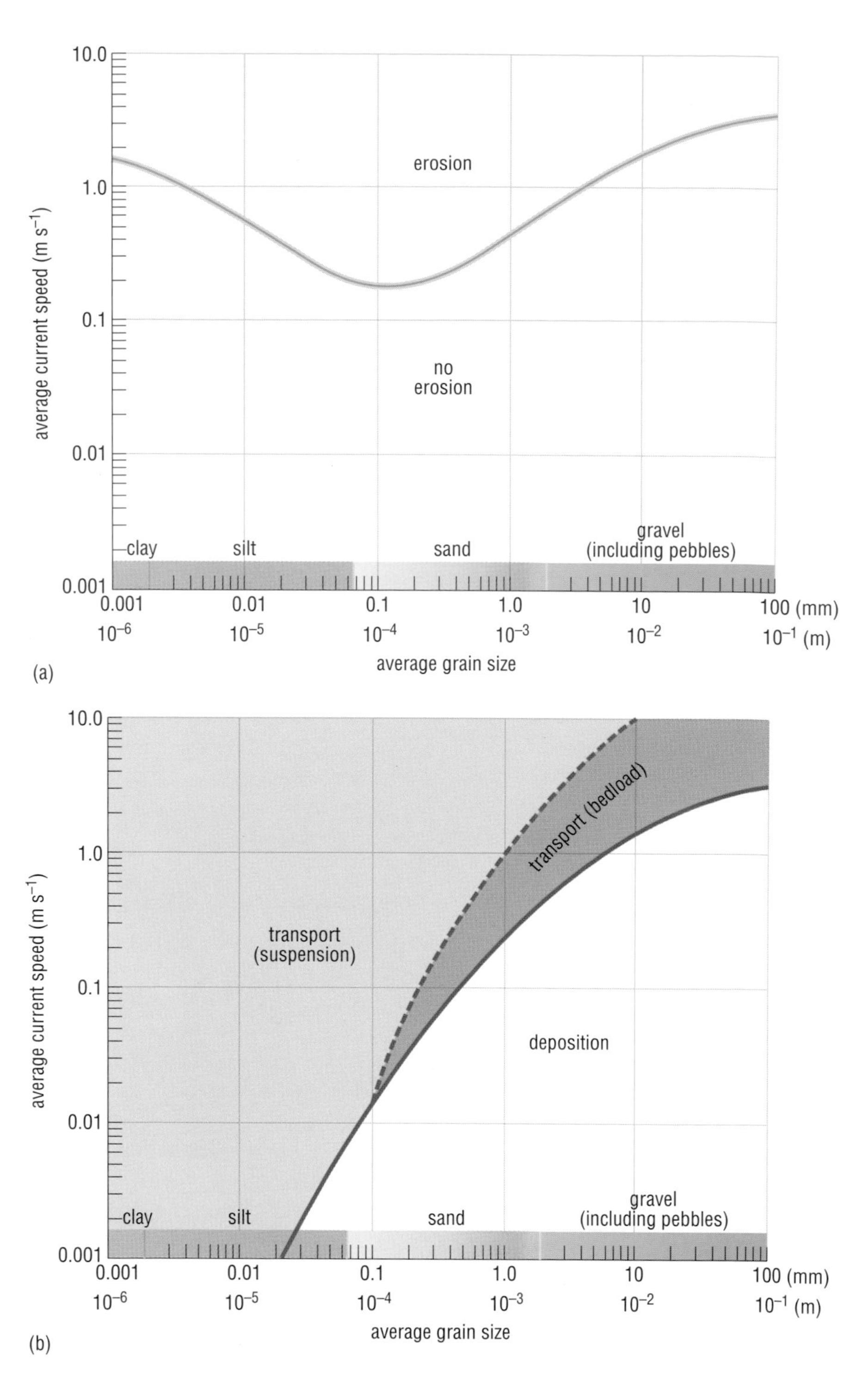

Figure 4.6 (a) Diagram showing the range of average current speeds at which sediment particles of different sizes are eroded, i.e. set in motion. The curve for sediments finer than about 0.1 mm is for relatively uncompacted silts and muds (see also Section 4.2.1).
(b) Diagram showing the range of average current speeds at which sediment particles of different sizes are transported, in suspension or as bedload, and below which they are deposited. The broken line indicates the transition between bedload and suspension transport.

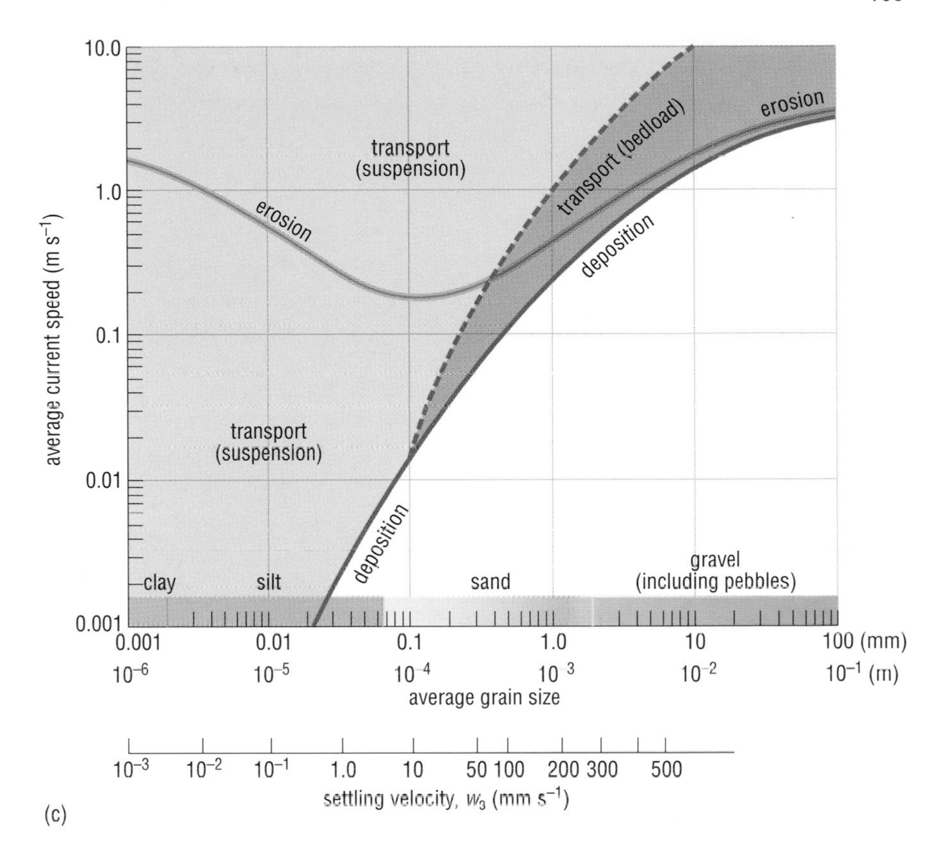

(c) The curves in (a) and (b) are combined in this diagram. Note that particles can be transported *both above and below* the 'erosion' curve, and that deposition of fine-grained sediments occurs only at very slow current speeds. Average settling velocities (w_s) for particles of different sizes deposited from suspension in still water are also shown. (Diagrams compiled from various sources; all scales are logarithmic.)

4.2.2 THE CONCEPT OF SHEAR VELOCITY

It can be shown theoretically that shear stress at the sea-bed is directly dependent on the viscosity of the water and the average velocity gradient in the overlying water:

$$\tau_0 = (\mu + \eta) \times \frac{d\bar{u}}{dz} \quad (4.2)$$

where τ_0 is shear stress on the sea-bed as before;

μ (Greek '*mu*') is the molecular viscosity (see below);

η (Greek '*eta*') is the eddy viscosity (see below); and

$\frac{d\bar{u}}{dz}$ is the time-averaged velocity gradient directly above the bed.

(Note that the symbol A is sometimes used to denote eddy viscosity. Note also that viscosity units are N s m^{-2}, u is in m s^{-1} and z is in m, so τ_0 will again be in N m^{-2}.)

The **molecular viscosity** of a fluid is a measure of its resistance to flow in a laminar fashion (Figure 4.4(a)), and results from the forces that tend to 'glue' molecules together, e.g. motor oil and treacle are more viscous than water and flow less readily. Molecular viscosity is a constant physical property for any fluid at a given temperature, and in general increases with decreasing temperature, i.e. liquids flow less readily when cooled. The viscosity of water in polar regions is about twice that in equatorial regions, and you may have noticed that water in rivers appears almost 'oily' in cold weather, when it flows more sluggishly.

In the boundary layer, however – and indeed throughout the oceans – turbulent flow (Figure 4.4(b)) predominates, and it is the **eddy viscosity** that matters. In turbulent flows, a random multidimensional component is added to the overall (or net) flow direction (Section 4.1.1), and parcels of water (rather than individual molecules) are constantly interacting with one another. The intermolecular forces are still present but are greatly exceeded by resistance to motion brought about by randomly moving (eddying) parcels of water. As a result, the eddy viscosity is several orders of magnitude greater than the molecular viscosity and is highly variable.

QUESTION 4.2 From these descriptions, how would Equation 4.2 be simplified for (a) laminar and (b) turbulent flow?

Looking at Equation 4.2 in connection with Question 4.2, you may have realized that it is not a straightforward exercise to determine the shear stress at the sea-bed. That is partly because the eddy viscosity continually varies in turbulent conditions, and partly because the rate of increase of current velocity with height above the bed (the velocity gradient) in the boundary layer is not constant (Figure 4.2).

However, it turns out that at least in the lowermost couple of metres of the boundary layer, flow velocity increases logarithmically with height above the bed (Figure 4.7(a)). Accordingly, if we measure current velocities at various heights above the bed within the lower part of the boundary layer and plot the data on a graph using a log scale for height above the bed, the velocity gradient is a straight line (Figure 4.7(b)).

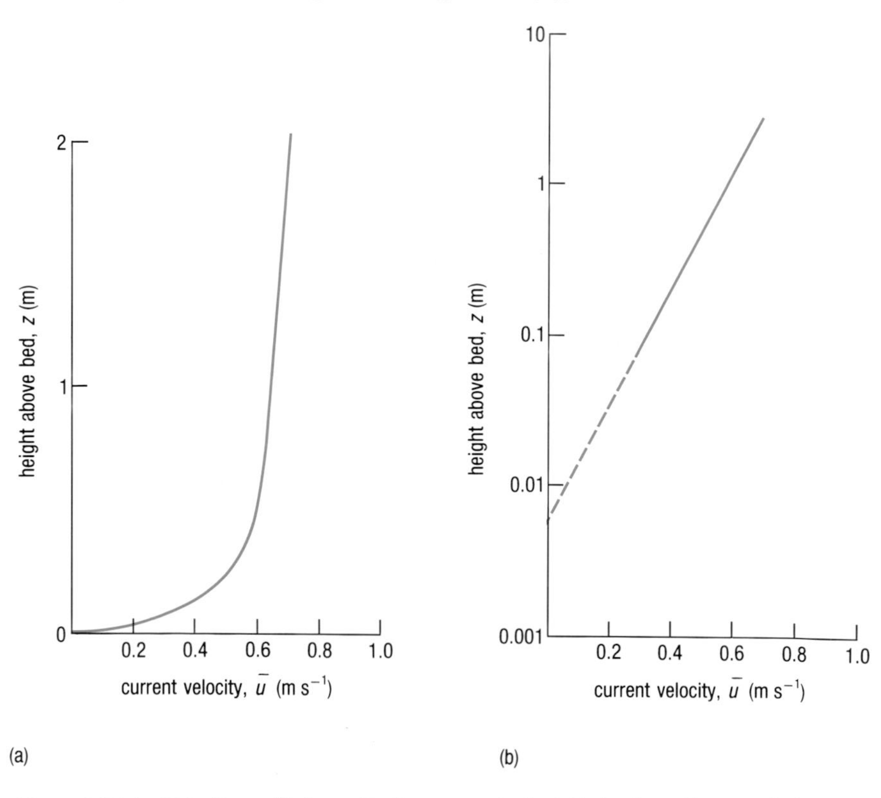

Figure 4.7 (a) Velocity profile for water flow over a bed, plotted using a linear scale for both the horizontal and vertical axis (i.e. for both velocity and height above the bed).
(b) The same velocity data as in (a), plotted using a $\log_{10}$ vertical scale (height above the bed) and a linear horizontal scale (current velocity).

The straight line in Figure 4.7(b) is dashed below about 0.05 m (5 cm) because current meters cannot be placed any closer to the bed. The dashed line intersects the vertical (depth) axis at a finite distance above the bed, for reasons we shall explain shortly. There is a simple relationship between the gradient of the straight line (Figure 4.7(b)) and a quantity known as the **shear velocity**, u_*, from which the shear stress can readily be calculated. Note that the shear velocity is a *derived* quantity, which means that it cannot be directly measured by suspending current meters in a water flow. More important, *to obtain the shear velocity we use the* ***inverse*** *of the gradient of the graph in Figure 4.7(b).*

The reason for this lies simply in the way the graph is plotted, *not* in any mathematical 'magic'. Conventionally, when plotting a graph, the variable that can be 'fixed' is plotted on the horizontal (x) axis, and the variable that is to be measured is plotted on the vertical (y) axis. However, if we were to plot depth (the 'fixed' variable) horizontally, and current speed vertically, we would lose the visual effect of change in velocity with height above the bed. That is why we normally plot height (or depth) vertically, following the same convention as in velocity profiles (e.g. Figure 4.2).

The shear velocity is related to the *inverse* gradient of the graph by the simple empirically determined equation:

$$u_* = \frac{1}{5.75} \times \frac{d\bar{u}}{d\log z} \qquad (4.3)$$

An example of how the calculation is made using a logarithmic velocity profile is shown in Figure 4.8 (overleaf), and Figure 4.9 (also overleaf) is a graph plotted from actual measurements. Note that in Equation 4.3, $\bar{u}$ has units of velocity (m s^{-1}) and log z is simply a number, so u_* also has units of m s^{-1} (which is why it is called the shear *velocity*). In general, shear velocities (calculated) are about an order of magnitude *smaller* than the mean velocities (measured) of the real currents involved.

The relationship between shear velocity and shear stress is given by:

$$u_* = \sqrt{\frac{\tau_0}{\rho}} \qquad (4.4)$$

or

$$\tau_0 = \rho u_*^2 \qquad (4.5)$$

where ρ is the density of the water, and is effectively constant for most practical purposes (cf. Equation 4.1). Note once again that shear stress is proportional to the *square* of a velocity term, and that its units are those of force per unit area: kg m^{-3} m^2 s^{-2} = kg m^{-1} s^{-2} = N m^{-2}.

QUESTION 4.3

(a) Using the value of u_* calculated in the box in Figure 4.8, work out the value for the shear stress at the bed (τ_0) beneath the flow. (Assume the density of water is 10^3 kg m^{-3}.)

(b) How does the value of u_* compare with the actual time-averaged current speed 1 m above the bed in Figure 4.8?

(c) Is the shear velocity calculated using the depth interval 1 m to 10^{-1} m on Figure 4.8 the same as that between 10^{-1} m and 10^{-2} m?

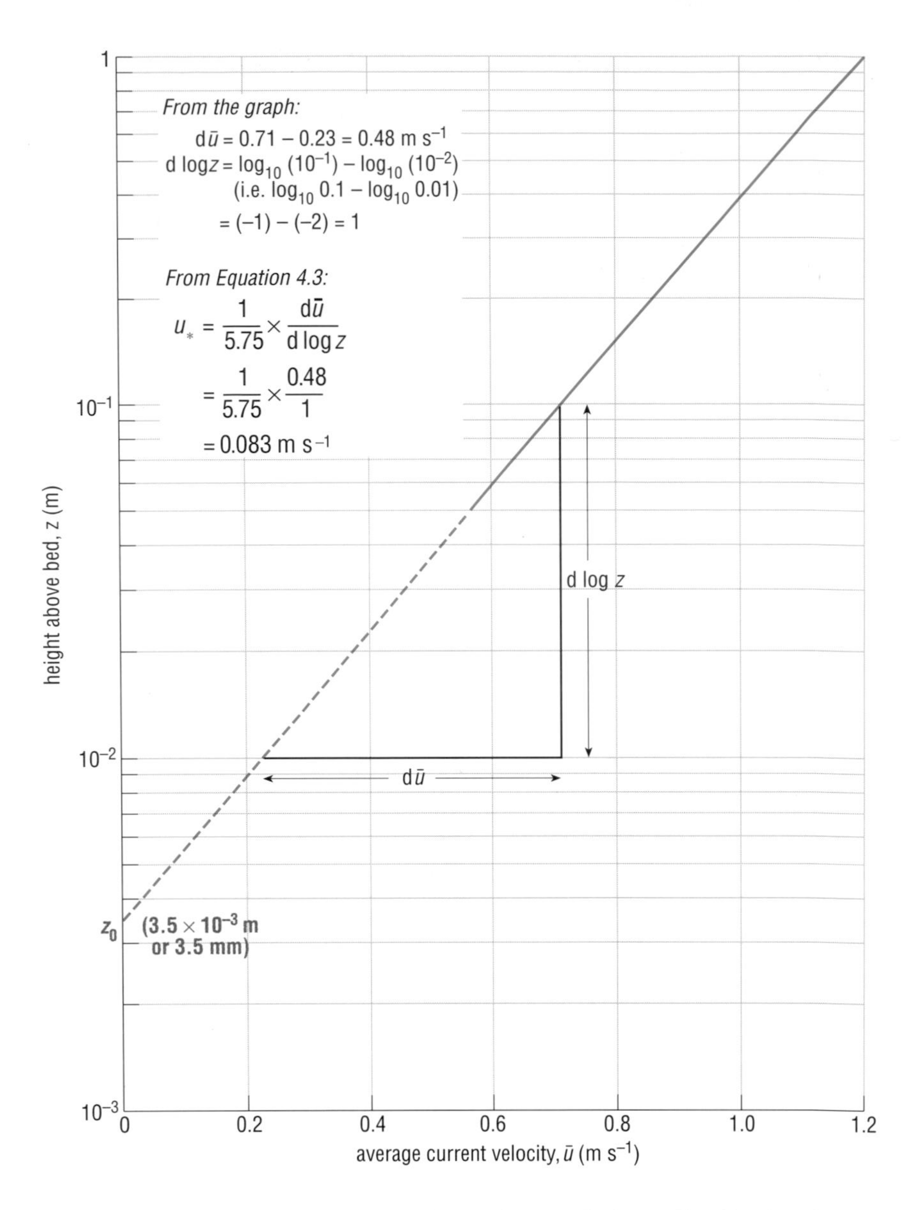

Figure 4.8 The calculation of shear velocity (u_*) from a velocity profile with a logarithmic ($\log_{10}$) vertical scale. The significance of z_0 (the roughness length) is explained in the text.

The main difficulty with this method is that the theory is strictly valid only for steady flow, and marine currents are rarely steady. We discuss these issues again in Section 4.2.4, but first we consider the intercept of the straight line with the vertical axis at z_0 on the vertical scale in Figure 4.8 (cf. Figures 4.7(b) and 4.9). At first sight, the intercept suggests that current velocity is zero at some finite height above the bed, which is plainly unrealistic and is an artefact of the underlying theory. However, it does have some basis in physical reality: the value of the intercept z_0 is called the **roughness length**, because its value provides an indication of how rough the sea-bed is.

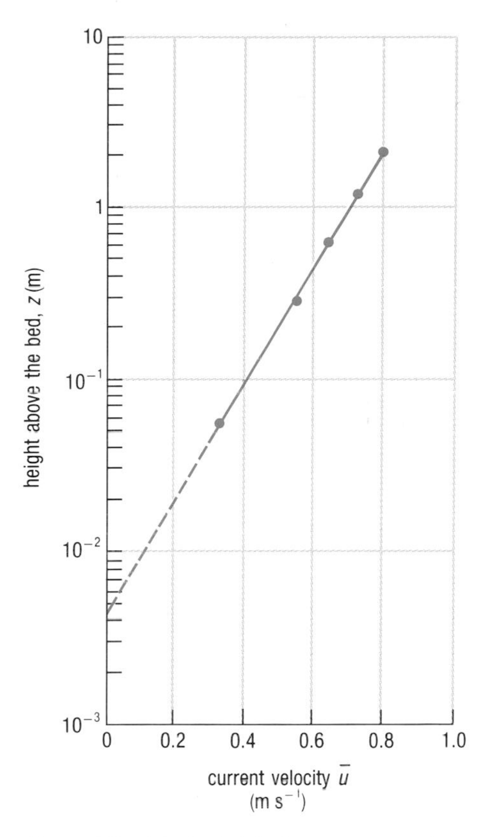

Figure 4.9 A typical plot of average current velocity, $\bar{u}$, measured at different heights, z, above the sea-bed in a tidal current. The vertical scale is logarithmic as in Figures 4.7(b) and 4.8.

Roughness length is related to the grain size of the sediment and/or to upraised features on the bed, such as sand ripples, where these occur (see Section 4.4). Its value increases as the grain size of the sediment on the bed (or the size of ripples in the sediment) increases, and so the velocity profiles constructed from current meter data can also provide an idea of the roughness of the bed beneath the flow. However, values of roughness length obtained from graphs such as Figures 4.8 and 4.9 should be taken only as a general indication of bed roughness, not as an accurate measure of particle size.

QUESTION 4.4

(a) How would you expect the slope of the graph in Figure 4.7(b) to change if:

(i) the current increased in speed, but continued to flow over a bed with the same roughness;

(ii) the current remained steady, but flowed over a bed with much coarser grain size?

(b) What are the implications of these changes for the value of τ_0 and erosion at the bed?

A generalization arising out of Question 4.4 is that the more nearly vertical the line on graphs such as Figure 4.9 the slower the current and hence also the smaller the shear velocity and corresponding shear stress. The theory that gives us the shear velocity, u_*, is not always strictly valid for flow very close to the bed, which we consider in the next Section.

4.2.3 THE VISCOUS SUBLAYER

Flow is not invariably turbulent throughout the full thickness of the boundary layer. Experiments have shown that under some circumstances there is a very thin layer adjoining the bed, in which flow is essentially laminar. This is the **viscous sublayer**, no more than a few millimetres thick, which develops only when current velocities are not excessive and where the bed is sufficiently even – i.e. the grain size of sediment particles and the size of upraised features such as sand ripples are not too great. Within the viscous sublayer, the velocity profile is effectively linear, so the velocity gradient is constant (Figure 4.10(a)). Since flow in the sublayer is laminar, the shear stress there is given by Equation 4.2, and is the product of the molecular viscosity and the gradient of the velocity profile (which is linear in the sublayer, cf. Question 4.2(a))*.

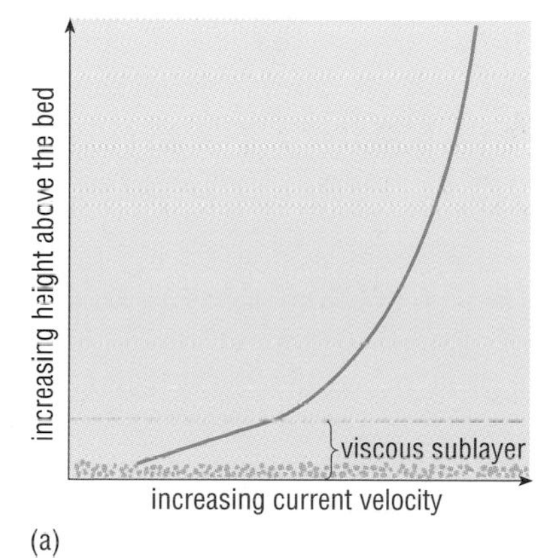

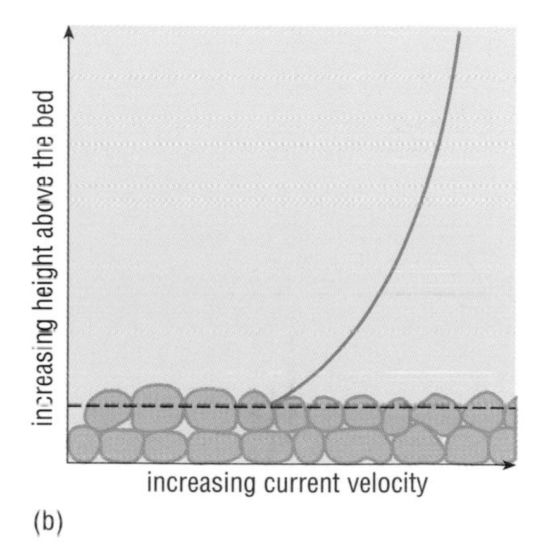

Figure 4.10 (a) Highly schematic velocity profile for smooth flow over a fine-grained bed. The profile is virtually linear within the viscous sublayer close to the bed.
(b) Highly schematic velocity profile for rough flow over a coarse-grained bed. The viscous sublayer (potential thickness is indicated by the dashed line) is disrupted by the coarse grains. Turbulent flow (the 'logarithmic layer', where d $\bar{u}$/d log$z$ is constant) extends very close to the bed.

* If you are mathematically inclined, you may be interested to learn that in mathematical treatments of the sublayer, the kinematic viscosity, (ν) (Greek '*nu*'), which is molecular viscosity divided by density) becomes important. For example, the sublayer thickness is proportional to ν/u_*.

The thickness of the viscous sublayer is inversely proportional to current speed, and hence also to shear velocity, u_*.

Does the thickness of the sublayer increase or decrease with rising current speed, other things being equal?

As its thickness is inversely proportional to the speed of flow, the viscous sublayer becomes thinner as the current speed increases. This means that a given bottom material may be 'smooth' under a slow current, so that flow is effectively laminar at the bed; but it will become 'rough' under a fast current, so that flow is turbulent right down to the bed.

From Equation 4.2 and your answer to Question 4.2, in general would you expect the thickness of the sublayer to be greater or less in cold than in warm water?

Molecular viscosity increases with falling temperature, so the viscous sublayer should be thicker where the water is cold (e.g. in polar regions), and thinner where it is warm (e.g. equatorial regions).

In naturally occurring turbulent boundary layers, much of the unevenness of the bed is provided by the sediment particles forming the bottom. If the average grain diameter is less than one-third of the thickness of the viscous sublayer, the sublayer remains intact. The main body of the turbulent flow above is 'unaware' that these grains exist and does not readily move them (Figure 4.10(a)). The near-bed flow is dominated by viscous forces and the flow can be described as *hydraulically smooth* (smooth boundaries are afforded by most mud beds and by the finer grades of sand, Table 4.1).

If the current flows over sediment with grain diameters greater than about one-third of the thickness of the viscous sublayer, the grains begin to disrupt the flow in the sublayer so that flow conditions become transitional between laminar and turbulent (Figure 4.10(b)), and the flow can be described as *hydraulically rough*. Once grain diameters have reached about seven times the notional thickness of the sublayer, they protrude so far into the turbulent layer that the sublayer breaks down altogether and fully turbulent flow conditions extend right down to the bed. Eddies are now able to 'reach' down between the sediment grains and there is much greater potential for sediment movement.

These changes in flow conditions near the bed are schematically shown in Figure 4.11, but it is important to remember that it is not only the grain size of the underlying sediment that determines whether smooth or rough flow will occur: the flow speed is also important. To give you some idea of the numbers involved, when u_* is 0.01 m s^{-1} (actual current speed *c.* 0.1 m s^{-1}) and the bed is smooth (very fine sand or smaller, Table 4.1), the sublayer is about 1.2 mm thick; it begins to break down when the average grain diameter reaches 0.4 mm, and ceases to exist at grain diameters of a few millimetres (very coarse sand or fine gravel, Table 4.1).

Would you expect the thickness of the viscous sublayer to be greater or less if u_* were greater?

It would be less, since sublayer thickness is *inversely* proportional to shear velocity: for a shear velocity of 0.04 m s^{-1} (actual current speed *c.* 0.4 m s^{-1}), the viscous sublayer over a smooth bed would not be thicker than about 0.3 mm.

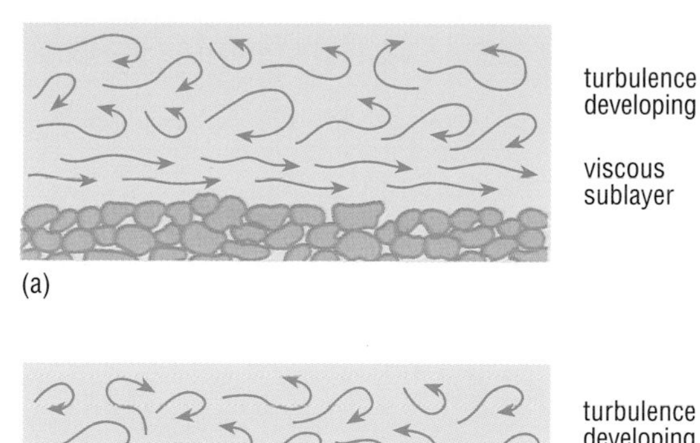

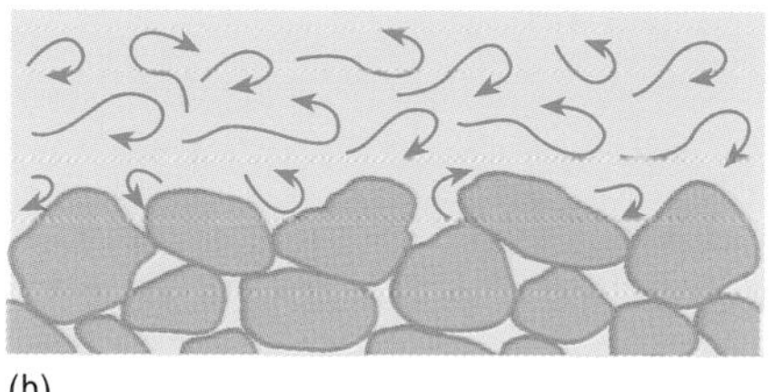

Figure 4.11 (a) A viscous sublayer, with laminar or quasi-laminar flow, over a smooth bed, cf. Figure 4.10(a).
(b) Turbulent flow occurs where the bed is rough, cf. Figure 4.10(b).

Considering for the moment *only non-cohesive sediments*, when the grain size is small and the flow is sufficiently slow the sediment grains are protected by the viscous sublayer and no movement occurs. As the shear velocity increases with increasing flow speed, the viscous sublayer starts to break down and the smallest particles are taken directly into suspension, while coarser grains begin to roll or occasionally to slide or bounce across the bed as bedload. Eventually, with increasing flow speed the shear velocity increases enough to randomly lift grains into suspension and temporarily redeposit them (i.e. the grains move by saltation, Figure 4.1). At still higher shear velocities, the grains are lifted permanently into suspension to be transported as suspended load, provided that eddies have vertical upward velocities exceeding the settling velocities (w_s) of the grains (see Section 4.3.2). In the case of cohesive sediments, the particles are typically so small that a viscous sublayer will develop when current speeds are low. However, for the majority of cohesive sediments, there can be no sublayer at the relatively high current speeds required to erode them (Figure 4.6(c)).

Shear velocities in shelf seas are generally in the range 0.01 to 0.04 m s^{-1}, so a viscous sublayer is not likely to develop where the bed is of gravel or where there are sand ripples. A sublayer can develop over a flat surface of silty sediment, but where there are flat sands it is likely to be more or less disrupted (depending on particle size), so flow close to the bottom will be transitional between laminar and turbulent. In the deep ocean, current velocities (and hence shear velocities) are generally about an order of magnitude less than in shelf seas, and so the sublayer is thicker (between about 3 and 12 mm).

Wherever a sublayer is developed and flow at the bed is smooth, particles can become trapped in the sublayer and be deposited. As mentioned above, sediment particles transported in suspension are kept there only so long as upward velocities in the eddies exceed the settling velocities of the particles. Once the particles get into the sublayer, where the flow is smooth, they can be deposited, even if the mean current speed is well above the 'deposition' curve of Figure 4.6. However, only a small proportion of the total suspended load would be deposited, because the sublayer is generally only a few millimetres thick. Where a current flows over a bed of gravels in which the particle size exceeds seven times the (notional) sublayer thickness, flow is turbulent right down to the bed, and silt and clay particles remain in suspension, because there is no sublayer in which they could be trapped. Of course, were the water then to flow at the same speed over a bed of very fine sands, flow at the bed could become smooth, with development of a sublayer in which suspended particles could become trapped and deposited.

4.2.4 VELOCITY PROFILES IN THE SEA

Currents in the sea continually vary in direction and speed. Tidal currents, in particular, change direction with time and accelerate from what may be effectively zero speed at slack water towards a maximum speed and then decelerate again (Section 2.4.1, Figures 2.19 and 2.21). The result is that the *logarithmic* velocity profile is not straight (as it is in Figures 4.8 and 4.9) for either flow direction, but curved, and the shear stress at the bed and values of both u_* and z_0 can be underestimated for accelerating currents, and overestimated for decelerating currents. However, for most situations in the sea, tidal current acceleration and deceleration occur near to slack water when the potential for sediment movement is low anyway. Furthermore, marine currents can have long 'memories', so when a current flowing over the sea-bed encounters a surface with a different roughness, it takes time for the velocity profile in the turbulent boundary layer to adjust. Similarly, changes in current velocity at the surface take some time to be 'transmitted' to the turbulent boundary layer near the bed.

In the real turbulent ocean, moreover, the shear stress is constantly varying, because of the eddies and vortices that occur on all scales from metres to centimetres. As you may imagine, sediment transport in natural flows is a difficult area to research, but observation and experiment combine to suggest that one of the most important processes involves what are known as burst–sweep cycles associated with smaller vortices near the bed. In brief, downward *sweeps* of high velocity water penetrate lower velocity layers near the bed (including the viscous sublayer, where present) exerting instantaneously high shear stresses at the bed and displacing both sediment and the immediately overlying lower velocity water as *bursts* upward into the main flow. Burst–sweep cycles are considered to be largely responsible for initiating sediment transport and also for the development of silt and sand ripples, which can significantly increase bed roughness.

The sea-bed is usually covered with both small-scale and large-scale sedimentary features known as *bed forms* (see Section 4.4). Probably most familiar are the small-scale sand ripples we have already mentioned, commonly seen on beaches and in estuaries, and formed by waves and currents. Burst–sweep cycles are likely to help such smaller scale bed forms to persist, once they have formed, because increased bed roughness leads to increased vertical and decreased horizontal water movements (and sediment transport). In other words, the ripples provide a partial obstruction to the flow, reducing its capacity to move sediment; this resistance to the flow is sometimes called *form drag*, In general, the shear stress available to move sediment is less than that calculated from the logarithmic velocity profile, because the transfer of energy and momentum from the moving water results *both* in movement of sediment *and* in formation of sediment ripples.

Finally, we should not forget that current flows in the sea can consist of both water and sediment, especially close to the bed, where concentrations of suspended sediment are likely to be greatest. The density of the fluid (water + sediment) thus increases downwards towards the bed, and this density gradient damps down the turbulence, as well as making the flow more sluggish (because the energy required to transport the particles extracts momentum from the fluid). It is more difficult for the turbulent eddies of burst–sweep cycles to penetrate to the bed and move the denser fluid upwards, and shear stresses at the bed can be lower than calculated from the current velocity.

QUESTION 4.5 Why might Equations 4.1 and 4.5 suggest that, for a given current speed, sediment-laden water is likely to have more erosive power than sediment-free water? Higher density of sediment laden wa
= higher shear stress

Waves and sediment movement

So far, we have considered only the effects of currents, but you know that waves can also move sediment on the sea-bed, at least in shallow water and along coasts – and the bigger the waves, the deeper their influence will extend (see Figure 4.16). The determination of shear stresses caused by wave motions is rather complicated as we have to deal with to-and-fro movements rather than with the generally more unidirectional flow of currents. We shall defer consideration of this topic to Chapter 5, but meantime it is worth noting that the basic principles are the same, and you should be aware that shear stresses due to waves *are proportional to the square of the orbital velocity* of the wave motion. You need also to bear in mind that waves and currents do not occur in isolation from one another, but interact.

4.2.5 SHEAR VELOCITY AND THE BEHAVIOUR OF NON-COHESIVE SEDIMENTS

Figure 4.12 summarizes shear velocities appropriate to the erosion, transport and deposition of *non-cohesive* sediments of various grain sizes. Because Figure 4.12 applies to *non-cohesive sediments only*, it departs increasingly from reality for particle sizes less than about 100 μm (0.01 mm, 10^{-4} m). In real sediments, the smaller the particles the greater the proportion of clay minerals among them; and we have seen that clay minerals are not rounded but plate-shaped and tend to stick together (they are cohesive materials). Furthermore, as in the case of Figures 4.7 and 4.8, for diagrams such as Figure 4.12 the flow is assumed to be steady and uniform, with a fully developed turbulent boundary layer over a flat bed of evenly sized particles. As we have seen, very few natural sediments even approach such a narrow size distribution, nor is the sea-bed generally flat; and marine currents are not normally steady and uniform. Thus, although Figure 4.12 can be used to describe the behaviour of non-cohesive sediment consisting of equidimensional quartz grains, for particle sizes smaller than about 0.2 mm, it diverges increasingly from the more realistic situation summarized in Figure 4.6.

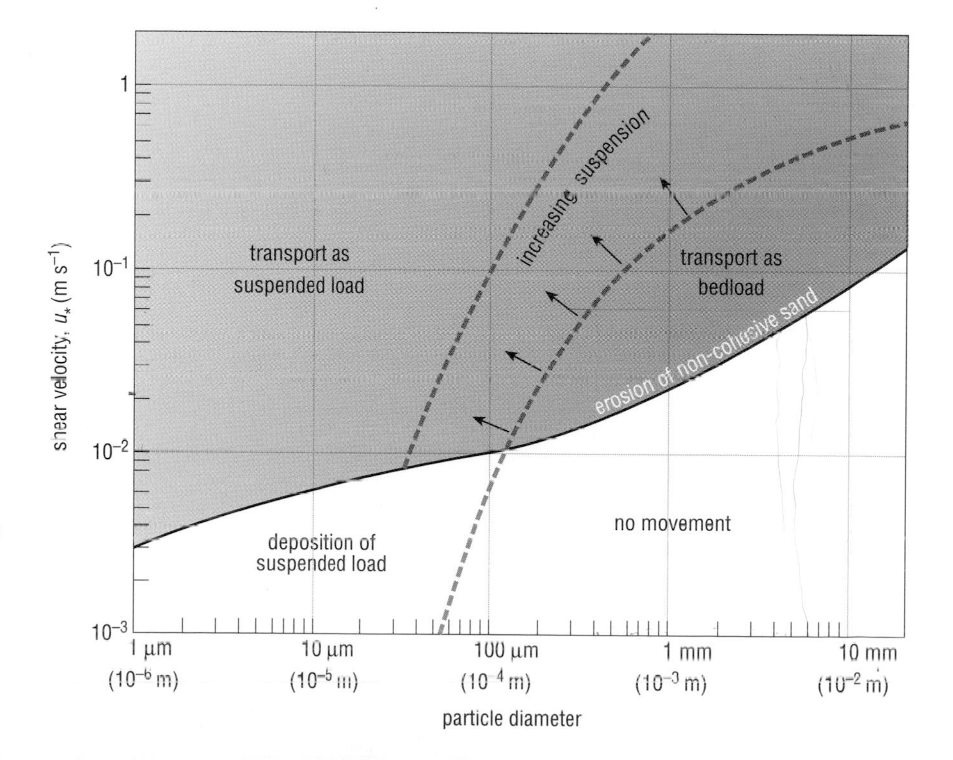

Figure 4.12 Empirically determined diagram showing shear velocities at which *non-cohesive* sediment particles of different grain sizes are eroded, transported and deposited. Broken lines represent gradational boundaries. Scales are logarithmic. The diagram is for silt, sand, and gravel-sized particles and the parameters used in its compilation are: density of particles (quartz) 2.65×10^3 kg m^{-3}; density of seawater 1.025×10^3 kg m^{-3} at 15 °C ; molecular viscosity 10^{-3} N s m^{-2}, and $g = 9.8$ m s^{-2}.

In addition, Figure 4.12 implies that current speeds for erosion and deposition are the same. This is quite different from the situation in real sediments (Figure 4.6) and again arises because only non-cohesive particles with narrow size ranges have been used to compile the diagram. Nonetheless, we can still make use of Figure 4.12 to obtain actual values for critical shear stresses.

QUESTION 4.6 Use Figure 4.12 and Equation 4.5 to calculate (a) the critical shear stress for a bed consisting of 1 mm particles, and (b) the minimum shear stress at the bed required to keep 1 mm particles in suspension. (Assume that the density of seawater is $10^3\,\text{kg m}^{-3}$.)

Information about shear velocities and shear stresses at the bed cannot be obtained from graphs such as Figure 4.6, chiefly because the curves are compiled from observations of *natural* sediments and provide no information about velocity gradients. So, for example, from Figure 4.6 we can merely *estimate* that a current averaging about $0.4\,\text{m s}^{-1}$ will set particles averaging 1 mm in size in motion as bedload, while average current speeds in excess of $1\,\text{m s}^{-1}$ are required to keep such particles in suspension.

4.2.6 RATES OF SEDIMENT TRANSPORT

The rate of sediment transport is the mass of sediment that is moved past a given point or through unit area of the water column in unit time. It is also called the *sediment flux*. To calculate the total sediment transport rate, we have to consider both the bedload flux and the suspended sediment flux. In general, the amount of suspended sediment will increase with increasing current speed, as progressively coarser grains are lifted permanently into suspension.

The bedload transport rate

Although it is difficult to make predictions about bedload transport in the marine environment, experiment and theory suggest that the rate of bedload transport (q_b) is proportional to the cube of the shear velocity, i.e.:

$$q_b \propto u_*^3 \qquad (4.6)$$

provided that the shear stress at the bed is greater than the critical shear stress. In addition, since shear velocity is itself related to average current velocity, the rate of bedload transport (the bedload flux) is also proportional to the cube of average current velocity measured at a fixed height above the bed (usually 1 m).

The relationship summarized in Equation 4.6 arises because the rate of bedload transport depends on the *power* of the current, which is the rate at which it supplies energy, i.e. its rate of doing work in moving sediment on the sea-bed (cf. discussion in Section 1.5 in relation to Equations 1.12 and 1.13). Since the energy of the current is proportional to (velocity)^2, and power is given by (energy) × (velocity), then the power must be proportional to (velocity)^2 × velocity, i.e. to (velocity)^3, as in Equation 4.6. The relationship in Equation 4.6 is important because it means that even very small changes in current speed or bed roughness can have significant effects on the rate of bedload transport. We can illustrate this with a practical example.

Figure 4.13(a) shows the changes in average tidal current velocity measured at one metre above the bed during a complete tidal cycle at a location in the North Sea. The south–south-westerly currents flow for slightly longer and attain slightly higher velocities than the north–north-easterly currents. As a result of the relationship between q_b and current velocity, there are appreciable differences between the amounts of sediment that can be transported in each tidal current direction (Figure 4.13(b)).

QUESTION 4.7

(a) What is the difference between the maximum current speed of the NNE tidal current, and that of the SSW current?

(b) Estimate the difference between the sizes of the shaded areas in Figure 4.13(b), and hence suggest whether there is likely to be significant net transport of sediment in one particular direction at this location.

(c) Examine the dashed horizontal lines and shaded areas in Figure 4.13(b). What can you conclude about the particle size(s) of sediment being transported at the current velocities corresponding to the shaded areas?

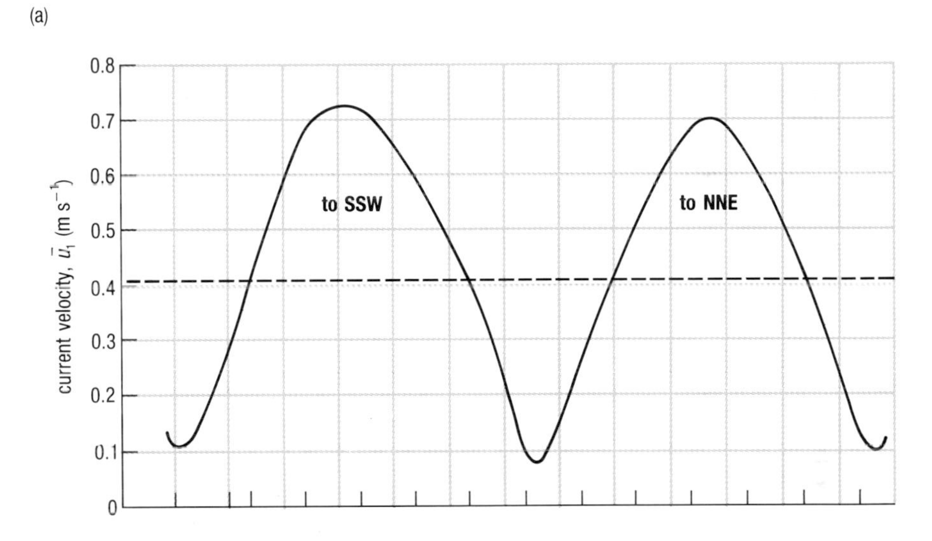

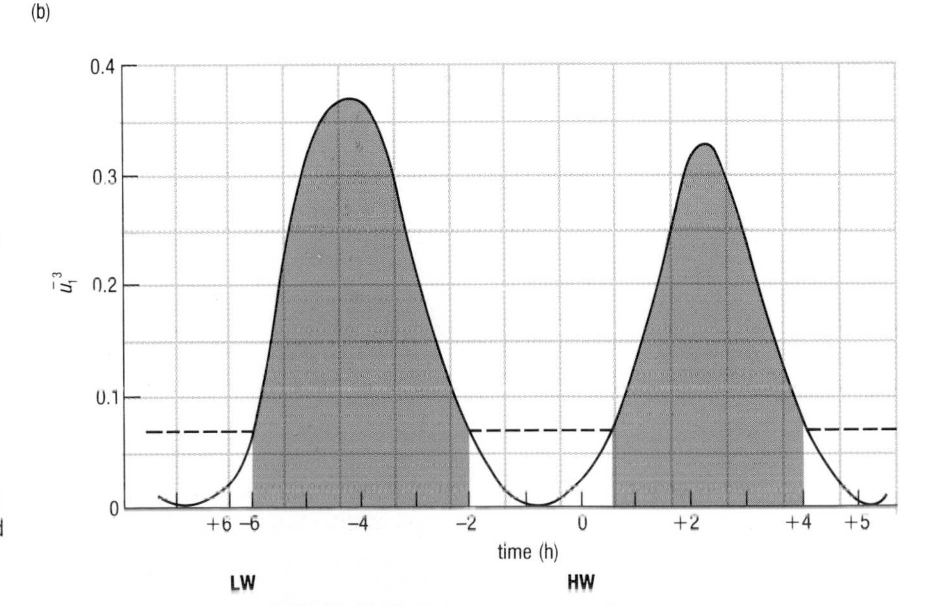

Figure 4.13 (a) The changes in average tidal current velocity at 1 m above the sea-bed $(\bar{u}_1)$ during a complete tidal cycle at a location in the North Sea. The small gap in the time-scale between +6 and −6 hours is because the tidal cycle is about 12.5 hours long (Chapter 2). The dashed line is the threshold velocity required to move sand grains of 0.3 mm diameter in the bedload.
(b) The changes in $\bar{u}_1^3$ with time during the same tidal cycle. Because q_b is proportional to $\bar{u}_1^3$, the shaded areas are proportional to the amount of sediment transported. The horizontal dashed line represents the cube of the threshold velocity shown in (a). LW and HW are low and high water respectively.

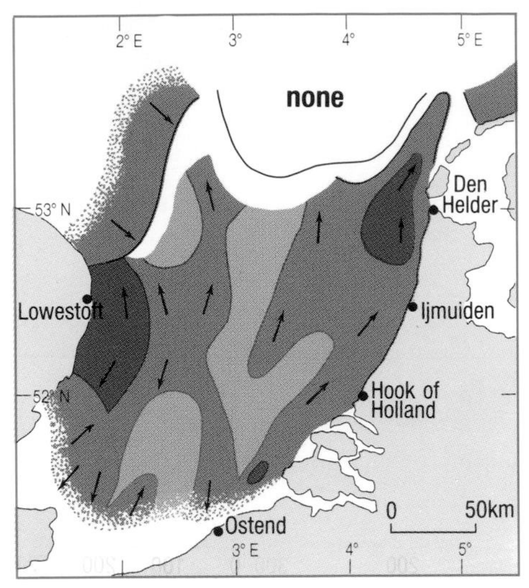

theoretical net sand transport

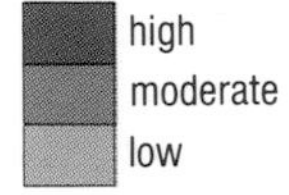

Figure 4.14 Theoretical net transport of sand by tidal currents in the southern North Sea using threshold velocities for grain diameters of 0.2 and 0.3 mm. Arrows show direction of net transport.

Figure 4.14 has been compiled by repeating the procedure shown in Figure 4.13 for many other locations in the North Sea. It is a fair approximation to the observed pattern of bedload sediment movement in the southern North Sea. The actual rate of bedload transport in the sea is difficult to measure directly, because by definition the movement takes place at, or very close to, the interface between the sea-bed and the overlying water. If we try to sample the bedload, the likelihood is that we shall also sample some of the suspended load, and even sediment from the bed beneath the moving bedload.

One method of measuring bedload transport rates uses the movement of sediment ripples at the sea-bed. These migrate as the sediment moves. So, if we know the size of the ripple, we can calculate the mass of sediment that is moving for a given length of the ripple crest, and then work out the transport rate from how fast the ripple is migrating. To do this, a rod, about one metre long, is held horizontally in the flow about ten centimetres above the bed. A light is shone obliquely onto the rod from one side so that the shadow of the rod on the bed appears as a zig-zag line where it is distorted by the ripple. The progression of the sand ripple is measured photographically by recording the changing size and shape of the shadow. As the method allows for only a few ripples to be measured, the results may not be representative of the overall pattern of sediment transport in an area.

Another approach has been to detect the noise produced by grains colliding during sediment movement. The intensity of the sound is related both to the amount of sediment and to the grain size that is moving. This method is most suitable for larger grain sizes but there are problems in calibrating the acoustic signals, chiefly because the erratic motions of turbulent current flow mean that the bedload is moved intermittently – mostly during the burst–sweep cycles described in Section 4.2.4.

Rate of transport of the suspended load

The determination of the rate of suspended load transport is straightforward by comparison with measurement of the rate of bedload transport, but it still has its problems. Current speeds (Figure 4.15(a)) and sediment concentrations (Figure 4.15(b)) are measured throughout the water column and then the suspended sediment flux (through unit area at right angles to the current), q_s, is calculated by multiplying together the two sets of data. Traditional methods for determining the concentration of suspended sediment by direct sampling of the water column are laborious and expensive and often inaccurate. Another technique uses electromagnetic flow meters in conjunction with fast-response sensors (or transmissometers) which record the impact of sediment grains. Nowadays, it is common to use sensors which can determine suspended sediment concentrations from the intensity of backscattered light or sound (the higher the frequency of sound used, the smaller the grain sizes that can be detected). Perhaps best known of these devices is the Acoustic Doppler Current Profiler (ADCP), which can measure both current speed and suspended sediment concentration from a vessel underway.

QUESTION 4.8 What can you conclude from Figure 4.15(b) and (c) about the way in which sediment concentrations, and the suspended sediment fluxes for grains of different sizes, vary through the water column?

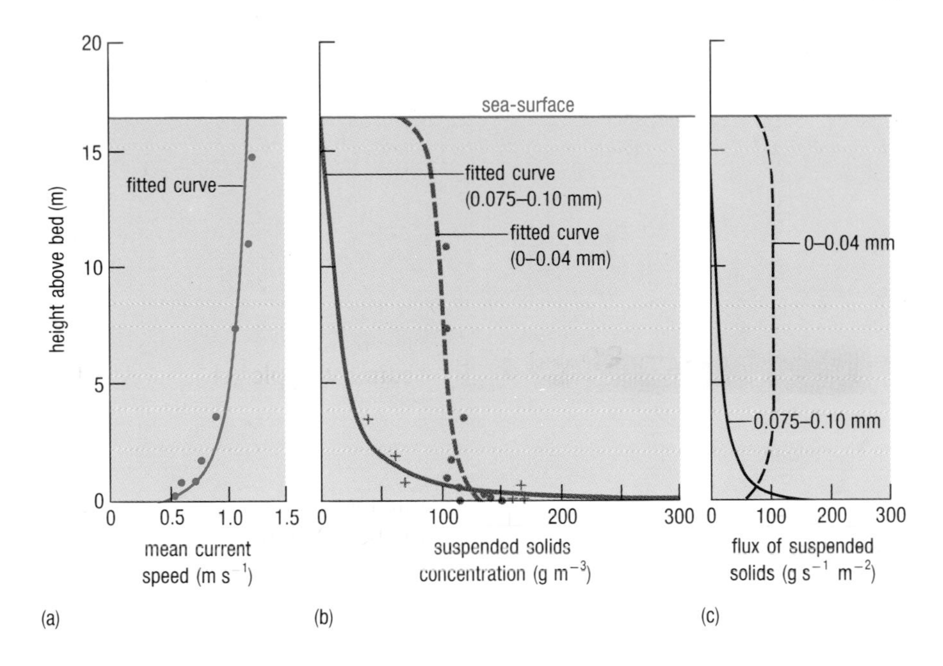

Figure 4.15 Profile of (a) mean current speed, and (b) concentrations of sediment grains with diameters in two different size ranges, multiplied to give (c) the sediment fluxes throughout the water column. Note that (m s^{-1}) × (g m^{-3}) gives g s^{-1} m^{-2}, so the flux is transport per unit (vertical) area. Part (c) does not show the total suspended flux, only the fluxes for the particle sizes represented in (b).

Rates of erosion and transport in relation to current speeds

As you know, in shallow shelf seas such as the North Sea, the influence of waves and currents at the surface commonly extends right to the bottom.

Should we therefore expect sediment erosion and transport (both in bedload and suspended load) to be disproportionately large during storms?

Yes. Wind-driven currents will tend to flow faster during storms, and where conditions at the sea-surface can affect the benthic boundary layer, sediment transport would be greater at such periods, since doubling the current speed quadruples the shear stress (Equation 4.1), which is the principal factor responsible for sediment erosion. In addition, the cube relationship of Equation 4.6 means that doubling the current speed leads to an eight-fold increase in bedload flux.

Faster currents also keep larger amounts of sediment in suspension (Figures 4.6 and 4.13). In addition, the orbital motions of large waves generated by storm winds are more likely to reach the sea-bed (Figure 1.8), stirring more sediment up into suspension and keeping it there for longer. Thus, in many areas, storms can determine the overall magnitude and direction of sediment transport, even though they may be relatively rare and of short duration. This is well illustrated in Figure 4.16(a) (overleaf), part of a record of observations made during a stormy period in the eastern North Sea. It shows that the amount of sediment taken into suspension, and available for transport by tidal currents, is greatly increased when waves are large; Figure 4.16(b) suggests that the greater the wave height, the larger the particles that can be lifted into suspension.

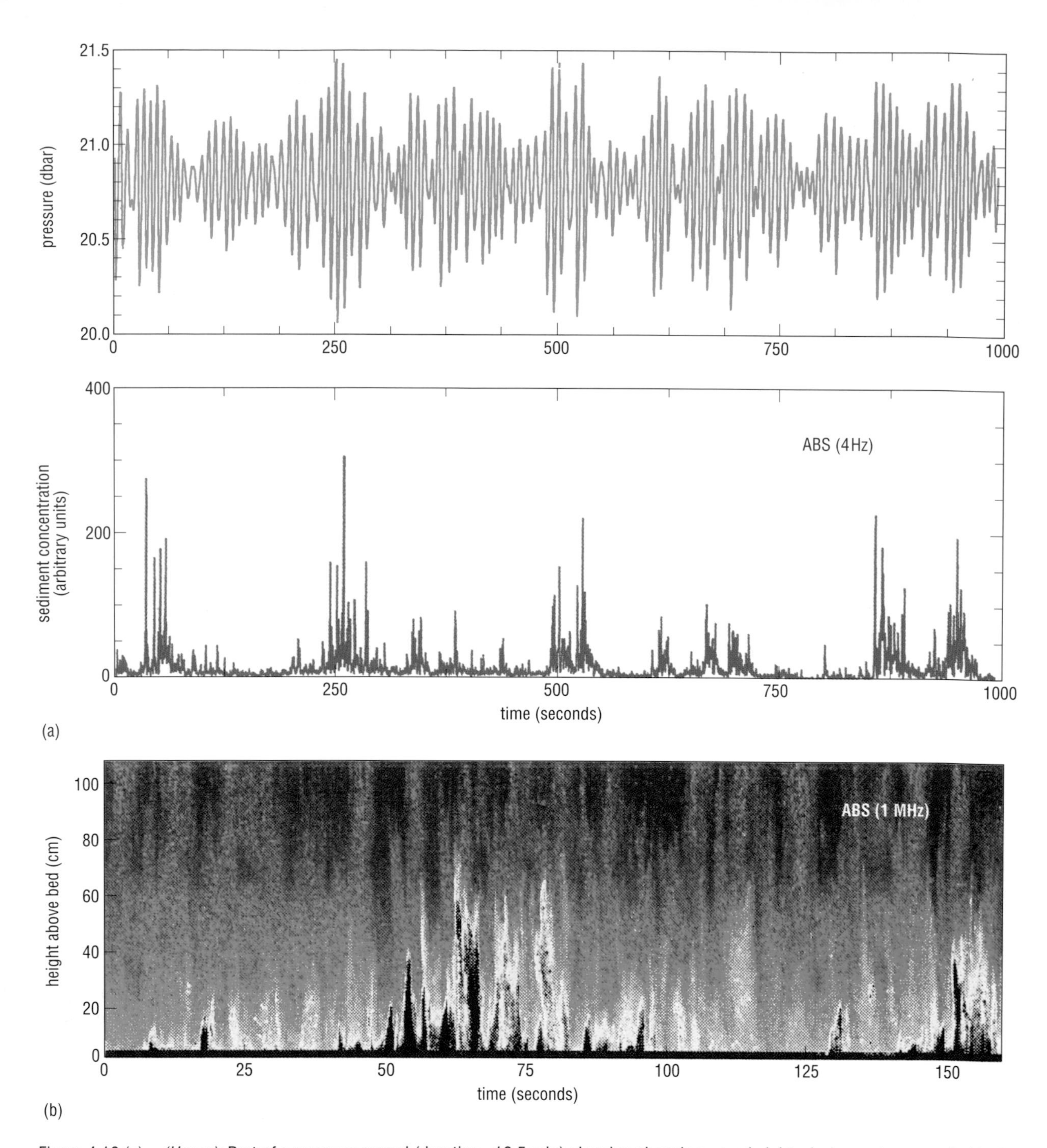

Figure 4.16 (a) *(Upper):* Part of a pressure record (duration ~16.5 min) showing changing wave heights during a stormy period in the eastern North Sea, where the average water depth was ~21 m (dbar = decibar; 1 dbar = 1 m). In shallow water, pressure variations at the sea-bed due to the passage of waves are not simply determined by hydrostatic pressure. In this case, the maximum sea-bed pressure variation of 1.35 dbar 'converts' to a maximum crest–trough wave height at the surface of ~2.7 m.
(Lower): Acoustic backscatter (ABS) record of sediment concentration above the bed at the same location. Note that the relatively small waves cause little resuspension into the water column, but that resuspension increases significantly in response to the groups of larger waves; also that the first couple of waves during a group of larger waves apparently have little effect, but serve to 'fluidize' the bed, so that subsequent large waves put large amounts of sediment into suspension.

(b) Concentrations of suspended sediment recorded using acoustic backscatter, above a sea-bed of mainly coarse sand (in the same part of the North Sea as (a)); the ABS frequency used is 1 MHz, which is sensitive to relatively large grain sizes. The vertical axis shows height above the bed, and concentrations are indicated by colour, ranging from dark blue for lowest concentrations (1 mg l^{-1}) to dark red for highest (10 g l^{-1}). Dark red areas corresponding to high concentrations of coarse sand are only seen in association with the larger disturbances, such as those resulting from waves of ~4 s period towards the middle of the plot (i.e. larger grains are lifted only by larger waves).

4.3 THE DEPOSITION OF SEDIMENT

As flow conditions in a shallow marine environment change, it is often useful to know the rate at which sediment is likely to be deposited if current speeds decrease, and the rate at which it will be eroded if speeds increase. The relationship between erosion, transportation and deposition for *natural* sediments was summarized in Figure 4.6, which takes account of the increasing cohesiveness of sediments as particle size decreases, and of the greater current speeds required to erode such sediments (Section 4.2.1).

4.3.1 DEPOSITION OF THE BEDLOAD

Only coarser sediment (grains larger than about 0.1–0.2 mm diameter) is transported as bedload. For a given grain size, these particles will stop moving when the bed shear stress (τ_0) is only a little less than the critical shear stress (Section 4.1.1) that was needed to start them moving in the first place – as illustrated by the proximity of the 'erosion' and 'deposition' curves in the right-hand half of Figure 4.6(c).

Some idea of potential deposition rates from the bedload can be gained if the rate of bedload transport is known. As $q_b \propto u_*^3$ (Equation 4.6), we may assume that the transport rate will decrease (and so the rate of deposition will increase) in proportion to the reduction in u_*^3 and hence also in proportion to the reduction in the cube of the average actual current velocity $\bar{u}_1^3$.

4.3.2 DEPOSITION OF THE SUSPENDED LOAD

Particles in suspension will begin to settle towards the bed as soon as gravitational forces exceed buoyancy forces, but grains larger than about 0.1 mm will continue to move as part of the bedload, perhaps being taken intermittently back into suspension by eddies. Particles smaller than about 0.1 mm do not go through a stage of bedload transport and are deposited directly from suspension (Figures 4.6 and 4.12). Moreover, as a current slows down, suspended particles of a given grain size do not all reach the bed at the same time because they will be distributed at different depths in the water column (Figure 4.15(b)); hence the rate at which suspended sediment is deposited depends on more than just the decrease in current speed. The time that particles take to settle will depend to a large extent on their settling velocities, w_S (Figure 4.6(c)), and on the degree of turbulence in the water column – and while the particles are settling, they continue to be transported in the direction of net current flow.

Since very small particles settle significantly more slowly than large ones (especially if they are flaky), they will eventually reach the bed some distance from where they began to settle, i.e. there is a settling *lag*. Also, grains of slightly different sizes may settle at very different rates because (as you read in item 3 of the discussion of Figure 4.6), settling velocities of small particles are proportional to the *square* of the diameter. So, for particles in the clay to very fine sand range, a very small decrease in grain size results in a significant change in settling velocity (Figure 4.6(c)), and settling lag thus increases dramatically with decreasing particle size. Conversely, for grains coarser than about 2 mm, settling velocity depends on the *square root* of the diameter, so even quite large changes in diameter result in only small variations in settling velocity, and there is less separation in sediment grain size during deposition.

In short, the smaller the particle size, the slower the current can flow while continuing to transport sediment, the greater the settling lag, and the slower the rate of deposition. As you will see in Chapter 6, for muds to be deposited at all, either the water must be completely still for long periods, or the particles must somehow be brought together into larger aggregates with greater settling velocities. Other factors affecting rates of deposition of suspended sediment include: the degree of bed roughness (which generates near-bed turbulence); the extent of resuspension of sediment by marine organisms; and occasional brief episodes of erosion during an overall period of net deposition. We need to recall also that once sediment settles into the viscous sublayer (where present), it is effectively trapped there (Section 4.2.3).

QUESTION 4.9 Given a settling velocity close to 10 mm s^{-1} (Figure 4.6(c)), calculate roughly how long it would take for most of the sediment 0.1 mm in diameter shown in Figure 4.15(b) to be deposited, assuming there is no turbulence to slow the settling of the grains.

4.4 BED FORMS

The sea-bed is rarely flat: it is usually covered with both small- and large-scale sediment features. The most familiar of these are probably the small-scale sediment ripples formed by waves and currents, commonly seen on sandy beaches and in the silts and muds of estuaries and tidal flats (Figure 4.17). These features are known as **bed forms** and they range in size from small ripples up to the large sand-waves and sand-banks beneath shelf seas, which may rise more than 15 m above the sea-floor.

Figure 4.17 Sediment ripples in the muddy sands of an estuary. The wavelength is a few cm.

The oscillating water flows in wave motions make sediment ripples with generally symmetrical cross-sections, but current-produced bed forms are usually asymmetrical with the steeper slope facing downstream. During active sediment transport, sediment is moved mainly as bedload up the shallower slope, and redeposited down the steeper slope. In this way, both the sediment and the bed form migrate across the sea-bed. You may recall that the rate of migration of ripples has been used to estimate the rate of transport in the bedload (Section 4.2.6). It is not unusual to find the bed forms produced by, say, a flood tidal current partially eroded and flattened by the following ebb tide. After transport ceases, bed forms become static and, if buried by further layers of sediment, they may become preserved within the sedimentary record, although not always in a perfect state.

Current ripples commonly develop where relatively slow currents move over sediments finer than about 0.6 mm grain size, and water flow is only slightly disturbed above the ripples (Figure 4.18(a)). It seems they can form only where the viscous sublayer is not completely broken down by turbulence and grain roughness (Figure 4.10). Since the thickness of the sublayer decreases as the speed of a current increases, it follows that ripples are unlikely to develop at fast current speeds or where there is coarse sediment on the bed.

At higher current speeds, and in coarser sediments, somewhat larger bed forms known as *megaripples* are produced (Figure 4.19(b)). These are up to a metre or more in height, and often have wavelengths of several metres or even tens of metres. The water flow above a megaripple is often disturbed up to the surface, producing ‘boils’ (Figure 4.18(b), cf. Section 4.1.1).

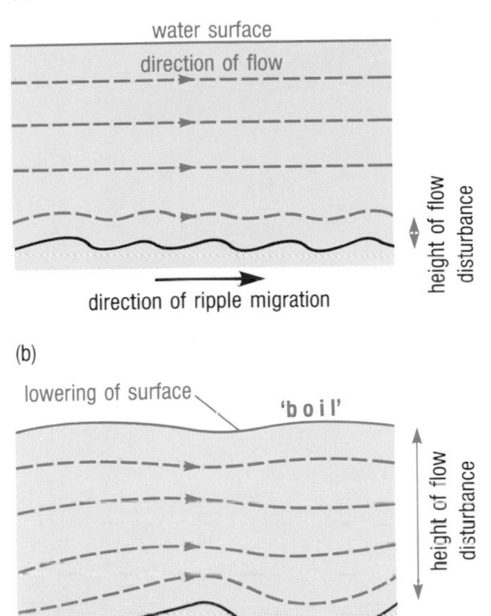

Figure 4.18 The flow of water (dashed lines) over ripples and megaripples and the direction of migration of the bed forms (solid arrows).
(a) Ripples: there is only minor disturbance of the flow by the bed forms.
(b) Megaripples: flow can be disturbed up to the surface where 'boils' may develop.

The shapes of bed form crests are also related to flow conditions. Where flows are relatively slow and/or the water is deep, bed forms are linear with long straight crests, like the ripples shown in Figure 4.17. At higher speeds or in shallower water, the crests become progressively more indented until eventually they are broken up into short, curved sections, like the ripples shown in Figure 4.19(a). Fluctuating flows can lead to the superposition of smaller bed forms on larger ones (e.g. ripples on megaripples, Figure 4.19(b)).

At current speeds of up to about 0.75 m s^{-1}, *sand-waves* can occur extensively. For example, they cover an area of 15 000 km^2 off the Netherlands coast. Sand-waves may reach 18 m high and have wavelengths of nearly a kilometre, though most are much smaller (Figure 4.20, overleaf). Individual sand-waves may migrate in the direction of the mean or residual current, at rates of between 10 and 150 m per year. Their amplitude is enhanced by the oscillatory nature of tidal flows, and the largest sand-waves stand well above the level of the charted sea-bed and may constitute a hazard to shipping in shallow water. Hydrographic surveys of many shelf seas were completed when most vessels had draughts of less than 17 m, and many of these charts are still in use. As vessels have increased progressively in tonnage and draught, there is greater risk of grounding on large uncharted sand-waves. Large bed forms can be detected on SAR (*S*ynthetic *A*perture *R*adar, Section 1.7.1) images taken from satellites, but the risks remain, as the stopping distance for large vessels is several kilometres.

(a)

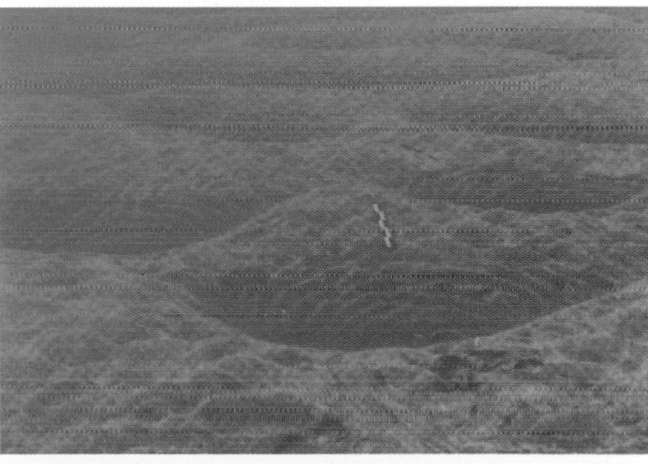

(b)

Figure 4.19 (a) Current-formed ripples with curved crests. The matchbox provides scale.
(b) Megaripples with superimposed, curve-crested current ripples. The steeper slopes face towards you, suggesting that the current flow was from top to bottom of the scene in the photograph. The 12-inch ruler provides scale.

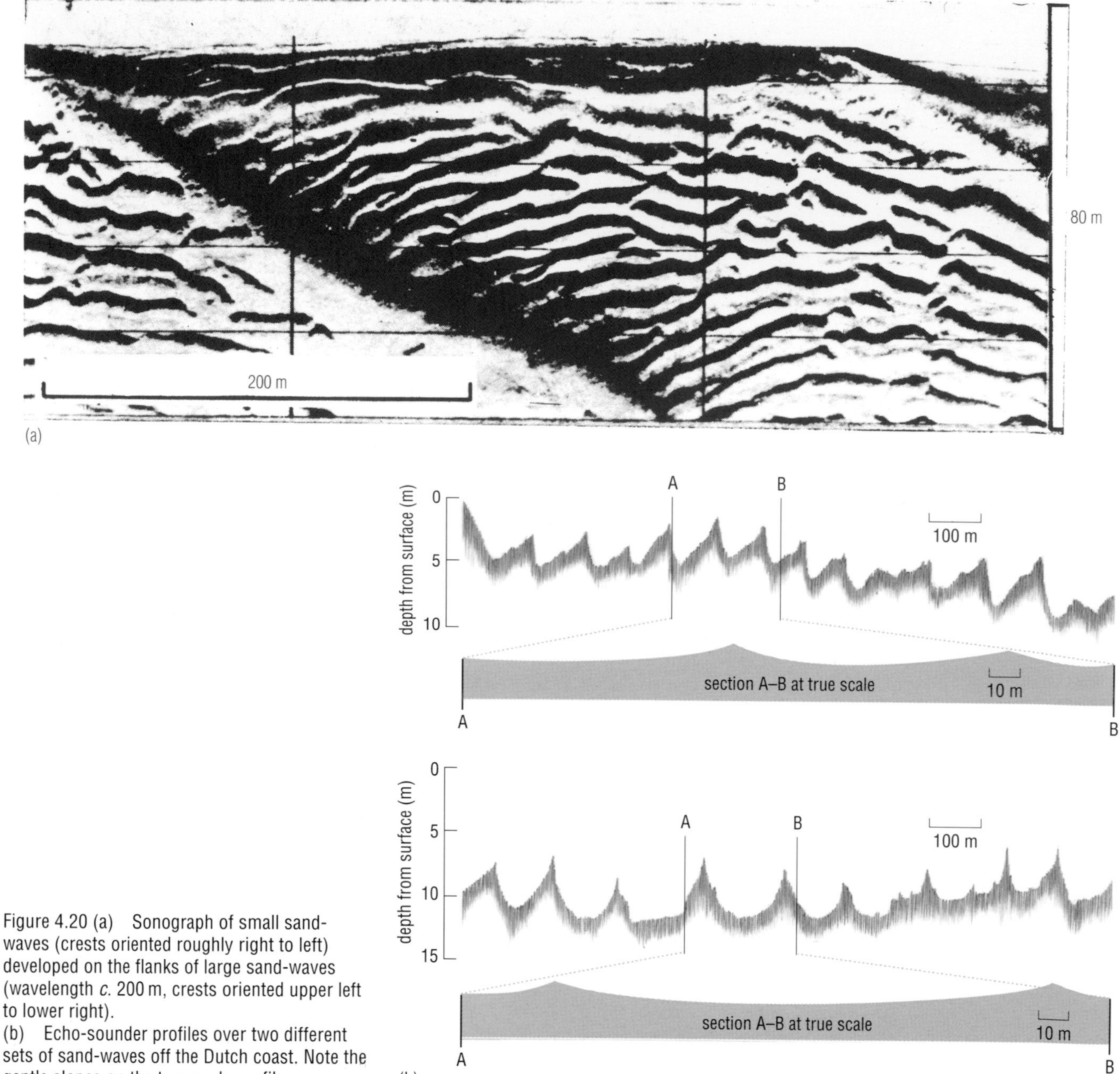

Figure 4.20 (a) Sonograph of small sand-waves (crests oriented roughly right to left) developed on the flanks of large sand-waves (wavelength *c.* 200 m, crests oriented upper left to lower right).
(b) Echo-sounder profiles over two different sets of sand-waves off the Dutch coast. Note the gentle slopes on the true-scale profiles.

At slower current speeds, larger sand-waves pass into smaller sand-waves and megaripples, which frequently have current-formed ripples on their backs (Figure 4.19(b); cf. Figure 4.20(a)). It is possible that, in some circumstances, the smaller current ripples form first and may even initiate formation of megaripples. Once formed, the current ripples increase bed roughness, so that turbulence extends right down to the sea-bed, generating the megaripples without erasing the current-rippled surface. All of these bed forms are transverse to the current flow and develop where surface current flows do not exceed about 0.5–0.75 m s^{-1}. At greater current speeds, bed forms become linear and approximately parallel to the flow direction.

Where there is abundant sand on the sea-bed and mean tidal current flows are more than about 0.5 m s^{-1}, *sand-banks* can be expected to develop. These large submarine features are found on many continental shelves, particularly in shallow water, and may reach 80 km in length, 3 km in width and tens of metres in height, oriented parallel to the tidal currents or converging in the down-current direction (Figure 4.21). Like large sand-waves, they can be a hazard to shipping: the Goodwin Sands, for example (north-east of the Straits of Dover), are particularly notorious.

Where near-surface peak current speeds reach 1.5 m s^{-1} or so, bottom currents are such that sand is winnowed away completely, either exposing bed rock or leaving a residue of gravel with linear bifurcating *furrows*, which extend down-current for several kilometres and can be up to 1 m deep and more than 10 m wide (Figure 4.22). The furrows are probably passageways for removal of sand, as linear *sand ribbons* often extend far down-current from them, reaching 15 km in length and 200 m in width, but only about 1 m in thickness. Where current speeds exceed about 1.5 m s^{-1}, *scour hollows* may develop. The elongate depression known as the Beaufort Dyke, oriented in a north–north-westerly direction in the North Channel of the Irish Sea, may be an elongate scour hollow, and others occur at the entrance to San Francisco Harbour and in the Hayasui Strait between the islands of Kyushu and Shikoku in Japan (Figure 4.23).

Sand-banks and other current-parallel bed forms are probably produced by secondary helical eddies with axes parallel to the direction of the current – such eddies are analogous to the helical vortices of **Langmuir circulation**, which develop with their axes parallel to the wind direction in well-mixed surface waters and form 'windrows'– more or less regular lines of smooth water and/or streaks of foam.

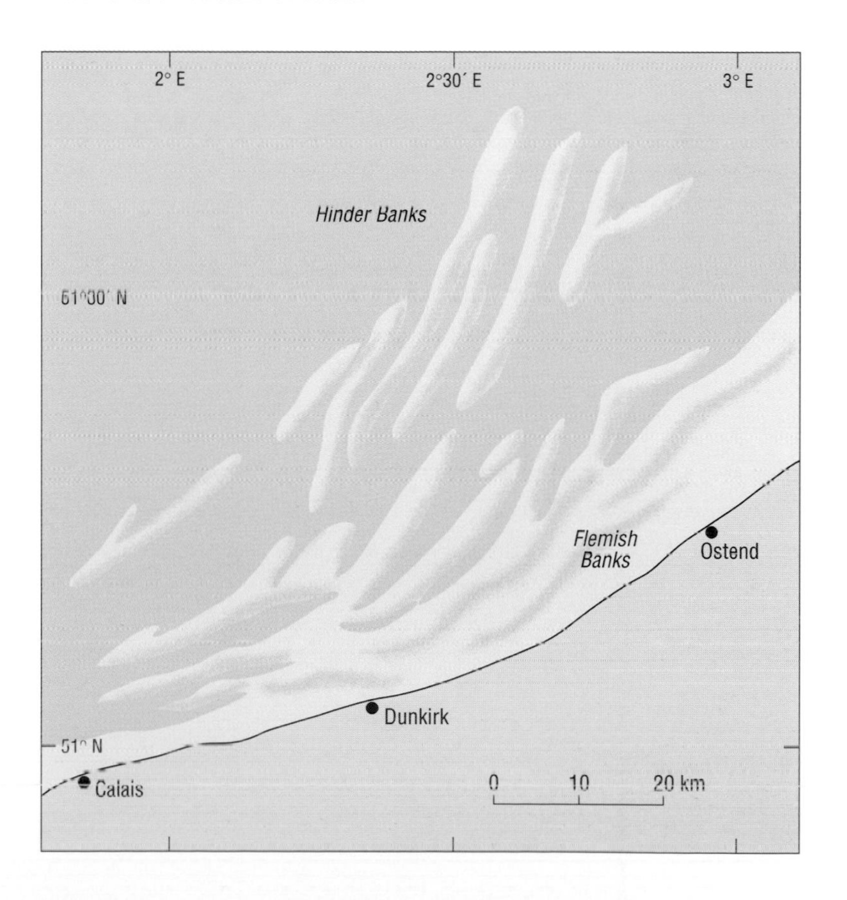

Figure 4.21 Sketch map of sand-banks in the southern North Sea, illustrating the contrast between the generally straight and parallel Hinder Banks and the often convergent Flemish Banks. This picture represents a snapshot in time – the pattern today would probably look somewhat different.

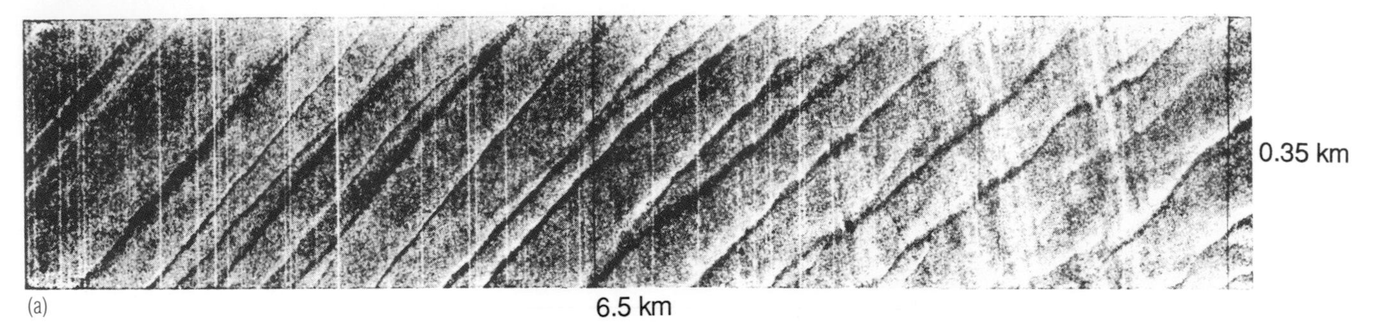

(a)

(b)

Figure 4.22 (a) Sonograph showing longitudinal furrows in gravels in the English Channel. (b) Sea-bed photograph of sand ridges (with superimposed ripples) and furrows near the edge of the Malin Shelf off north-western Scotland, in water *c.* 200 m deep. A hard pavement of cemented sand and gravel is visible in the furrows, which are about half a metre across. The prevailing flow is due to the slope current along the shelf edge, and is towards the NNW (from upper right to lower left) (direction from compass, lower right). The boulder in the foreground has feathery bryozoans on its upcurrent side and the calcareous tubes of polychaete worms on its downcurrent side.

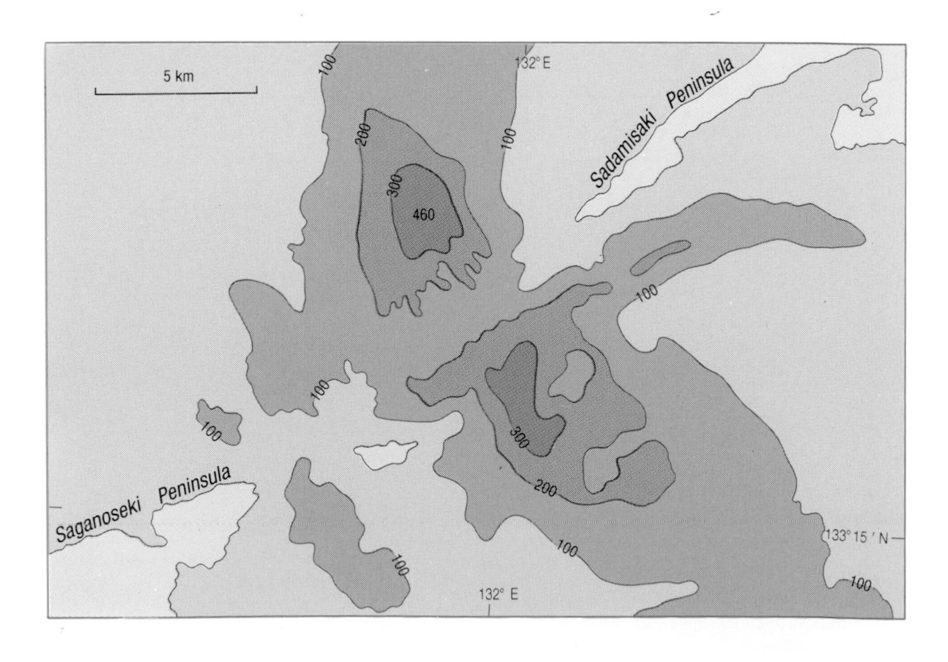

Figure 4.23 Deep scour hollows, Hayasui Strait, Japan. Contours show depth below average level of sea-bed (m).

This concludes our exploration of the basic principles and processes that characterize the movement of sediments in shallow marine environments. In succeeding Chapters, we shall look at some of those environments in greater detail, to apply the general principles in particular cases. We begin with the near-shore regions, partly because they are accessible and well-studied, and partly because they are the most influenced by human activities.

4.5 SUMMARY OF CHAPTER 4

1 Friction between flowing water and the sea-bed generates a boundary layer in which turbulent flow dominates, except very close to the bed. Movement of sediment (erosion) occurs when the shear stress generated by the frictional force of water flowing over the sediment overcomes the force of gravity acting on the sediment grains and the friction between the grains and the underlying bed. Shear stress is proportional to the square of the mean current speed (and to the density of the water). Movement of grains of a given size begins when the shear stress at the bed reaches a critical value (critical shear stress). Particles larger than about 0.1–0.2 mm in diameter are initially moved as bedload, then lifted intermittently into suspension, and as current speeds increase further they are finally lifted permanently into suspension. Grains smaller than about 0.1 mm are lifted directly into suspension as soon as the critical shear stress is reached.

2 Cohesive sediments contain a high proportion of fine-grained clay minerals and are more difficult to erode than non-cohesive sediments, which consist mostly of quartz grains. For cohesive sediments, the smaller the particle size, the greater the current speed required to erode them. The resistance of muds to erosion is assessed by their yield strength. Once in suspension, clay particles can be transported for long distances by currents that would be much too weak to erode them.

3 Shear stress is proportional also to the velocity gradient in the boundary layer and to the viscosity of the water. When current speed is plotted against the height above the sea-bed (as the vertical axis) on a log–linear graph, the inverse velocity gradient $d\log z / d\bar{u}$ is linear. The slope of the line is used to calculate the shear velocity, and the intercept of the line with the depth axis gives a measure of the bed roughness length (z_0) which increases as the sediment grain size increases; roughness length will also be greater if there are bed forms such as sand ripples

4 When water flows over a smooth (very fine-grained) bed, the lowermost water layer appears to flow in a laminar fashion, forming a viscous sublayer only a few millimetres thick, which decreases in thickness as flow speed increases (and as viscosity decreases with increasing temperature). Suspended particles which settle into the sublayer are subjected only to laminar flow and are soon deposited. When water flows over a coarse-grained bed, or at high current speeds, grains protrude through the viscous sublayer, breaking it down, and turbulent flow extends to the bed, giving rise to greater potential for sediment movement. In the marine environment, values of shear velocity (and hence of bed shear stress) may be overestimated or underestimated because of: decelerating or accelerating tidal currents (respectively); the time taken for a velocity profile to adjust to a new bed roughness; and the concentration of suspended sediment close to the bed. Erosion and transport of sediment is initiated mainly by cycles of downward sweeps and upward bursts that result from turbulent motions in the boundary layer

5 The rate of bedload transport is proportional to the cube of the shear velocity ($q_b \propto u_*^3$) and hence also to the cube of the average current speed, but it is difficult to measure directly in the sea. The rate of suspended load transport can be calculated from the product of the current velocity and the sediment concentration.

6 Deposition of the bedload begins when the current speed falls so that the shear stress at the bed is only a little below the critical shear stress required to start the sediment moving. The rate of deposition of the bedload is proportional to the reduction in the cube of the average current velocity, $\bar{u}_1^3$. The rate of deposition of the suspended load varies according to sediment grain size. The settling lag of fine suspended particles means that they may reach the bed well after they began to settle from suspension. The rate of deposition of grains of a given size in the suspended sediment load depends both on the vertical distribution of sediment concentration above the bed, and on the settling velocity of the grains.

7 Raised sediment features on the sea-bed are called bed forms. Bed forms produced by waves are symmetrical, those formed by currents are asymmetrical. The type of bed form depends mainly on current speed and sediment grain size and partly also on water depth. Small-scale current ripples form at relatively slow current speeds and where sediment is finer than about 0.6 mm grain diameter. As current speeds increase or where sediments are coarser-grained, larger-scale megaripples are formed; and sand-waves develop where sand is abundant. At current speeds of more than about 1 m s^{-1}, bed forms such as sand-banks and sand ribbons develop parallel, rather than transverse to, the current flow. At still higher current speeds, erosional features such as furrows and scour hollows develop.

Now try the following questions to consolidate your understanding of this Chapter.

QUESTION 4.10 This question is based upon Figure 4.9 (Section 4.2.2), plotted from actual measurements in a tidal current flow over the sea-bed.

(a) Calculate the shear velocity of the tidal current and the shear stress on the bed beneath the flow. (Assume the density of seawater is 10^3 kg m^{-3}.) How does the value of u_* compare with the actual current speed one metre above the bed?

(b) Why may the values for u_* and τ_0 that you have calculated in (a) be reliable only for a limited period of time?

(c) Use Figure 4.12 to deduce the approximate range of sizes of sediment particles that could be moved only as bedload, while the current is flowing as represented in Figure 4.9. Which types of sediment (Table 4.1, p. 96) does this range include?

(d) According to Figure 4.12, what is the maximum size of particles that could be transported entirely in suspension by the current represented in Figure 4.9? Which type of sediment (Table 4.1) does this represent?

QUESTION 4.11 How might inspection of bed forms on the bed of the southern North Sea help to confirm the net transport paths for sand shown in Figure 4.14?

QUESTION 4.12 Figure 4.19(b) shows a series of large, current-formed megaripples with superimposed, small-scale, current-formed ripples. Were the smaller ripples formed by currents flowing slower or faster than those which formed the megaripples?

CHAPTER 5 BEACHES

'The waves massed themselves, curved their backs and crashed.
Up spurted stones and shingle.'

From *The Waves* by Virginia Woolf.

'... I gain the cove with pushing prow,
And quench its speed in the slushy sand.
Then a mile of warm-scented beach ...'

From *Meeting at Night* by Robert Browning.

For many people, the word 'beach' means simply that stretch of sand or shingle beyond the reach of all but the highest tides, where they can lay out their sun beds, erect the deckchairs and keep a wary eye on the rising tide while the children paddle or build sandcastles. Their interest lies mainly in those sandy or pebbly areas of the shoreline which are exposed at some stage in the tidal cycle. This is the **intertidal zone**, sometimes also called the *littoral zone*, between the high and low tide levels. In this Chapter, however, we consider also the broader region beyond the intertidal zone, whose landward boundary is generally cliffs or sand dunes and/or the seaward limit of land plants. The seaward boundary, on the other hand, is well offshore, where sediment is not disturbed by wave action during fair weather conditions – i.e. around 10 m to 20 m water depth at low tide. This region can lie several kilometres offshore where beach gradients are low (and sometimes it is incorporated into definitions of the littoral zone, together with the intertidal zone).

Beaches bordering continental areas in temperate climatic regions tend to be formed of pale yellow to brown quartz-rich sands, the most common solid products of weathering and erosion. Pebble or shingle beaches also occur, in which the pebbles are usually derived from some fairly local source such as an adjacent line of cliffs. Around volcanic islands, beaches often consist entirely of lava fragments and may be black in appearance. The green sands round parts of Hawaii contain a high percentage of crystals of the mineral olivine, derived from the surrounding volcanic rocks. In tropical regions, where biological productivity may be high and land-derived sediment scarce, many beaches are comprised of brilliant white sand and gravel-sized fragments of coral and shells, and even carbonate grains which have been precipitated inorganically from seawater (Figure 3.2(b)). Some beach sediments may be artificially derived. For example, beaches near coal mining districts often contain a high proportion of sand-sized coal fragments, and the pebble beach at Lynmouth in Devon contains pieces of wave-rounded bricks and tiles, debris resulting from the 1952 flood disaster, during which entire houses were washed down the Lyn valley into the sea.

Changes in sea-level during the Pleistocene Ice Age have largely determined the present distribution of beach sediments. Sands of many present-day beaches migrated away from the shelf edge as sea-level rose following the last glaciation. There are several examples along the central eastern coast of North America, where at present there is little sediment supply to form beaches.

5.1 THE MORPHOLOGY OF BEACHES

No two beaches are the same, and beach widths can range from a few tens to many hundreds of metres, depending upon factors such as slope, tidal range, prevailing winds, sediment type (which is related to local geology, see above), and so on. In general, most beaches display features that can be identified by reference to one or other of the two extremes illustrated in Figure 5.1(a) and (b).

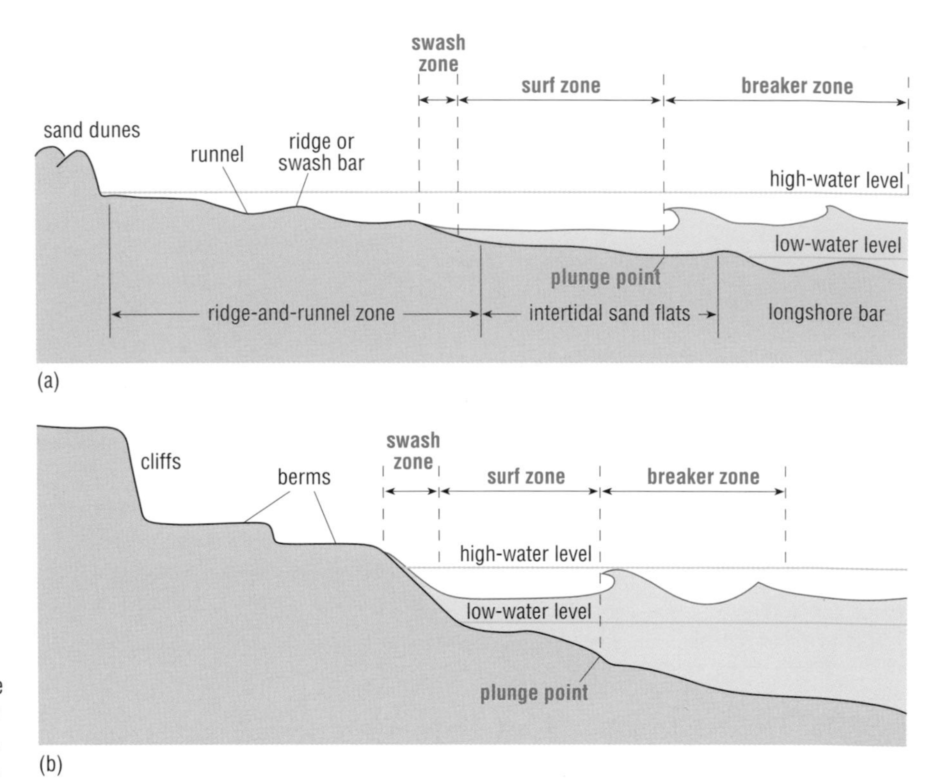

Figure 5.1 Diagrammatic beach profiles showing (a) a beach of shallow slope and wide intertidal zone, with ridge-and-runnel structure, and (b) a steep beach with narrow intertidal zone and berms. The plunge (or break) point is where most waves break (especially plunging breakers, Figure 1.18). For Figure 5.1(c)–(f), see opposite.

Beaches are accumulations of loose sand or pebbles. Much of the energy of waves breaking upon the shore is dissipated in moving sediment, though some is always reflected back out to sea, the proportion depending on the beach slope (Section 1.5.2), and beach profiles can change quite rapidly in response to changes in wave energy (determined chiefly by prevailing wind and tide conditions).

Broadly speaking, on most beaches, three zones of wave action can generally be identified (Figure 5.1). As waves approach the shore, they become unstable and break, forming a *breaker zone* which can be quite broad, because the waves have a range of wave heights and do not all break at the *plunge point* (see Section 5.1.1). Landward of the breaker zone is the surf *zone*, generated by the breaking waves; and finally there is a **swash zone**, where smaller waves are projected up the beach slope, which is alternately covered by the upsurge of water (the *swash*) and exposed as the water retreats in the *backwash*. On steep beaches, the breaker zone may extend to the swash zone without an intervening surf zone (and, in some texts, the breaker zone is not recognized separately, but is included as part of the surf zone). The sloping part of the beach between high and low water is sometimes called the *beach face*.

Figure 5.1 (c) A beach of shallow slope, with a breaker zone passing into a surf zone inshore.
(d) Sandy beach of moderate slope (intermediate between (a) and (b)), with a berm (see also Figure 5.12).
(e) Aerial view of the ridge-and-runnel zone on a gently sloping beach backed by dunes, seen at low tide.
(f) A ridge (swash bar) and runnel in the intertidal zone.

QUESTION 5.1 Would you expect the wave zones described above, (a) to maintain a constant position during a tidal cycle, and (b) to have comparable widths off beaches of steep and shallow slope? (c) Would you expect the plunge point (Figure 5.1(a) and (b)) to change position with the tidal cycle?

Shallow sloping beaches may be characterized by a series of low broad sandy ridges or bars (sometimes known as *swash bars*) separated by linear depressions or *runnels* which are parallel to the shoreline (Figure 5.1(e) and (f)). These features are formed by sediment movement in the surf and swash zones and a whole series of swash bars and runnels may develop, especially in the upper part of the intertidal zone, above the intertidal sand flats where present (Figure 5.1(a)). When the tide is rising, it can be dangerous to walk down to the sea across a beach with a well-defined **ridge-and-runnel zone.** The runnels fill with water first (Figure 5.1(e, f)) and it is easy to find yourself stranded on a swash bar, separated from the next bar or the shore by a stretch of often relatively deep, fast-flowing water. Seawards, beneath the breaker zone, a *longshore bar* may develop in the winter season, when more sediment is moved offshore (see overleaf).

Beaches with moderate to steep slopes and relatively narrow intertidal zones are often characterized by **berms**, flat-topped ridges at the upper limit of wave action (Figure 5.1(b) and (d)), usually built by the migration of sand or shingle bars up a beach under the influence of low-amplitude waves in calm weather – although the steep seaward slopes of berms often result from erosion by storm waves.

There are nearly always waves from distant storms (i.e. swell, Section 1.4.3), but their effect is swamped, especially in winter, by steep waves generated during local storms. In general, where there is a seasonal difference between fair weather, swell-dominated waves in summer, and steep storm waves in winter, sediment is moved up the beach slope during summer to build swash bars and/or berms, whereas sediment is removed from the beach during winter and transported seawards, where it may build a longshore bar (Figure 5.1(a)). During the following summer, when swell waves prevail again, sediments are moved back up the beach to reconstruct the swash bars and/or berms. Small, low-amplitude gentle waves and swell thus tend to build up beaches whereas high steep storm waves tend to tear them down (Figure 5.2).

Figure 5.2 (a) A sandy beach in autumn becomes (b) stripped of sand during winter storms, so that eventually (c) underlying rocks and boulders are exposed. During summer (d), the beach becomes sandy again.

Figure 5.3 is a series of beach profiles illustrating the progressive onshore movement of a sand-bar under the influence of low-amplitude waves in the Northern Hemisphere summer. The shoreward slopes of the ridges are steeper than the seaward slopes, a situation that is likely to be reversed under the erosive action of higher steeper waves in winter.

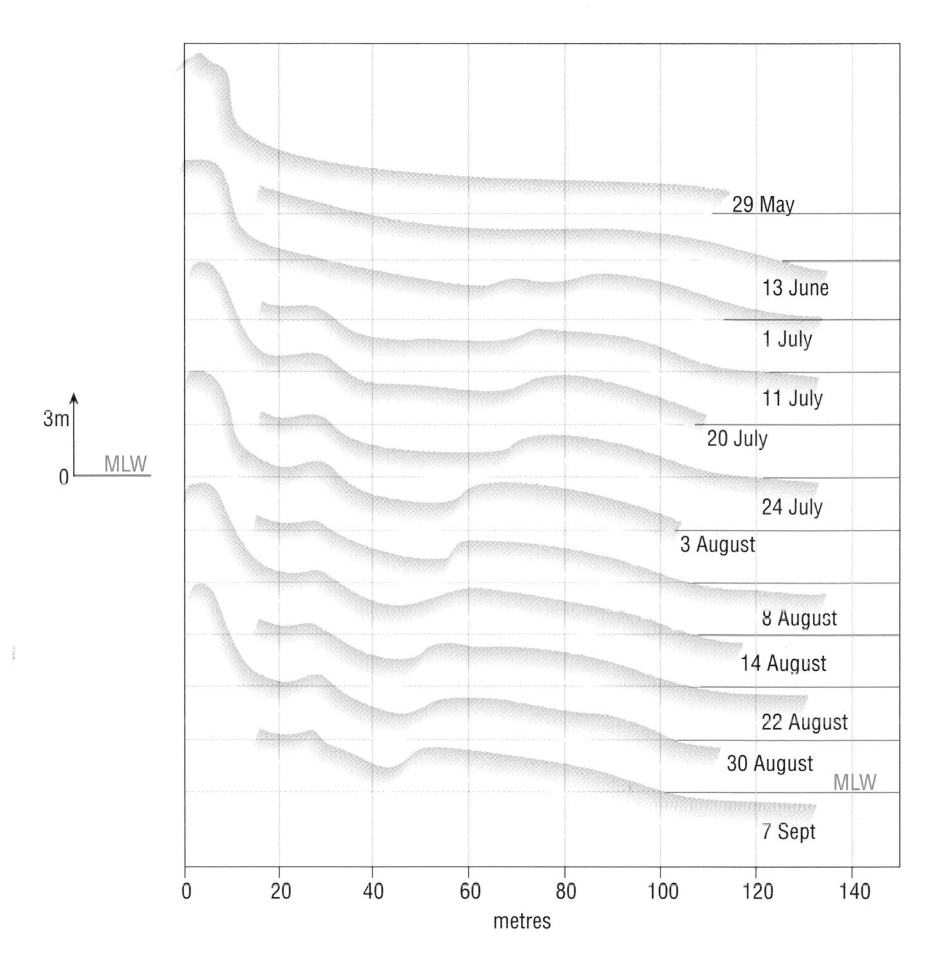

Figure 5.3 A series of profiles (starting at the top) showing the growth of a beach by onshore movement of a sand-bar at Crane Beach, Massachusetts (USA), during the summer of 1967. MLW = Mean Low Water.

Sand dunes

Beach sands above the normal high tide level, as well as the sand dunes that form the landward limits of some beaches, are built up by onshore winds transporting fine to medium sands (i.e. in the 0.1–0.5 mm (100–500 μm) size range, Table 4.1), originally supplied to the beach by waves. Dunes will only develop where winds are favourable and beach sediments fall in the appropriate size range. As with bedload transport by currents (Equation 4.6), the rate of transport of wind blown sediments is proportional to the cube of the wind speed (but winds need to be much faster than currents to move the sand any distance, because of the lower density of air). Sand dunes tend to migrate inland and can travel long distances if wind conditions allow. They can be thought of as migrating bed forms resulting from wind action (cf. Section 4.4), and they can be stabilized by vegetation, which traps and binds the sand.

5.1.1 BEACH PROFILES IN RELATION TO GRAIN SIZE AND WAVE STEEPNESS

If you visit the coast frequently, you have probably noticed that beaches made of shingle or pebbles are steeper than those made of sand, so sediment grain size must affect the beach profile. When a wave breaks on the shore, sediment is transported up the beach by the swash, and deposited as the swash peters out. Some of it is then re-eroded and dragged back down by the backwash. Partly because of the dynamics of shallow-water waves (see Section 5.2.2), and partly because some of the water brought in on the wave is lost by percolation into the beach sediments, the backwash is weaker than the swash. As a result, there is a net onshore movement of sediment up the beach, until the slope reaches a state of dynamic equilibrium, and as much sediment is being moved landward as is being returned seaward. The rate of percolation into the beach sediments depends on grain size. Water percolates much more easily into shingles or pebbles than into sands, and so the backwash over pebbles is proportionately weaker than over sands.

Wave steepness is also an important factor controlling the dynamic equilibrium of beach slopes. When steep waves break onto the beach (as spilling or plunging breakers (Figure 1.18), their energy is dissipated over a relatively narrow area. The swash does not move far up the beach, so less energy is lost in moving sediment up the beach face, and there is less opportunity for percolation to occur. Consequently, the backwash is strong and a lot of material can be moved seawards. Plunging breakers are particularly erosive, for they tend to scour the beach near the plunge point (Figure 5.1(a) and (b)), lifting large quantities of sediment into suspension, to be removed in the backwash.

Where less steep waves break on the beach (collapsing or surging breakers, Figure 1.18), the swash is strong and moves a good deal of sediment up the beach. The backwash is relatively weak because much energy has been used in moving sediment up the beach, and there is more opportunity for water loss by percolation.

QUESTION 5.2 As you saw in Figure 1.18, the angle of a beach slope determines whether waves of given height and steepness will form plunging, spilling, collapsing or surging breakers. Would you expect spilling and plunging breakers to be commoner than collapsing and surging breakers on (a) shallow or (b) steep beaches?

Figure 5.4 summarizes relationships between the three variables we have just been considering, namely predominant grain size, prevailing wave steepness, and beach slope.

QUESTION 5.3

(a) Use the curves in Figure 5.4 to describe how the influence of wave steepness on beach slope varies as the sediment grain size decreases.

(b) For waves of steepness 0.02, what angle of beach slope would you expect to find where the sediments are predominantly (i) fine sand, (ii) fine gravel?

An empirical relationship has also been established between beach slope and the depth of water where the waves break. A combination of theory and systematic observations over many years suggests that the ratio of wave height to water depth at the break point (H/d) is constant for a particular

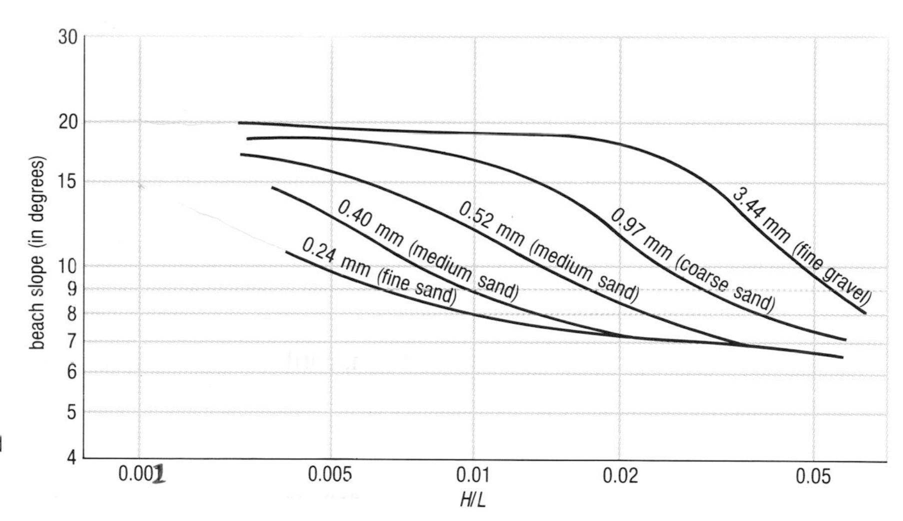

Figure 5.4 Relationships between beach slope (measured in degrees), wave steepness (*H/L*) and average grain size. Note that the scales for both beach slope and wave steepness are logarithmic.

beach slope, and that the ratio decreases with increasing beach slope. Put simply, this means that all waves of a particular height, approaching a beach of given slope, will break at about the same depth, such that for a particular slope, *H*/*d* is effectively constant, irrespective of whether they are storm waves or swell. It also means that waves of a given height will break in shallower water (i.e. *H*/*d* will be greater) over a gently shelving beach than over a steeply sloping one. So, on beaches whose slope increases towards the high water mark, the value of *H*/*d* will decrease as the tide rises. Values of *H*/*d* appear generally to be in the range 0.7–0.8, and lie at the upper end of this range for gently sloping beaches, at the lower end for steeper ones.

If all waves had the same height, therefore, the plunge point or break point (Figure 5.1(a, b)) would be a very narrow zone, migrating back and forth with the tides. However, as waves approaching the shore have a range of wave heights, they will break at different depths and distances offshore, so there is a relatively broad breaker zone (Figure 5.1(c)). The plunge point, lying within the breaker zone near its landward end, is a somewhat narrower zone where the majority of the waves will break, for any particular combination of tidal conditions and sea state.

Careful analysis of wave records has also revealed that, in addition to the familiar waves we see when we visit the coast, there are longer-period oscillations in the wave field. These *infragravity waves* or long waves, with periods ranging from 20 seconds to 200 seconds (Figure 5.5, overleaf), can also influence beach processes and beach profiles; but the theoretical relationships are complicated and we do not consider them further here.

Before moving on, however, we should mention the phenomenon of *edge waves*, which can develop where waves approaching a shore obliquely are effectively trapped by a combination of refraction and reflection. As you saw in Figure 1.15, refraction causes oblique waves to become more nearly parallel with the shore as they move into progressively shallower water. If the waves are then reflected, their crests will be rotated *away* from alignment with the shore as they move into deeper water. With the right combination of water depth and wavelength, the reflected waves can then be refracted back towards the shore, to be reflected again. The waves are thus trapped, and interference between the incident and reflected waves gives rise to a sequence of edge waves (Figure 5.6, overleaf).

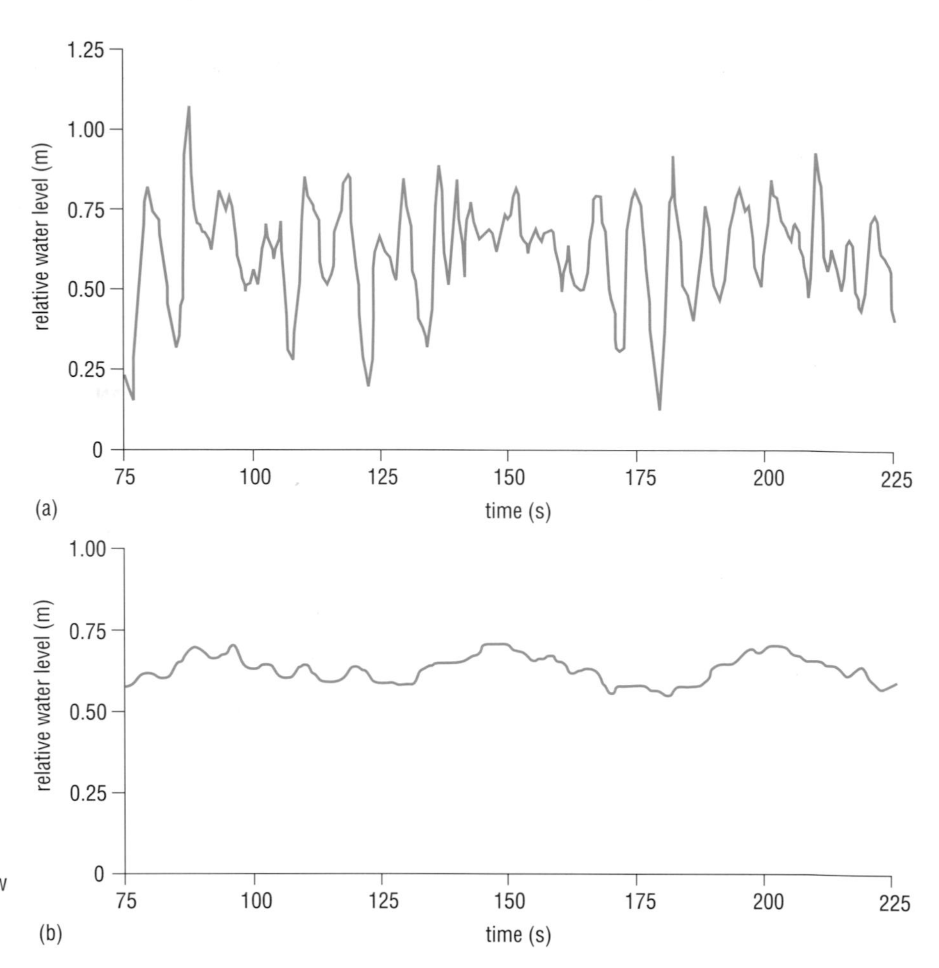

Figure 5.5 (a) A surface wave record made at Slapton Sands, Devon, statistically filtered to show (b) an underlying series of infragravity waves with period of about 50 s.

Edge waves are best developed where the boundary is steep (e.g. a cliff or harbour wall) and the water is relatively deep, so that the waves are reflected before they break. The phenomenon is much less easy to observe along beaches, where incident waves typically break before they can be reflected, but they are more likely to occur where beach slope is steep than where it is shallow.

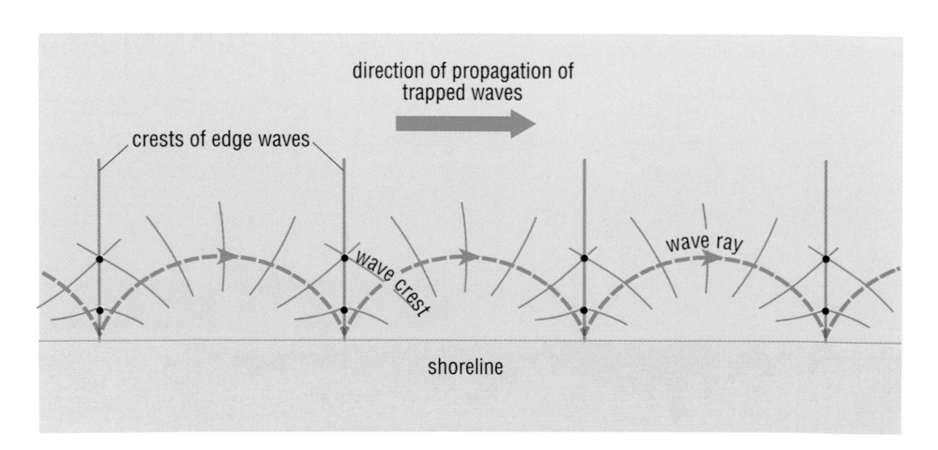

Figure 5.6 Development of edge waves. The blue broken line is a single wave ray (cf. Section 1.5.1); short curved blue lines are wave crests, combining where they intersect.

5.2 SEDIMENT MOVEMENT IN THE BEACH ZONE

Factors that control the rates of movement of sediment onto and off beaches must include the height, speed and steepness of waves, water depth, beach slope, settling velocity of sediment particles, magnitude and direction of local currents, and so on. As we saw in Chapter 1, water particles at the surface in a wave travelling over deep water follow virtually circular orbits that become progressively flatter ellipses in shallow water, until eventually there is simply a to-and-fro movement of water at the sea-bed (Figure 1.8).

5.2.1 ORBITAL VELOCITIES AND BED SHEAR STRESS

The speed at which the water moves in wave orbits is calculated from the length of time it takes a water particle to complete an orbit; i.e. for the particle to move from crest to trough, and back to the crest of the next wave as the wave-form passes. This speed is the *orbital velocity*, and it has a maximum horizontal *landwards* component immediately beneath the crest of a wave, and a maximum horizontal *seawards* component just beneath the trough.

For waves in water depths less than $L/2$ but more than $L/20$, the maximum horizontal orbital velocity near the bed (u_m) is related to the wave height H, water depth d, wavelength L, and wave period T, by:

$$u_m = \frac{\pi H}{T \sinh(2\pi d / L)} \tag{5.1}$$

Sinh is the hyperbolic sine, a mathematical function analogous to the hyperbolic tangent introduced in Equation 1.2, such that sinh $x \approx x$ when x is equal to or less than about 1. For shallow-water waves, $2\pi d/L < 1$ so sinh $(2\pi d/L) \approx 2\pi d/L$.

The important points to notice in Equation 5.1 are:

(i) that for waves of a constant period, increased wave height results in increased u_m; and

(ii) somewhat counter-intuitively perhaps, that for waves of a given height, u_m is *greater for short-period waves than for long-period waves*; i.e. there is an *inverse* relationship between u_m and T.

Does that mean we might expect swell waves to have smaller horizontal orbital velocities than steep storm waves?

Yes it does, because in general, steep storm waves have shorter periods than swell waves.

In water of depths less than $L/20$, waves behave as true shallow-water waves, and the long axes of the elliptical orbits, and hence the orbital velocity, remain constant with depth (cf. Figure 1.8(d)).

Since $\sinh(2\pi d/L) \approx 2\pi d/L$:

$$u_m = \frac{\pi H}{T \times (2\pi d/L)} = \frac{L}{T} \times \frac{H}{2d} = \frac{cH}{2d} \quad (5.2a)$$

$$\text{or} \quad u_m = \frac{H}{2}\sqrt{\frac{g}{d}} \quad (5.2b)$$

where H is wave height, c (= L/T) is wave speed, d is water depth and g is the acceleration due to gravity. Note that Equation 5.2b is obtained by substituting $\sqrt{gd}$ for c (Equation 1.4) in Equation 5.2a, and this has the advantage that only two variables, wave height and water depth, need to be measured in order to calculate u_m.

For reasons discussed earlier, as waves move into progressively shallower (shoaling) water they become higher, i.e. as the value of d decreases, the value of H increases. Both changes lead to an increase in the maximum orbital velocity of the wave. The shear stress at the sea-bed thus increases, and so does the potential for sediment movement.

QUESTION 5.4 In water of a given depth, what would you expect to happen to the shear stress exerted by waves at the sea-bed if the wave height were to double? It increases by x4

5.2.2 SEDIMENT MOVEMENT BY WAVES

Although there is a wave boundary layer analogous to the current boundary layer described in Section 4.1.1, it is much more complex than for unidirectional flow and we shall not discuss it in detail. The complexity arises because u_m, and hence the shear stress, reverses direction as a wave passes, so a boundary layer never becomes fully established. The wave boundary layer is also very thin, no more than a few millimetres to one or two centimetres, compared with up to 10 m or even more for the boundary layer produced by current movement in the sea (Section 4.1.1). However, as in the case of current flows, where medium-grained sands and finer sediment (i.e. sediments less than about 0.5 mm diameter, Table 4.1) occur on the sea-bed, water flow beneath waves is more or less laminar; but where the sediment is coarser-grained the flow is typically turbulent.

Influence of wave characteristics

As the maximum horizontal orbital velocity is (inversely) related to wave period (Equation 5.1), it is possible also to relate wave period and the size of particle that can be moved, as illustrated in Figure 5.7; while Figure 5.8 shows the combinations of wave heights and water depths necessary for a wave with period 15 s to move grains of the sizes represented, on a flat sea-bed.

A number of important points arise from Figures 5.7 and 5.8:

1 Figure 5.7 shows that the horizontal orbital velocity required to move a grain of a given size *increases* as the wave period increases. For example, u_m must be about 0.25 m s^{-1} in order to move a quartz grain 1 mm in diameter beneath a wave of 1 s period, whereas it needs to be about 0.4 m s^{-1} to move the same grain beneath a wave of 15 s period. This inverse relationship (discussed earlier in relation to Equation 5.1) results from the rapidity with which water particles accelerate to their maximum horizontal velocity. The acceleration is greater for shorter- than for longer-period waves, and hence so is the frictional (shear) stress at the bed.

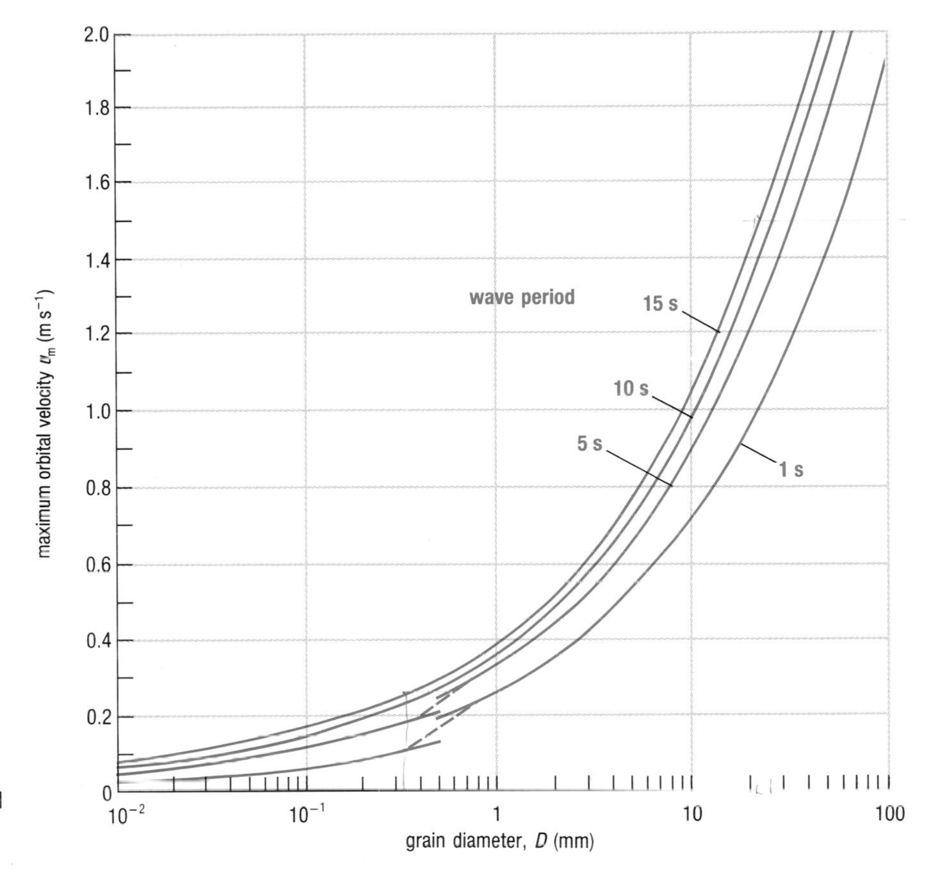

Figure 5.7 Empirically determined curves showing the near-bed horizontal orbital velocity needed to move sediments of different grain size under waves of different periods. The curves are constructed specifically for quartz grains with a density of 2.65×10^3 kg m^{-3}. Breaks in the 1 s and 5 s curves are the result of the wave boundary layer behaving in a laminar fashion for grain sizes less than about 0.5 mm diameter, and in a turbulent fashion for larger grain sizes.

2 As you might expect, comparison of Figures 5.7 and 4.6(c) shows that maximum near-bed horizontal orbital velocities of wave motions are of the same order as the near-bed current speeds required to move sediment particles (e.g. of the order of 0.3 m s^{-1} for 1 mm diameter grains).

3 Figure 5.8 demonstrates that large storm waves are capable of moving sediment at considerable depths on the continental shelf (as you might have inferred from Figure 4.16).

4 Equation 5.1 shows that there are many combinations of wave period, water depth and wave height which could produce the horizontal orbital velocity necessary to move grains of a given size.

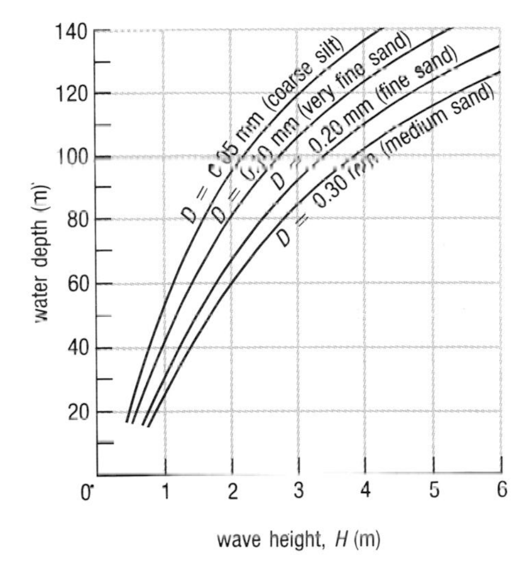

Figure 5.8 Empirically determined heights of a wave of period 15 s required to move particles of various sizes on a flat bed in different water depths.

QUESTION 5.5 For waves with $H = 3$ m in water depths such that $d = L/10$, calculations with Equation 5.1 give values of u_m close to 1.5 m s^{-1} and 1 m s^{-1} for wave periods of 10 s and 15 s, respectively.

(a) What are the maximum particle sizes that could be moved by these waves, according to Figure 5.7?

(b) How could you infer (i) from Equation 5.1 that coarser particles can be moved by shorter-period waves; and (ii) that the 10 s period waves are probably both steeper and in shallower water than the 15 s period waves?

(c) If the wave height decreased (e.g. after a storm had passed), but water depth remained the same, would you expect the sediment grains to remain in transport?

(d) If the wave height remained the same, but water depth increased (e.g. on the incoming tide), would you expect the grains to remain in transport?

The answers to Question 5.5 show that waves of quite long period can move coarse sediment particles even in relatively deep water (cf. Figure 5.8), so it is hardly surprising that when the waves behave as true shallow-water waves, pebbles and boulders can easily be shifted.

For instance, according to Equation 5.2b, what would be the orbital velocity of 3 m high waves in water, say, 2 m deep?

If we substitute these numbers into Equation 5.2b, we get:

$$u_{\mathrm{m}} = \frac{3}{2}\sqrt{\frac{g}{2}}$$

which works out to about 3.3 m s^{-1}, well off the vertical scale of Figure 5.7. That is quite sufficient to move even small boulders on the sea-bed.

Onshore and offshore movement of sediment by waves

As we saw in Section 5.1.1, for any particular set of wave conditions, there must generally be net transport of sediment up the beach until the slope reaches a state of dynamic equilibrium and as much sediment is moved landwards as is returned seawards. However, although the *amounts* of sediment shifted in each direction may be the same, the distribution of particle sizes in each direction is *not* the same.

The maximum horizontal orbital velocity is attained twice as a wave passes: once on the forwards stroke of the wave as the crest passes, and secondly on the backwards stroke as the trough passes. This means that sediment should be moved landwards beneath the crest and seawards again beneath the trough. So, why is there a net movement of sediment either landwards or seawards?

On the backwards stroke of the wave, when a water particle is in the trough, it is brought closer to the sea-bed than when it is at the crest of the wave, and frictional retardation at the bed is greater. This means that water particle speeds are not the same in both directions. They are greater during onshore movement, but are maintained for only a short period of time, when coarser sediment is moved shoreward as bedload and finer sediment as suspended load. Speeds are less during offshore movement, so while most of the suspended load and finer bedload are returned seawards, some of the coarser bedload material is left behind. As the slower offshore movements are maintained for longer periods of time (Figure 5.9), there is net offshore movement of fine material, and net onshore movement of coarse material.

In addition, as outlined in Section 5.1.1, the offshore flow of water (backwash) from breaking waves is also weaker than the onshore flow (swash) because of percolation into the beach sediments – the coarser the sediments, the greater the percolation and the weaker the backwash relative to the swash. Moreover, as you saw in Figure 4.6(c), flow speeds required to move (erode) even non-cohesive sands and gravels are greater than those at which such sediments are deposited.

In general, then, waves of moderate size can move sand (and even shingle) shorewards, and remove finer sands and muds seawards. The overall effect is to build up the beach. High steep waves can transport large pebbles (and even boulders) shorewards, but they will remove smaller gravel particles and sand seawards (again partly in suspension). However, especially during winter storms, the beach tends to be saturated with water, there is less percolation and the backwash is proportionately stronger, so the overall effect is to erode the beach (cf. Figure 5.2).

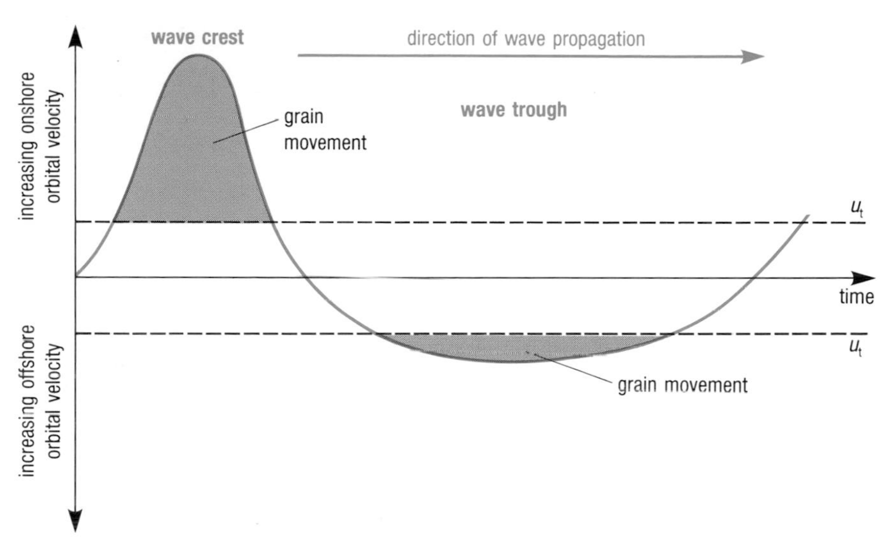

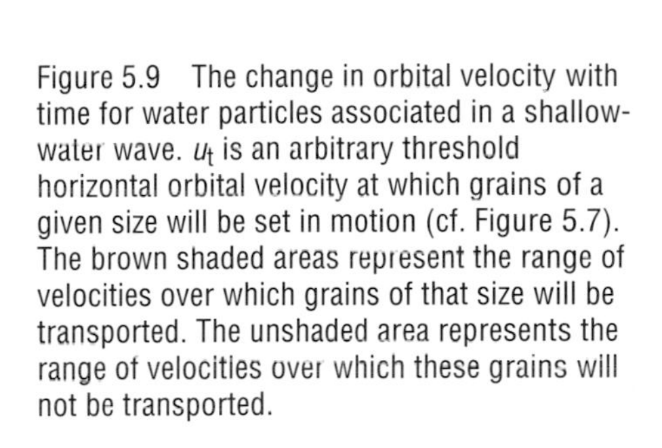
Figure 5.9 The change in orbital velocity with time for water particles associated in a shallow-water wave. u_t is an arbitrary threshold horizontal orbital velocity at which grains of a given size will be set in motion (cf. Figure 5.7). The brown shaded areas represent the range of velocities over which grains of that size will be transported. The unshaded area represents the range of velocities over which these grains will not be transported.

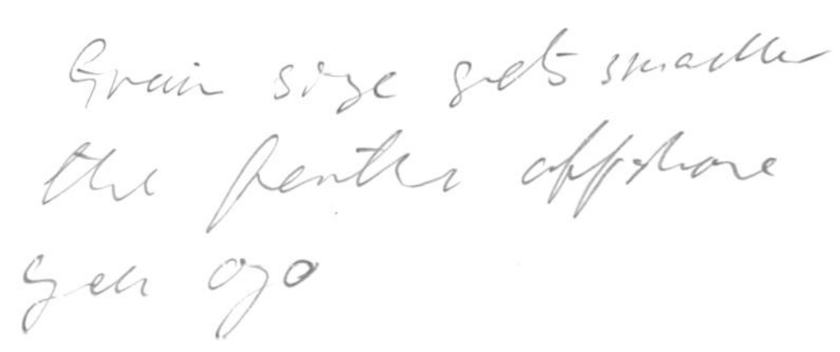

QUESTION 5.6 In light of what you have just read, summarize how these inequalities of flow affect the seawards distribution of sediment grain sizes in the beach zone.

Sedimentary features (bed forms) on beaches

One of the most common bed form features seen on the beach are wave-generated ripples, which are symmetrical and have long, straight crests that occasionally bifurcate. These features distinguish them from the asymmetrical ripples produced by unidirectional currents. Wave-formed ripples are symmetrical because sediment is moved towards the ripple crest on both sides: on the seaward-facing side as the wave crest passes and there is horizontal movement of water particles in a landwards direction; and on the landward side as the wave trough passes and the water particles move seawards.

Wave-generated ripples begin to form as soon as the threshold orbital velocity for grain movement is reached. As orbital velocities increase, the height of the ripples decreases until the sediment moves as a suspended 'sheet flow', backwards and forwards over the sea-bed.

Occasionally, the ripple marks found in runnels (Section 5.1) are a fascinating combination of straight-crested wave-generated ripples interspersed with asymmetrical current-formed ripples, trending roughly at right angles to the wave ripples (Figure 5.10, overleaf). The resulting patterns are known as *ladder-back ripples*. Sometimes, the current ripples give an indication of the longshore current direction, but in the case of Figure 5.10 they are probably formed by water draining from a runnel on the ebb tide. Another common feature you have probably seen on sandy beaches is the rhomboid pattern formed by fast-flowing backwash (Figure 5.11, overleaf).

A typically somewhat larger kind of structure commonly seen on beaches are *cusps*, which are among the most familiar of beach features, being commonest and best developed around high tide at the top of the swash zone. They are series of regularly spaced shallow embayments that are concave seawards and join at sharp seaward pointing 'horns', and are most obvious on shingle beaches (Figure 5.12, overleaf), though not confined to them. Typical cusp 'wavelengths' are from a metre or two to several tens of metres, and they may be associated with the edge waves that can result from the motions of oblique waves being reflected from the beach face (Section 5.1.1). Once formed, the cusps tend to remain in place, perhaps because wave crests are slightly higher and break sooner at the 'horns' than in the embayments.

Figure 5.10 Ladder-back ripples: wave-formed ripples with linear, bifurcating crests, in between which are smaller, current-formed ripples running at right angles to the wave-formed ripples. The scale is given by the camera lens cap.

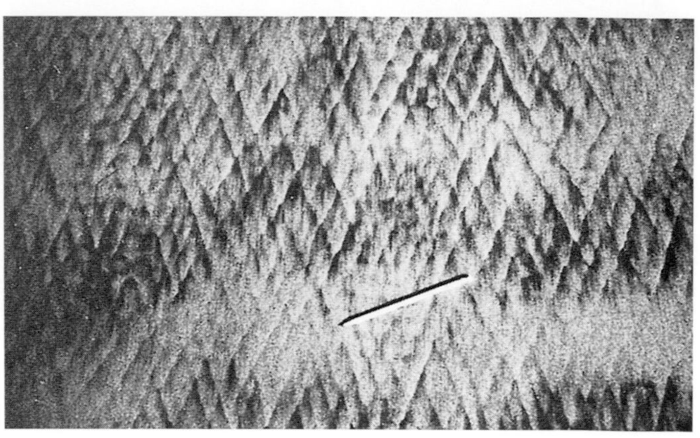

Figure 5.11 Rhomboid marks in beach sands. The scale is given by the pencil. The acute angles of the rhomboids point in the direction of flow (towards the bottom of the picture in this case).

Figure 5.12 A line of cusps bordering a well-developed berm at the top of a steeply sloping shingle beach.

5.2.3 LONGSHORE SEDIMENT TRANSPORT BY WAVE-GENERATED CURRENTS

When waves approach a straight coastline at an oblique angle, as commonly happens, a **longshore current** is established which flows along the shoreline with speeds between about 0.3 and 1 $m\,s^{-1}$. The speeds of such longshore currents are proportional both to the maximum orbital velocities of waves in the breaker zone, and to the angle that the wave-fronts make with the shoreline as they approach it. Longshore currents are best developed along straight coastlines and are an important way in which sediment is moved along shorelines where there are gently sloping beaches.

On steep beaches, longshore transport by the swash and backwash of waves is more important than that by longshore currents. When a wave breaks obliquely to the shoreline, the swash drives sediment up the beach face at an angle to the shoreline while the backwash drags the sediment down the beach at right angles to the shoreline – so successive waves move the sediment along the beach face in a zig-zag pattern.

The longshore movement of sediments in the beach zone is often referred to as *longshore drift*, and it can lead to the formation of *spits* of sand or shingle, elongated in the direction of longshore transport. Spits are generally linear extensions of beaches, sometimes terminating in the open water of bays and estuaries (Figure 5.13), sometimes lying relatively close inshore, forming long narrow coastal lagoons (see Section 6.3).

(a)

(b)

Figure 5.13 (a) Spurn Head at the mouth of the Humber (see Figure 6.5) is formed by longshore drift southward along the Humberside (Holderness) coastline (we are looking south, so east is to the left, west to the right). From time to time, the narrowest part of the spit is severed by winter storms
(b) Closer view of Spurn Point (the lighthouse is about half-way along), showing megaripples and tidal flats (see Section 6.3) on its western shore.

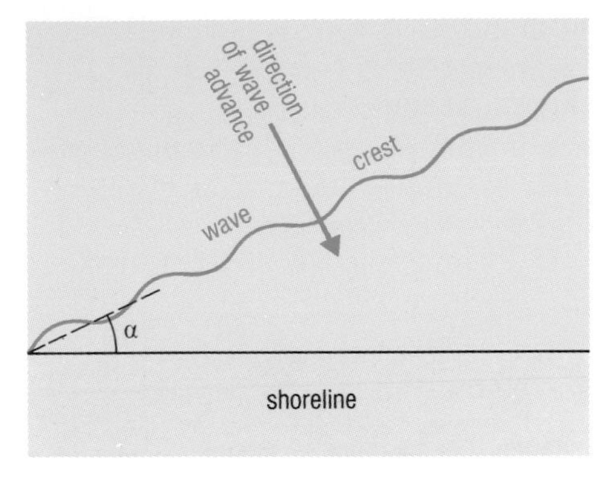

Figure 5.14 Diagram of a wave crest approaching a shoreline obliquely at an angle α.

The longshore sediment transport rate

In order to quantify longshore drift (the longshore transport rate), we need to know the longshore wave power (P_l) available. This depends on the height of the waves at their break point (H), their group speed (c_g), and angle (α) between the advancing wave crest and the shoreline at the break point (Figure 5.14). The relationship can be expressed by:

$$P_l = c_g(\frac{1}{8}\rho g H^2)\sin\alpha\cos\alpha \qquad (5.3)$$

where ρ is the density of the seawater and g is the acceleration due to gravity.

You should recognize the expression in brackets as defining wave energy (E; Equation 1.11) and c_g as the speed with which the energy is arriving (i.e. group speed). The sine term in Equation 5.3 is needed in order to determine the *longshore* component of wave power (per metre crest length of wave). To be meaningful, however, this must be converted into power per metre length of shoreline, which is the reason for using the cosine term in Equation 5.3.

If you were to stand on a beach and watch the breaking waves, you could use Equation 5.3 to arrive at a first-order estimate of wave power in the beach zone by making a few simple observations and estimates of basic wave characteristics. Conveniently, group speed is the same as wave speed in shallow water (Section 1.3), and you would need to estimate the average wave height at the break point (plunge point) of the waves.

QUESTION 5.7

(a) Given wave group speed = 0.5 m s^{-1} and average wave height = 1 m, and using $\rho = 10^3$ kg m^{-3} and $g = 9.8$ m s^{-2}, calculate what the longshore wave power would be if wave crests approached the shoreline at an angle of 30°. Pay attention to units and note that your answer should be in joules per second (watts) per metre length of shoreline.

(b) What would you expect the longshore wave power to be when the crests of these same waves approach parallel to the shoreline?

The wave power for waves breaking directly on the beach is, therefore, simply:

$$P = c_g E$$

i.e.
$$P = c_g(\frac{1}{8}\rho g H^2) \qquad (5.4)$$

Having calculated the wave power, the longshore rate of sediment movement for sand-sized grains, q_l, can be determined from the empirically derived equation:

$$q_l = \frac{0.77P_l}{g(\rho_s - \rho)\,0.6} \qquad (5.5)$$

where q_l is measured in m^3 s^{-1}, ρ_s and ρ are the densities of sediment and water respectively, 0.77 is a coefficient of efficiency relating to loss of water due to percolation through the sediments, and 0.6 is another coefficient which represents the average proportion of the bulk sediment occupied by particles, rather than pore space. The longshore transport of sediment is often quite substantial. For example, along parts of the south-eastern coast of the USA, the net southwards transport rate may be as much as 0.5×10^6 m^3 per year.

Sediment transport by combined waves and currents

Although we have confined our discussion in preceding Sections to the influence of wave action, in practice waves and currents act together in the coastal zone. Most net sediment transport is likely to occur where movement by currents is enhanced by wave motions, because waves are very effective at stirring up sediment on the sea-bed (cf. Figure 4.16). Once waves have lifted sediment into suspension, it can then be transported by currents which would be unable to lift the sediment off the sea-bed by themselves. These currents include both longshore currents and rip currents (see below), as well as tidal currents, and even the slight landwards movement of water caused by the wave drift described in Section 1.2.1.

5.2.4 RIP CURRENTS

Rip currents are strong, narrow currents with speeds up to 2 m s^{-1} which flow seawards from the surf zone (Figure 5.15, overleaf). They are potentially very dangerous because a swimmer caught in a rip current may be swept out to sea quite rapidly and drown after becoming exhausted by trying to swim back to the shore against the current. The best means of escape is to swim parallel to the shore for a few metres, away from the rip current, before trying to swim shorewards. However, experienced surfers are quite happy to exploit these currents by riding them out to sea. Beach anglers also know that rip currents can be productive areas in which to fish.

The longshore current generated where waves break obliquely on the shoreline is confined largely to the surf zone landward of the breaker zone. As more and more waves add their longshore components on breaking, the longshore current tends to increase with distance along the shore. This clearly cannot happen indefinitely and the requirements of continuity (Section 2.4.1) dictate that the discharge of water from waves into the longshore current must be balanced by an offshore flow, i.e. at intervals there must be transport of 'excess' water out through the surf zone. This is one way in which rip currents develop (Figure 5.15(a)).

Rip currents also develop where waves are moving directly onshore. As you know, an increase in wave height along a wave crest occurs when the wave enters shallow water and begins to slow down (Section 1.5). Variations in wave height along a wave crest will occur if one section of the crest encounters shallow water before another. This happens either when a wave crest approaches an irregular coastline and shallow water is encountered first off seaward projections, or where the nearshore submarine topography is irregular for other reasons (cf. Figure 1.16). In short, the location of the rip currents is determined by the offshore bathymetry. Just shorewards of the breaking waves, i.e. in the surf zone (Figure 5.1), the average water level falls, and then rises continuously towards the shore, an increase in level known as **wave set-up** (Figure 5.15(b)) – which can also occur with obliquely breaking waves. Where wave heights are greatest along the wave crest, the wave set-up is also greatest and horizontal pressure gradients are established between regions of higher and lower wave set-up (i.e. higher and lower average water level). Water therefore tends to flow from positions of higher to lower wave height, generating longshore currents which move towards each other in the surf zone, turning seawards at the convergence as rip currents. Rip currents generated in this way thus form part of a cell-like circulation of water (traced out by the arrows in Figure 5.15(c)). The strength of rip currents depends upon the nature and size of the waves that generate them, and may vary with the state of the tide.

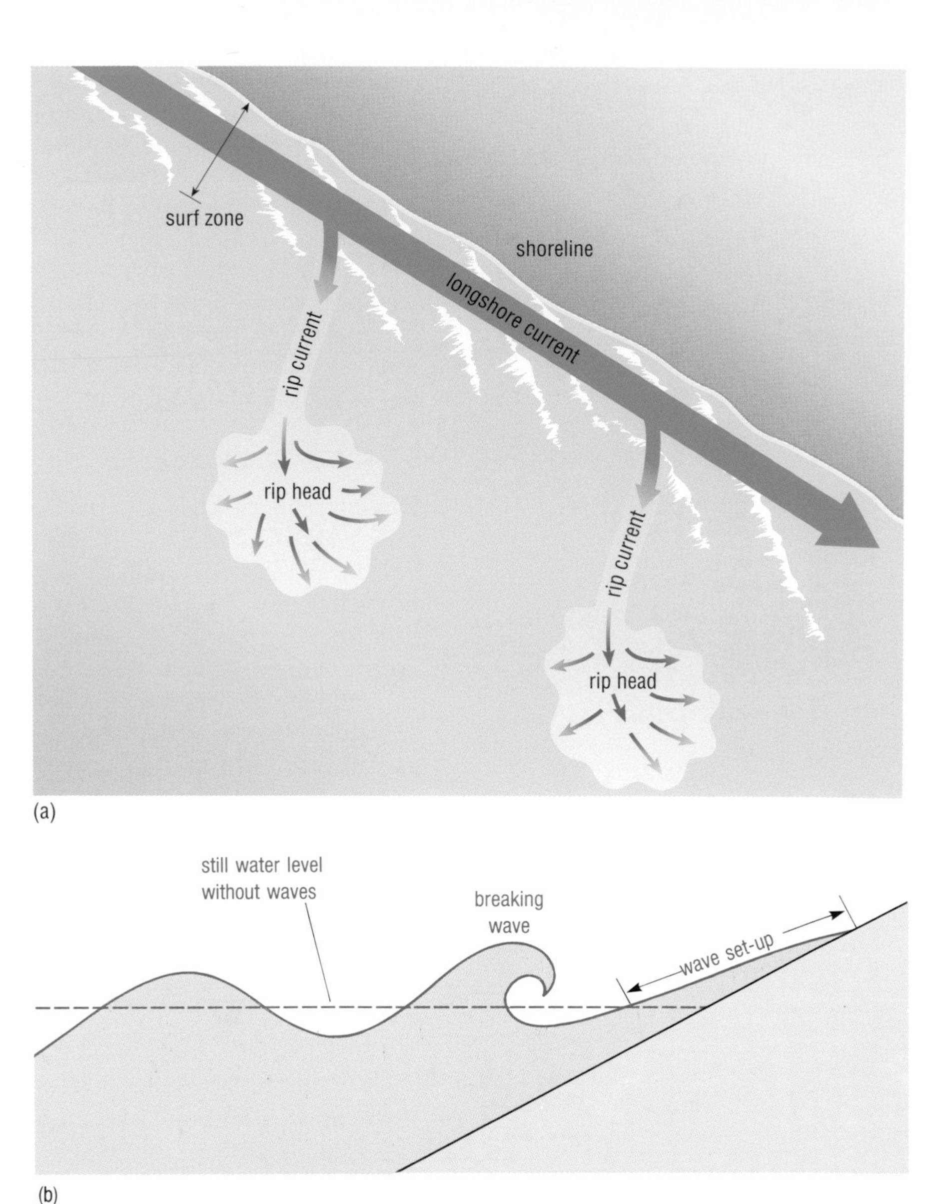

Figure 5.15 (a) Plan view of a section of coastline showing the development of rip currents from a longshore current generated by obliquely breaking waves.
(b) Wave set-up: the rise in the mean water level inshore of a breaker. For (c)–(d), see opposite.

Some of the sediment being transported by longshore currents will be carried seawards by rip currents (Figure 5.15(d)), which thus scour out and maintain the channels in which they flow. Rip current circulation can also be an important means of renewing water in the nearshore zone and flushing out sewage and other pollutants discharged in coastal regions.

5.3 DYNAMIC EQUILIBRIUM OF SEDIMENT SUPPLY AND REMOVAL

If estimates can be made of the amount of sediment moved by different means in the beach zone, a *sediment budget* can be drawn up of material imported (e.g. from rivers or cliff erosion) and material exported (e.g. by longshore or offshore transport), and the balance used to determine whether the beach at any one place is undergoing active deposition or erosion

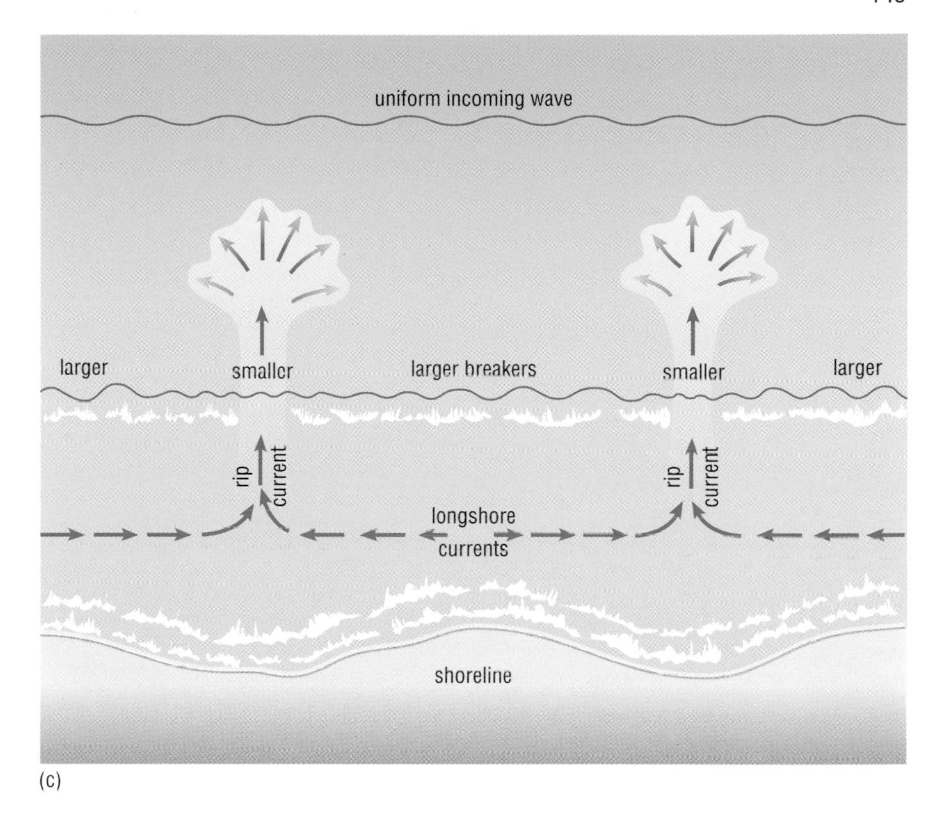

(c)

(d)

Figure 5.15 (c) Plan view of a section of coastline showing the formation of rip currents due to variation in wave height along wave crests.
(d) The presence of rip currents seawards of the breaker zone is indicated by a brownish foamy patch (right-hand side of photograph) showing sand being transported offshore. This situation corresponds to that shown in part (a), with waves coming onshore obliquely.

(Figure 5.16(a), overleaf). In practice, this procedure can be difficult, given the uncertainties and assumptions in making calculations based on equations such as Equations 5.3 to 5.5. However, the *likelihood* of beach erosion or deposition at a particular location can be estimated, and this is especially useful where the dynamic equilibrium of the coastal zone is likely to be (or has been) disturbed by, for example, the construction of a harbour or breakwater (Figure 5.16(b)). Sediment budgets can also be used to quantify the inputs and outputs for larger-scale *sediment transport cells*, which are adjacent but discrete segments of coastline where sediment sources and sinks are in overall long-term balance (Figure 5.17(a) and (b)).

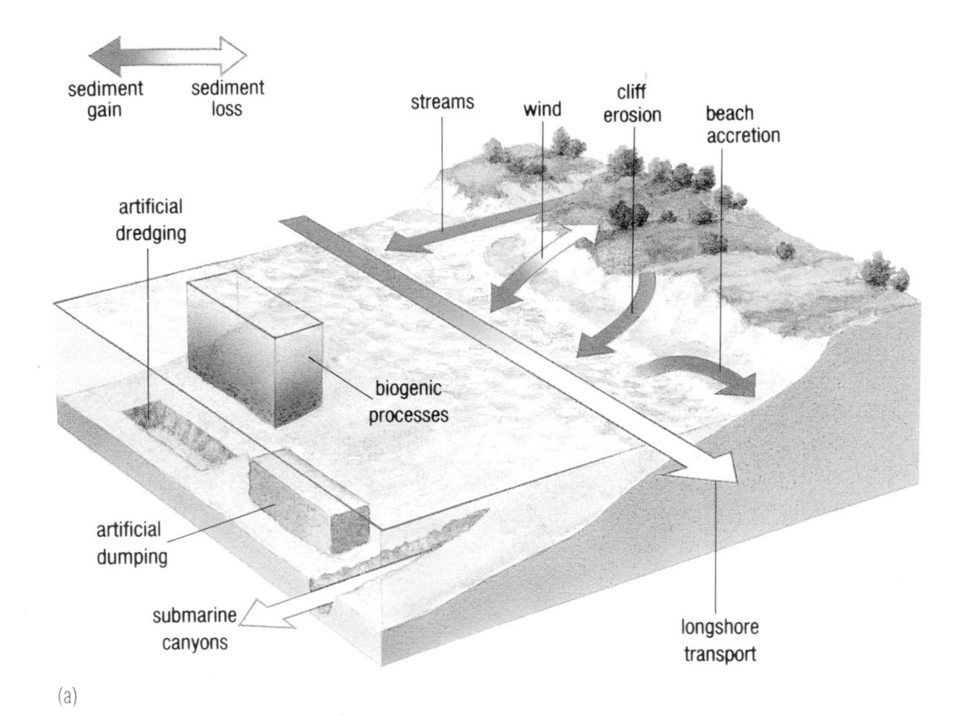

(a)

(b)

Figure 5.16 (a) The components of a coastal sediment budget.
(b) Whitby harbour and mouth of the River Esk. The harbour blocks the south-eastward longshore drift and little sediment is deposited along the shore beyond.

Figure 5.17(a) (opposite) illustrates an important general point: dynamic equilibrium in the coastal zone does not necessarily mean a situation in which erosion is balanced by deposition. It can equally well correspond to a situation of net erosion or net deposition – and frequently does.

Would you say that there is net erosion or net deposition of sediment along the section of coastline represented in Figure 5.17(a) (opposite)?

As the sediment is being transported offshore in each of the sediment transport cells, there must be net sediment erosion along that particular stretch of coastline.

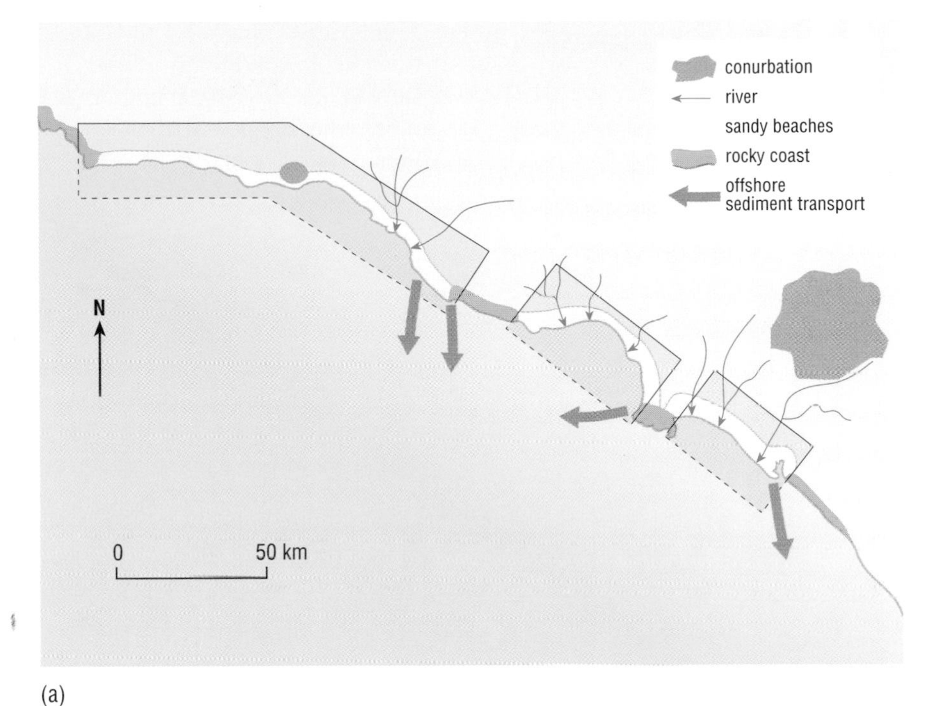

(a)

(b)

Figure 5.17 (a) The sediment transport cells identified along part of a coastline in California. Sand supplied to the coast by rivers and streams and coastal erosion is carried along the shore by waves and currents, and eventually lost to deeper waters offshore (brown arrows).
(b) The rocky headland at Scarborough provides a boundary between adjacent sediment transport cells.

However, sediment eroded from a shoreline in one area may be returned to it elsewhere and be deposited, and the resulting sediment transport cells can look quite different from those in Figure 5.17(a). For example, along the Holderness coast of eastern England north of Spurn Head (Figure 5.13), cliff erosion supplies great quantities of sediment to the North Sea. Much of the mud eroded from these cliffs is deposited within the Humber estuary (see Figure 6.5), some is transported further south and deposited on coastal mud flats and salt-marshes (see Section 6.1), and the rest is carried in suspension across to the eastern side of the North Sea, ending up on tidal flats and in estuaries there (see Figure 6.15). Sand eroded from the Holderness cliffs is transported both offshore to build sand-banks and along the shore to supply beaches further south, along

Spurn Head (Figure 5.13). However, there is limited evidence of sand transport beyond Spurn Head, i.e. across the mouth of the Humber estuary. Instead, it seems that incoming waves are refracted round Spurn Point and there is net longshore transport of sand northwards along the 'inside' of the peninsula, developing the megaripples seen in Figure 5.13(b).

Most sandy beaches are the product of erosion some way 'upstream' along the coast, so if cliff erosion is prevented, beaches further 'downstream' become starved of sediment. Attempts to slow down cliff erosion by building coastal defences therefore have the adverse effect of cutting off supplies of sediment to coastal sediment transport systems. The dynamic equilibrium of the systems is disturbed, and in consequence areas that were previously accreting (building up) cease to do so, while other areas erode more than before. Structures such as sea walls similarly interfere with the dynamics of wave and current action in such a way that after a few years the net effect may actually be to accelerate the rate of erosion. If cliffs are not interfered with, they erode at a fairly constant rate, but if the sea-bed gets deeper (as it will if sea-level continues to rise and/or if sediment supplies are cut off), the energy of the waves reaching the cliffs will be greater and rates of erosion will increase.

In contrast, reducing the depth of the sea-bed offshore causes incoming waves to interact more with the sea-bed (e.g. creating bed forms), and so lose more energy to it. 'Nourishment' of offshore bedforms can be used to slow erosion. For example, sand-bars (longshore bars, cf. Figure 5.1) off the Friesian islands in the eastern North Sea (see Figure 6.17) are migrating offshore, which results in net removal of sand from coastal beaches. Dumping large volumes of sand in troughs between the sand-bars has slowed the rate of bar migration and hence removal of sand from the beach. Beach 'nourishment', i.e. adding sand to beaches to keep them high, also slows rates of beach erosion. Another approach to coastline protection by absorbing wave energy is by means of artificial reefs and islands (Figure 5.18).

In the long term, however, it would seem prudent to let coastlines establish themselves naturally, eroding here and accreting there. Trying to work against natural forces by disturbing the dynamic equilibrium with often expensive applications of technology is likely to have only short-term benefits, i.e. on time-scales of a few years to a decade or two.

Figure 5.18 A beach in West Sussex, with artificial islands offshore. The photograph was taken in winter, and the shingle beach has a well-developed berm with cusps.

5.4 SUMMARY OF CHAPTER 5

1 Beach profiles are controlled by the influence of waves, tidal range, and sediment particle size. The wave zones are the swash zone, the surf zone and the breaker zone. Steep beaches are characterized by berms, shallow ones by swash bars and runnels. Longshore bars may develop seawards of the intertidal zone. Coarse-grained sediments lead to steep beaches because water is readily lost through percolation and the backwash is too weak to move much of the sediment that has been transported up the beach face by the swash. Conversely, fine-grained sediments lead to shallow beaches. Small gentle waves and swell waves tend to build up beaches and steep storm waves tend to tear them down and flatten them.

2 Water particles in shallow-water waves follow orbital paths which become progressively flattened towards the sea-bed. In shallow water, the maximum horizontal orbital velocity and shear stress at the bed increase as the wave height increases and as water depth decreases. The conditions which determine sediment movement for a given grain size may be achieved from many different combinations of wave height, wave period and water depth. The orbital velocity necessary to initiate sediment movement (threshold velocity) for a given grain size increases as the wave period increases.

3 Beneath a wave, sediment is moved landwards as the crest passes and seawards as the trough passes. Strong shoreward velocities move both coarser sediment (as bedload) and finer sediment (as suspended load) landwards. Weaker seaward velocities, of longer duration, move only the finer bedload and suspended load seawards. Coupled with the effects of percolation, this leads to a net movement of coarse sediment landwards and fine sediment seawards.

4 Straight-crested, symmetrical ripples form as a result of the oscillatory water movement beneath waves. Rhomboid patterns are formed by fast-flowing backwash. Larger sedimentary structures on some beaches include cusps formed at the high tide mark.

5 Wave-induced longshore currents are generated when waves break obliquely to the shoreline. These currents, and the zig-zag movement of swash and backwash on steep beaches, move sediment along the shoreline, and can also lead to the generation of rip currents. Rip currents develop also as a consequence of horizontal pressure gradients between regions having wave set-up of different heights. Convergences of resulting longshore currents lead to the return of water seawards in narrow fast-flowing (rip) currents.

6 The wave power available for longshore sediment transport can be calculated from the wave group speed, average wave height and the angle the wave crest makes with the shoreline. The rate of sediment transport along the shoreline can be estimated using the wave power. Most net sediment transport occurs when movement by currents is enhanced by wave action. Wave action lifts sediment into suspension where it is transported by currents which, by themselves, may be unable to lift sediment off the sea-bed.

7 Drawing up sediment budgets and identifying coastal sediment transport cells can help to quantify the dynamic equilibrium of the coastal zone, i.e. whether there is net erosion or net deposition (accretion) of a coastline. Such studies are useful to assess the likely or actual impact of coastal engineering or construction works. Attempts to alter the dynamic equilibrium along one stretch of coastline (e.g. by erosion-prevention measures) are likely to disturb the equilibrium elsewhere, resulting in accelerated (and unwanted) erosion and/or accretion elsewhere. It is probably wiser to let Nature takes its course.

Now try the following questions to help consolidate your understanding of this Chapter.

QUESTION 5.8

(a) The orbital velocity of waves of 15 s period is found to be 0.25 m s^{-1}. According to Figure 5.7, what is the maximum size of quartz particles that can be moved by such waves?

(b) According to Figure 5.8, in water 100 m deep how high would these same waves have been?

QUESTION 5.9 Many years ago, improvements were made to the main river channel of the St John's River, Florida, which had silted up so much that it could be used only by small boats. A deep channel was dug for ships and large jetties were built immediately to the north and south of the estuary mouth which extends eastwards into the Atlantic Ocean at Jacksonville. The predominant longshore drift is southwards. Predict the likely outcome of building these jetties on the sediment budget of the coastline immediately to the north and south of the estuary.

CHAPTER 6 ESTUARIES

'When low tides drain the estuary gold
Small intersecting ripples far away
Ripple about a bar of shifting sands'

North Coast Recollections, John Betjeman.

Estuaries are regions where rivers meet inlets of the sea, and most estuaries still retain the main features of river valleys, often having meandering courses and numerous tributaries. Their upper limit is generally considered to be the furthest point where the tidal rise and fall can be detected. Estuaries can usually be divided into three sections: a lower (or marine) estuary, in free connection with the open sea; a middle estuary, where most of the mixing between seawater and river water takes place; and an upper (or fluvial) estuary, dominated by freshwater influences but nevertheless subjected to daily tidal rise and fall, like the rest of the estuary (Figure 6.1).

Estuaries are ephemeral features on geological time-scales, having an average life of at most a few tens of thousands of years and generally much less. Most estuaries are geologically very young, for they developed during the latest post-glacial rise of sea-level, which inundated coastlines and drowned the mouths of river valleys (Figure 6.1(b)). So the world may be unusually rich in estuaries at the present time.

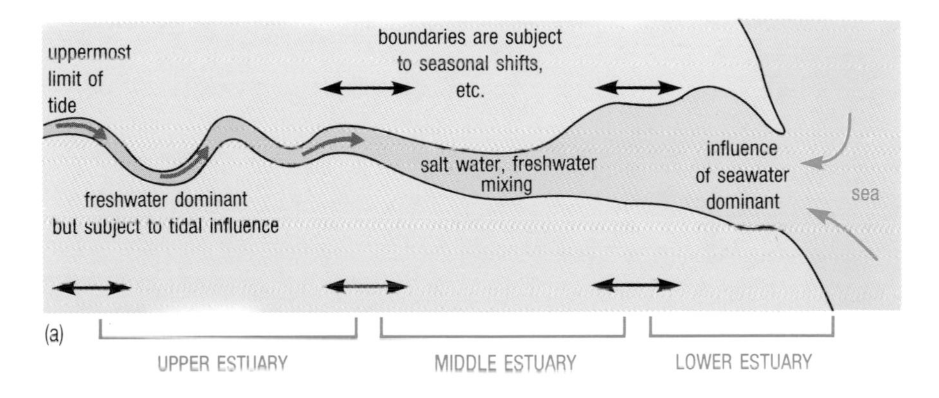

(b)

Figure 6.1 (a) A schematic map of a typical estuary showing the divisions into lower, middle and upper estuary. The boundaries are transition zones that shift according to the seasons, the weather and the tides.
(b) A river valley in South Devon drowned by sea-level rise after the last glacial period.

Estuaries occur at the mouths of rivers which transport relatively small amounts of sediment and discharge it into coastal waters where wave and tidal current action are sufficiently strong to disperse the sediment. An estuary is less likely to develop where sediment discharge is high, and a delta may grow seawards from the river mouth instead (see Chapter 7).

The physiographic setting of estuaries can vary enormously. In glaciated mountainous areas, for example, where river valleys have been deepened by glaciers, they end in fjords having a rock bar or 'sill' near their mouths, above which water depths can be as little as a few tens of metres. Inside the fjord, however, the water can be hundreds of metres deep and fjords may extend more than a hundred kilometres inland. Along low-lying coastlines, on the other hand, estuaries often develop as extensive shallow lagoons between rivers and the sea. Despite this wide range of possible settings, the processes controlling transport and deposition of sediments are much the same in all estuaries.

Estuaries have a global significance for continental shelf and oceanic processes because of the exchange of water and sediment with coastal seas. During transport through estuaries, the grain-size distribution of the sediment becomes altered by repeated deposition, re-erosion and transport, and some sediments become permanently trapped. Estuaries act as a sort of filter for sediment input to the oceans, and chemical reactions in estuarine waters can alter the character of some mineral particles, especially clays, which can thus influence pollutant transport. Suspended sediment concentrations are generally high, and the sediments are often richly organic, because of high biological production, both in the water column and in the sands and muds of the estuary bed.

Estuaries are also characterized by strong gradients of **salinity** (and hence of water density), of suspended sediment concentration, and of chemical and biological properties. There is considerable interaction between physical, chemical, biological and sedimentological processes.

6.1 SEDIMENT DISTRIBUTION IN ESTUARIES

Estuaries consist of one or more channels and **intertidal flats**, which are alternately covered and uncovered by the rise and fall of the tides. There is a progression in grain size from mud-dominated sediments at the high tide level to sand-dominated sediments at the low tide level. Intertidal flats typically have very low gradients, usually in the order of 1:1000. Figure 6.2 illustrates the zonation that commonly develops in estuaries in temperate regions.

1 The *main tidal channel* is the deepest part of the estuary, submerged for all or most of the tidal cycle and subjected to strong tidal currents and minor wave action. Sediments consist of well-winnowed sands (and sometimes gravels), which are partly exposed at low spring tides and may be locally colonized by mussels, especially where the underlying bed rock protrudes through the sediments.

QUESTION 6.1 What sort of bed forms might you expect to see in the main tidal channel? Asymmetrical wave formed ripples

Figure 6.2 (a) Intertidal zonation in a typical estuary (see text for explanation of zones). Vertical scale greatly exaggerated. (You may see alternative terminology used for the various zones in other books.)

(b) Photomosaic of part of a typical estuary at low tide showing the main channel, with its banks rising to the intertidal and high tidal flats and narrow salt-marsh, and (c) the same estuary at high tide.

2 The *intertidal flats* commonly form the widest zone, being submerged and exposed for roughly similar periods. As they are usually submerged during mid-tide, when tidal currents tend to reach their maximum speeds, sediment movement is mainly controlled by these currents; but wave action may also be a factor, especially when the water is shallow during low tide, and small bed forms (ripples) are commonly developed.

The sediments of this main zone of intertidal flats often consist of alternations of sands or silty sands and fine muds, in which the layering is commonly disturbed (bioturbated) by dense populations of burrowing organisms such as lugworms (*Arenicola*) and – where sands predominate – cockles (*Cerastoderma*) and other bivalve molluscs. In some estuaries, muds deposited on the lower part of the intertidal flats form a lower mud-flat zone, where the gradient steepens down to the main channel.

3 The *high tidal flats* are mud flats submerged only at high tide when current speeds fall close to zero, and are generally reached by waves of only low amplitude. Little bedload transport occurs, but during periods of slack water at the turn of the tide, muds settle out of suspension onto the mud flats. The transport of fine-grained silts and clays (i.e. muds) over the high tidal flats, and their subsequent deposition there, is encouraged by settling lag (Section 4.3.2). As the flood tide inundates the tidal flats and the current begins to slacken, these smallest particles begin to settle from suspension. However, they do not settle vertically through the water, but are carried into shallower water by the slowly moving current as they sink, eventually to be deposited some distance shoreward of where they began to settle.

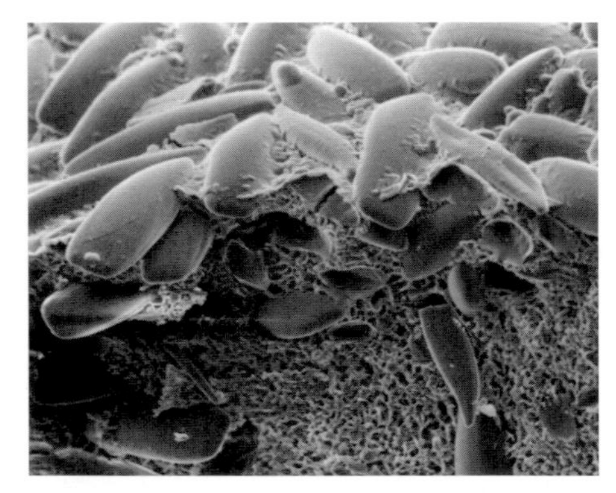

(a)(i)

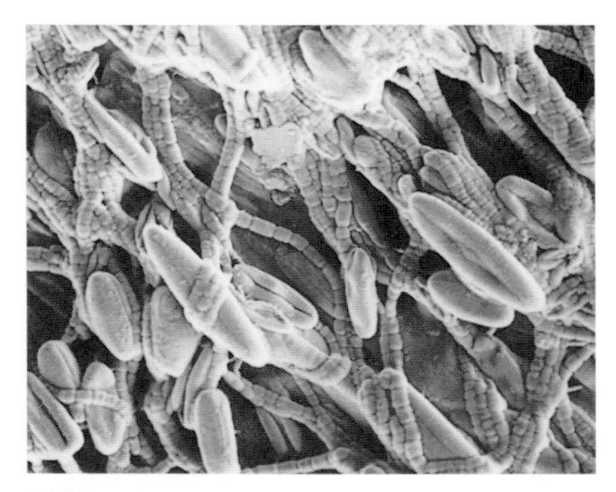

(a)(ii)

Figure 6.3 (a)(i) Scanning electron micrograph (SEM) showing a section through a dense mat of diatoms on a muddy sediment. The diatoms are *Scolioneis tumida* and are about 80 μm long. (ii) SEM of a mat from the sediment surface in an area of salt-marsh, showing strands of stabilizing cyanobacteria and a few individual diatoms.

Muds are cohesive sediments which, once deposited, are difficult to erode (Section 4.1.2). On tidal flats, therefore, the flow of water required to erode (or re-erode) fine sediment on the outgoing (ebb) tide is greater than that at which it can be deposited on the incoming (flood) tide. So suspended sediment will not be moved back downstream (with the ebb tide) as far as it has been moved upstream (with the flood tide).

An additional factor that promotes sediment accretion on tidal mud flats is binding of sediment particles by mats of filamentous algae, and/or biological 'glues' (known as extracellular polymeric substances or EPS) from various types of algae, especially **diatoms**, and/or bacteria (Figure 6.3(a)). In summer, the topmost layer of mud can contain extremely high concentrations of stabilizing organisms. In winter, these decline and wave activity during storms may lead to periods of erosion. However, high tidal flats are generally zones of sediment accretion, whose level rises as the muds accumulate, with the result that the depth and duration of submergence during high water progressively decreases.

4 Ultimately, the high tidal flat is exposed for sufficiently long periods for colonization by salt-tolerant higher plants to begin, leading to the development of a *salt-marsh*, flooded normally only during the highest spring tides (Figure 6.3(b)). The most common pioneering salt-tolerant plants in western Europe are *Salicornia* (the fleshy marsh samphire) and *Spartina* (the tough marsh cord grass). The plant roots bind the sediment and help prevent further erosion, while the plant stems retard the water flow and trap the mud, encouraging still further deposition. This helps to consolidate the sediment and build up the level of the salt-marsh, so that the older, higher parts are flooded less frequently. Erosion by waves and undercutting of the compacted sediments along the edge of the salt-marsh can locally form small 'cliffs' up to several tens of cm high.

The relative widths of zones 1–4 can vary widely, even within a single estuary, and the complete zonation is not always developed in all estuaries. For example, in relatively narrow estuaries and on shallow intertidal flats, where there is insufficient wave action to cause winnowing, there are mud-flats (Figure 6.3(c)) but the intertidal flats of Figure 6.2 are missing; this can also happen if little or no sand is being transported into the estuary. By contrast, in more exposed estuaries where wave action is greater, the high tidal mud flats may be poorly developed or even absent, and salt-marsh development may occur on a silty or sandy substrate. There will of course be no intertidal mud flats at all where the tidal range is zero.

In summary, there is a general progression in grain size, from the mud-dominated sediments of the high tidal flats to the sand-dominated sediments of the main channel. All intertidal flats are dissected by networks of tributary tidal channels flowing into the main one (Figure 6.2). The rising tide first fills the channels and then spills out over the intertidal flats. Conversely, as the tide falls, water from the flats drains into the channels, which are the last to empty. Over long periods of time, the salt-marsh gradually grows seawards as the estuary fills with sediment, but the network of drainage channels can persist long after the marsh has become dry land.

(b)

(c)

Figure 6.3 (b) A high tidal flat in a small estuary in southern Portugal. Clumps of *Salicornia* and *Spartina* pass into a fully vegetated salt-marsh on the right.
(c) Photograph of a typical estuarine tidal mud flat, in a tributary valley of the River Dart estuary, Devon.

6.1.1 AGGREGATION OF SEDIMENT IN ESTUARIES

QUESTION 6.2

(a) Approximately how long would it take sediment particles *c*. 2 μm in diameter to settle through 1 m of water to the bed of the estuary? Settling velocities for particles of this size are of the order of 5×10^{-3} mm s^{-1}, i.e. 5×10^{-6} m s^{-1} (cf. Figure 4.6(c)).

(b) Would you expect your answer to (a) to be greater or smaller if these were flaky clay particles?

Your answer to Question 6.2 indicates that if very fine-grained sediments are to be deposited, another process must be operating in the estuarine environment. This is the aggregation of the tiny grains to form larger ones which are deposited more rapidly. There are two principal ways in which aggregation can happen:

1 **Biological aggregation** is locally important in some estuaries. Clay particles are ingested by filter-feeding animals and excreted in faecal pellets up to 5 mm long, with settling velocities measured in centimetres per second, rather than millimetres per hour. There may also be 'fluffy' aggregates of dead and dying planktonic material, including bacteria. In estuaries without a great deal of biological activity, however, these processes are less important than:

2 **Flocculation**, which occurs as the result of the molecular attractive forces known as *van der Waals forces*. These forces are not particularly strong, but they vary inversely as the square of the distance between two clay particles and become important when particles are brought very close together. In fresh (river) water, flocculation does not take place because clay minerals normally carry a net negative charge and similarly charged clay particles repel one another. In seawater, however, the positively charged **cations** in solution neutralize these negative charges, so that when clay particles are brought sufficiently close together, the van der Waals forces dominate, and flocculation occurs.

Flocculation is thus an important process where freshwater and seawater mix, and it occurs in all estuaries. There are three main ways in which clay particles can be brought close together for van der Waals forces to take effect:

1 By wind- or current-generated turbulence in the water column.

2 By *Brownian motion*: very small suspended clay particles are continually buffetted by the random motion of water molecules.

3 By being **scavenged** by larger particles which sink rapidly through the water column, collide with smaller particles and 'capture' them.

Another process that occurs where freshwater and seawater mix is *cation exchange* between water and suspended clay mineral particles. The four main cations in seawater are sodium (Na^+), potassium (K^+), magnesium (Mg^{2+}) and calcium (Ca^{2+}), and they are the principal participants both in flocculation and in cation exchange. These same cations occur in clay minerals and the exchange reactions occur mainly with cations bound to particle surfaces by **adsorption**, and to a limited extent with cations inside the mineral structures themselves. The exchanges result in a net gain of Na^+, K^+ and Mg^{2+} ions by clay minerals from seawater, and a net loss of Ca^{2+} ions from clay minerals to seawater, although the balance of electrical change in both clay minerals and seawater remains unchanged.

Important though these reactions may be for the detailed chemistry of estuarine waters, their effect on the overall composition of both clay minerals and seawater is sufficiently small to be neglected for most practical purposes. However, in estuaries subject to contamination from industrial effluents, significant amounts of heavy metals can be removed from solution by adsorption on clay mineral particles, and then deposited in the sediments.

6.2 TIDAL CHANNELS OF ESTUARIES

The main channels of estuaries (Figure 6.2) are the principal conduits for both tidal and river flow and therefore control the transport and deposition of sediments. The magnitudes of tidal range and river discharge in the main channels enable all estuaries to be classified somewhere along a continuum between two extremes: highly stratified estuaries at one end, to well-mixed estuaries at the other, as illustrated in Figure 6.4 and described in the next Section.

6.2.1 THE ESTUARINE CONTINUUM

Highly stratified estuaries (Figure 6.4(a), overleaf) develop where rivers discharge into seas with a low tidal range (<*c*. 2 m, e.g. the Mediterranean and Black Sea). The less dense (more buoyant) river water flows over the surface of the underlying denser seawater which forms a **salt wedge** that penetrates and thins up-river. The extent of the salt wedge varies with the river flow. When the discharge is low, the salt wedge penetrates further up the estuary than when the discharge is high.

There are very sharp salinity and density gradients between the overlying freshwater and underlying seawater: a stable **halocline** develops, and the resulting strong density gradient (*pycnocline*) inhibits mixing between the two water masses. However, shear stresses at the interface between the flowing river water and the salt wedge generate internal waves (Section 1.1.1). Where these break, small quantities of salt water are injected into the overlying freshwater – in other words, salt water from below is **entrained** into the freshwater above, making it **brackish**. These are basic features of **salt wedge estuaries**, which develop only where the tidal range is small.

Where rivers discharge into a sea with a moderate tidal range (*c*. 2–4 m), the whole water mass moves up and down the estuary with the flood and ebb tides. Friction between the water and the estuary bed causes turbulence which mixes the water column more effectively than does simple entrainment at the freshwater/salt water interface. Salt water is mixed upwards, and freshwater is mixed downwards, so the **isohalines** are more steeply inclined, the halocline is less well defined, and the stratification is weaker (Figure 6.4(b)). There is a wide variety of such **partially mixed estuaries**, each with its own characteristics resulting from the combination of tidal range, river discharge, local topography and bathymetry, and climatic conditions; and as tidal ranges along coasts in most parts of the world are in the *c*. 2–4 m range, this is the commonest estuarine type.

Both tidal range and strength of tidal currents can fluctuate considerably between spring and neap tides. The stronger tidal currents of spring tides enhance turbulent mixing of salt water and freshwater, reducing the buoyancy of the surface layer and further weakening the stratification. By contrast, during neap tides and/or at times of high river flow, the stratification can be strengthened, i.e. the estuary may take on 'salt wedge' characteristics – though mixing at the interface is likely to be strengthened during times of high river flow.

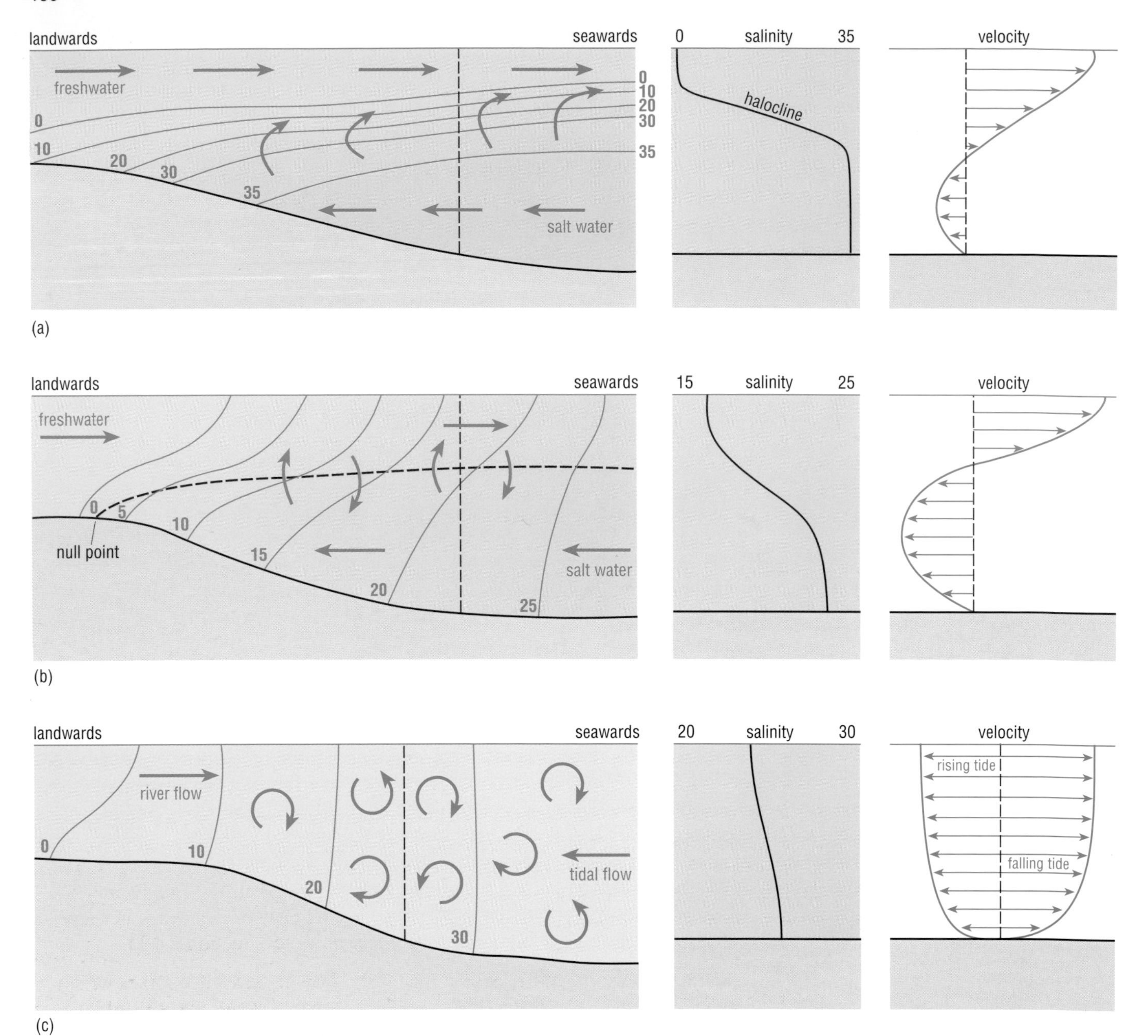

Figure 6.4 (a–c) Diagrammatic representations of water circulation, salinity distribution and velocity gradients within the continuum of estuarine types from salt wedge (a), through partially mixed (b), to well-mixed (c). The broken vertical line shows the position of the salinity and velocity profiles. Note the progressive weakening of the halocline from (a) to (c), a consequence of increasing tidal influence. In well-mixed estuaries (c), the salinity of the water column at any particular point in the estuary depends upon the state of the tide. Curved arrows on the longitudinal sections represent mixing.
Residual (net) flows in (a) and (b) (horizontal blue arrows) are seawards at the surface because of the river flow, landwards at the bottom because of vertical mixing and entrainment across the river water/seawater interface. Net flow in (c) is landwards on the flood tide, seawards on the ebb. In (b), the dashed sub-horizontal line on the longitudinal section shows the depth at which there is no horizontal residual flow either seawards or landwards, and its intersection with the bed near the head of the salt water intrusion defines the null point (see the text). Note that although shown only in (b) for clarity, a null point can be identified in any tidal estuary (see text, p.158).

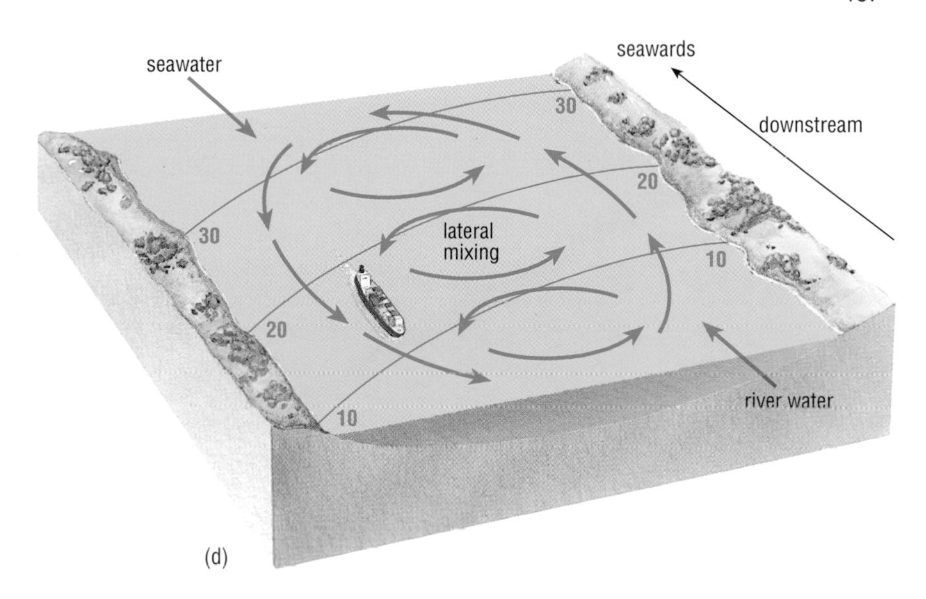

(d) Block diagram of middle and lower reaches of a well-mixed estuary (c), showing lateral deflection of seawater and river water flows by the Coriolis force, in this case in the Northern Hemisphere (see later text). Lateral mixing induces a horizontal residual circulation, leading to horizontal variations in salinity across the estuary, as indicated by the curved isohalines. Note that horizontal residual circulation is not confined to well-mixed estuaries; under appropriate conditions, it can also develop in partially mixed estuaries (see later text).

Residual flows and the null point: We have seen that in salt wedge estuaries, salt water is lost from the salt wedge by entrainment into the overlying freshwater with no corresponding gain of freshwater by the salt wedge. This has implications for *continuity* in the estuary.

What must happen in order to maintain the overall volume of water in the estuary?

The principle of continuity requires that the volumes of water flowing into and out of a given space in unit time must be equal (Section 2.4.1). Hence, water lost from the salt wedge must be replaced by a slow landward flow of seawater within the salt wedge itself. The principle of continuity also requires that, because of entrainment of salt water into the freshwater layer all the way from the apex of the salt wedge down to the estuary mouth, the total volume of water flowing out of the estuary in the upper layer in unit time (i.e. the net flow) must increase downstream. The slow landward flow within the salt wedge, and the 'extra' seawards flow in the freshwater layer, due to entrainment, are known as *residual flows* (cf. Section 2.4.1).

In partially mixed estuaries, the freshwater flowing seawards mixes with a large amount of salt water, so the total discharge of water via the surface layer can be an order of magnitude greater than the river discharge. Continuity requires that the salt water mixed into (and discharged by) the surface layer be replaced, and so the landward flow within the bottom salt water layer is significantly stronger than in salt wedge estuaries (cf. Figure 6.4(a, b)). Thus, in partially mixed estuaries, the vertical mixing between the upper (river water) and lower (seawater) layers produces larger landward and seaward residual flows. Although they vary from estuary to estuary, residual flows typically have less than 10% of the magnitude of the tidal and river currents superimposed on them, and they can occur even where mixing has reduced the salinity contrast between surface and bottom waters to little more than 1 (part per thousand).

The salt intrusion (Figure 6.4(b)) advances upstream as the tide rises, and retreats downstream as the tide falls, but the net (residual) flow within it is always *landwards* (upstream). Since net flow in the upper layer is always downstream because of the river flow, at any point in the tidal channel there must be a depth at which there is no net landward or seaward movement of water (where the velocity profile would cross the zero velocity line, cf. the profiles in Figure 6.4(a) and (b)). Where this depth coincides with the bed of the channel, there is a *convergence* (Section 1.2.1) where the landward and seaward flows along the bottom meet, and there is no net movement of water at the bed in either direction. This is known as the **null point** (Figure 6.4(b)) and it occurs near the head of the salt intrusion, where salinities are as low as 0.1 to 1 (part per thousand) in estuaries of low to moderate tidal range, up to about 5 (parts per thousand) or more where the tidal range is moderate to large.

Should we expect the null point to be fixed in position in any particular estuary?

No. The null point must move up and down the estuary with the tides, and over a greater distance during spring than during neap tides (though to a much smaller extent in salt-wedge estuaries, because the tidal range is small). In addition, there will always be some seasonal variation: the null point moves upstream when river discharge is low, and downstream when discharge is high.

A null point should thus be identifiable in any tidal estuary, even in **well-mixed estuaries** (Figure 6.4(c)), where the tidal range is large (> *c*. 4 m). Tidal currents are strong relative to river flow and the whole water column is mixed, so that salinity hardly varies with depth at all. The whole well-mixed water mass moves landwards and seawards with the tides, but (as in *all* estuaries) the average salinity decreases towards the upstream limit of tidal influence. Even where the estuary is otherwise well mixed, there will be some stratification in the upper estuary, with a residual (vertical) circulation and a null point (Figure 6.4(c)).

You need to bear in mind that nearly all estuaries are more or less funnel-shaped and widen seawards. In well-mixed estuaries, therefore, although the water column may be vertically homogeneous, there can be *horizontal* variations of both salinity and velocity across the width of the estuary, and a horizontal circulation can develop. This happens because the Coriolis force laterally deflects both the incoming tidal flow and the seaward-flowing river water. The result is that in the Northern Hemisphere seawater flows up-estuary on the left-hand side (facing downstream) and river water flows down-estuary on the right-hand side, while the reverse is true in the Southern Hemisphere. Mixing takes place laterally, and there is a *horizontal* residual circulation, as illustrated in Figure 6.4(d), rather than the vertical residual circulatory pattern that occurs in salt-wedge and partially mixed estuaries. This residual horizontal circulation is superimposed upon the main landward-and-seaward motion of the whole well-mixed water mass in the estuary, as it fills and empties with the tides.

QUESTION 6.3 In a well-mixed estuary, how would you expect the salinity of the water column at a particular point to change with time?

It is important to stress here, however, that the horizontal circulation illustrated in Figure 6.4(d) is not confined to the well-mixed end of the estuarine continuum, and can occur in stratified estuaries as well, if they are wide enough. Also, you need to be aware that the classification scheme outlined above is a general one. The pattern of water flow can vary between the extremes illustrated in Figure 6.4 even within a single estuary, depending upon conditions. For example, as we have seen, stratification tends to be enhanced when river discharge is high, and to be broken down by vigorous mixing during spring tides. Moreover, because tidal influences tend to dominate in the lower estuary, riverine influences in the upper estuary, an estuary may approach well-mixed conditions near the mouth and yet be quite well stratified near the upper limit of tidal action. An individual estuary may be well-mixed during spring tides, and partially mixed or even well stratified during neap tides. Major storms can disrupt these general patterns. If the river discharge is high enough it can push the salt intrusion out of the estuary altogether, and/or weaken (even destroy) any stratification; and gale force winds can also cause sufficient mixing to break down the stratification.

The size and overall cross-sectional form of the main channel of an estuary can also influence the water movements within it. Thus, the deeper and narrower the estuary, the more likely it is to be stratified (e.g. as in fjords), whereas broad shallow estuaries are more likely to be well-mixed and to have horizontal variations of salinity across them. However, a broad and moderately deep estuary could well be *both* stratified *and* have horizontal variations of salinity across it (e.g. as noted above, Figure 6.4(d) should not be seen to apply exclusively to well-mixed estuaries). In short, the patterns of water flow are unlikely to be the same for two estuaries of different shape, even where river discharge and tidal range are similar. No two estuaries are the same, each will have its own characteristics – but they can all be placed somewhere along the continuum represented in Figure 6.4.

Some examples of estuarine types

Only rivers discharging small amounts of sediment into a virtually tideless sea can form *salt-wedge estuaries*, and examples include rivers draining Texas and flowing into the Gulf of Mexico (e.g. the rivers Brazos and Sabine).

Partially mixed estuaries are common along the coasts of eastern America (e.g. the James River in Virginia), and north-western Europe (e.g. the Mersey and the Thames).

Well-mixed estuaries include the Severn estuary, the Firth of Forth in Scotland, the Gironde estuary opening into the Bay of Biscay, the Rio de la Plata, which opens into the South Atlantic, and the Humber estuary on England's North Sea coast. Large estuaries of this kind are commonly shallow and funnel-shaped: wide at the mouth and tapering inland (Figure 6.5, overleaf), and stratification is likely to develop at least some of the time.

QUESTION 6.4 Look at the estuary in Figure 6.6 (overleaf). The tidal range in this region varies from about 4–5 m on neap tides to about 7–8 m on spring tides.

(a) Whereabouts in the classification scheme of Figure 6.4 would you place this estuary?

(b) Why are the waves breaking slightly further offshore opposite the mouth of the main channel than opposite the beach to the right?

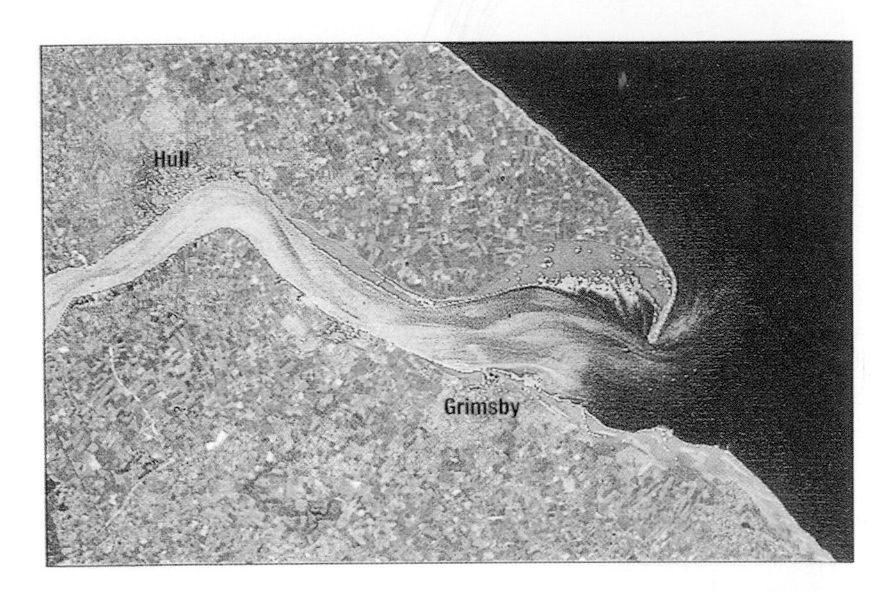

Figure 6.5 Landsat Multispectral Scanner image of the Humber estuary, on the east coast of England, approximately two hours before low water (taken in 1976 before the building of the Humber Bridge). Seawater and river water are black. The yellow coloration in the estuary is caused by high concentrations of suspended sediment and indicates the pattern of sediment distribution and movement in the estuary. Although most of the blue colour represents built-up areas, the blue bands bordering the north bank of the Humber as far as the spit at the mouth (Spurn Head, Figure 5.13), and extending seawards from the estuary mouth along the coast, are intertidal mud flats.

Figure 6.6 Aerial view of the seawards end of an estuary at low tide (Anglesey, North Wales). The main river channel hugs the left-hand side (looking landwards). The picture shows mostly the middle and lower estuary, with a combined length of about 5 km; the lower end of the much narrower upper estuary can just be discerned at the top of the picture

6.2.2 REGIONS OF FRESHWATER INFLUENCE

The answer to Question 6.4(b) provides a useful reminder that the influence of estuaries can extend well offshore. As the river water flows from the confines of the estuary mouth into the open sea, it spreads out over the surface of the seawater as a plume of brackish water. These buoyant plumes give rise to **regions of freshwater influence** (ROFIs for short), which can extend offshore for distances that range from a kilometre or two to several hundred kilometres, depending upon the magnitude of the river discharge. The ROFI of the Amazon – the world's largest river – can be detected more than 500 km from the mouth of its estuary (Figure 6.7).

Looking at Figure 6.7, would you expect the average salinity near the bed at the mouth of the Amazon to be greater or less than 0.1 (part per thousand)?

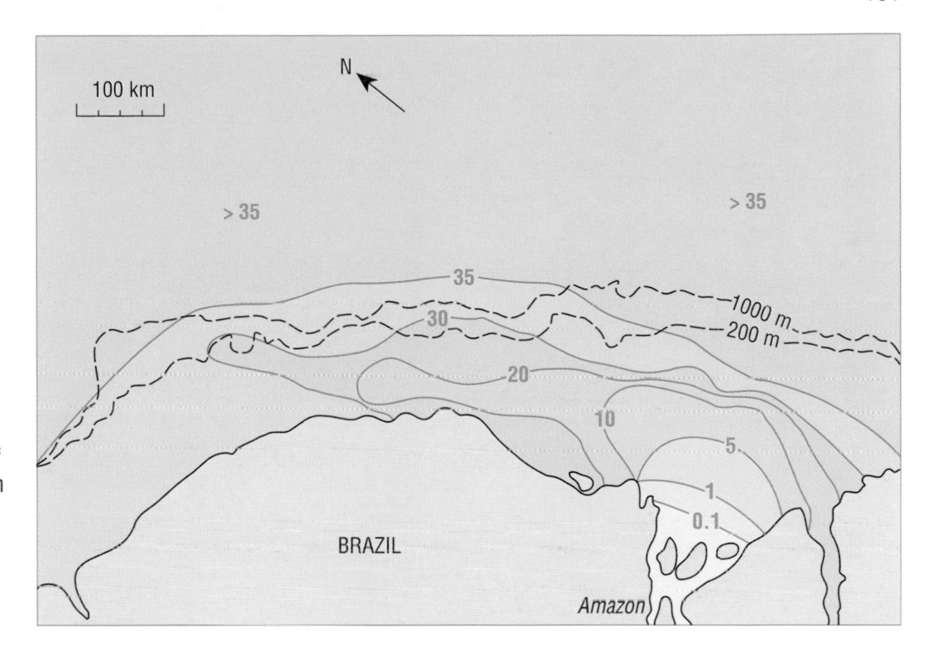

Figure 6.7 Contours of average salinity (parts per thousand) in surface waters off the mouth of the Amazon. Broken lines are isobaths (depths in metres); blue lines are isohalines. The plume of brackish water extends north-westward along the coast of northern Brazil because it is deflected by the equatorial current system of the Atlantic Ocean.

Even though this is a predominantly well-mixed estuary, we should expect some landward flow of seawater near the bed. The salinity near the bed should thus be greater than 0.1, because the water near the bed should be more saline than at the surface.

The degree of stratification and extent of mixing between river water and seawater in ROFIs depends upon both river discharge and tidal range, as well as upon weather conditions and the actual state of the tide (i.e. whether ebbing or flooding) – so it cannot remain constant with time, and Figure 6.7 is a time-averaged picture. As the plume of brackish water spreads out on leaving the river mouth, there is entrainment of seawater and mixing both at the base and along the margins, where there is convergence and mixing at *river plume fronts*. Because the fronts are surface convergences (cf. Figure 2.22), they are often delineated by froth or floating debris. River water is generally rich in nutrients washed off the land, so ROFIs are commonly regions of high primary productivity and hence of rich fisheries.

Where a ROFI does not extend very far offshore (i.e. the river flow is small), and where the tidal range is high and the mouth relatively narrow, the river plume can be forced back into the estuary by the strength of the flood tide, and a *tidal intrusion front* may form at the estuary mouth, sometimes in the form of a V pointing upstream. In some well-mixed estuaries, longitudinal fronts can also be observed at different stages of the tide, extending upstream for several kilometres from the estuary mouth. They show up either as areas of smooth or rippled water ('tidal smooths') or as irregular lines of froth or floating debris (Figure 6.8). These fronts are linear convergences resulting from transverse movement of surface water caused by lateral density gradients. Such gradients occur because tidal flows are strongest in mid-channel where the water is deepest, so that on the rising tide the water in mid-channel is slightly more saline (and therefore denser) than the water on either side – the situation being reversed when the tide falls. When such longitudinal fronts occur, the transverse surface currents can reach speeds of several cm s^{-1}. However, these are weak compared with tidal flows up and down the estuary, and they and their associated circulatory system have negligible effect on the transport and deposition of sediment.

Figure 6.8 Two views of the longitudinal front in the River Conwy (North Wales), resulting from cross-channel convergence of surface water. The front is only observed when the tide is rising (it disappears on the ebb), and is best developed during spring tides.

6.2.3 SEDIMENTATION IN ESTUARIES

In most estuaries, the null point is associated with a **turbidity maximum**, the region where concentrations of suspended material in the channel are greatest. It develops because material moves towards the null point from both upstream and downstream, and because turbulent mixing of river water with brackish bottom water near the null point leads to flocculation (Section 6.1.1). Some of the suspended sediment supplied by the river flocculates near the head of the salt intrusion in the upper estuary and settles into the lower layer. The rest is transported by the river flow further downstream, where some settles into the lower layer and is carried back up the estuary by the residual flow, along with suspended particles brought in from the sea (Figure 6.9). The rest of the suspended sediment escapes to the sea (including much biological material, some of terrestrial and riverine origin, some from the salt-marsh or mangroves – see Figure 6.12). Concentrations of suspended sediment of around 100–200 mg l^{-1} may occur in turbidity maxima of estuaries with a small tidal range, where turbulent mixing is generally weak. In estuaries with a large tidal range, on the other hand, where turbulent mixing is likely to be strong, concentrations in turbidity maxima can reach 10^3–10^4 mg l^{-1} (1–10 g l^{-1}). High concentrations of suspended matter in the turbidity maxima cut down the light available for photosynthesizing organisms. The particles also adsorb pesticides as well as heavy metals (Section 6.1.1). In addition, particulate organic matter provides sites of microbial activity. Concentrations of suspended particulate matter can be further increased in estuaries where there is heavy shipping traffic, which stirs up the sediments.

Coarser sediment is moved upstream by the residual landward bottom flow and deposited near the null point, along with the bedload and larger suspended particles supplied by the river (including coarser aggregates produced by flocculation near the null point). The residual circulation therefore acts as a sediment trap which impedes the escape of sediment to the open sea. In salt-wedge estuaries, where the river meets the salt wedge, at the head of the estuary, the freshwater leaves the bedload behind as it flows over the salt water, and a coarse sediment bar may build up close to the tip of the salt wedge.

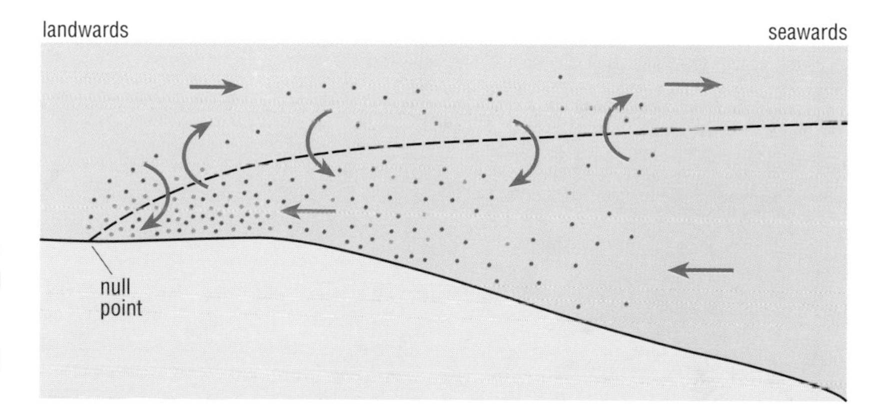

Figure 6.9 Schematic diagram illustrating formation of the turbidity maximum in a partially mixed estuary. Horizontal arrows = landward and seaward residual flows; curved arrows = mixing and particle movement. The broken line shows the depth at which there is no residual horizontal flow; cf. Figure 6.4(b).

As you might expect, the ebb and flow of the tide causes the turbidity maximum to shift up and down the estuary (Figure 6.10, overleaf). It is furthest upstream at high tide and starts to move seawards when the tide falls. Suspended sediment concentrations within it increase as sediment eroded from the bed joins sediment brought down by the river. At low tide, the turbidity maximum is near the mouth and, during slack water, near-surface concentrations decrease as some sediment settles from suspension. After the tide turns, the turbidity maximum moves back upstream and intensifies again, as sediment is eroded from the bed and turbulent mixing by the flood tidal currents increases. As the slack water of high tide approaches, there is re-deposition of suspended sediment, mainly in the middle estuary and especially on the high tidal flats (Figure 6.2); and suspended sediment concentrations in the turbidity maximum decrease again. In general, the turbidity maximum is better developed during spring than during neap tides. That is because tidal currents are weaker during neap tides, so less sediment is re-suspended and there is more opportunity for deposition during periods of slack water.

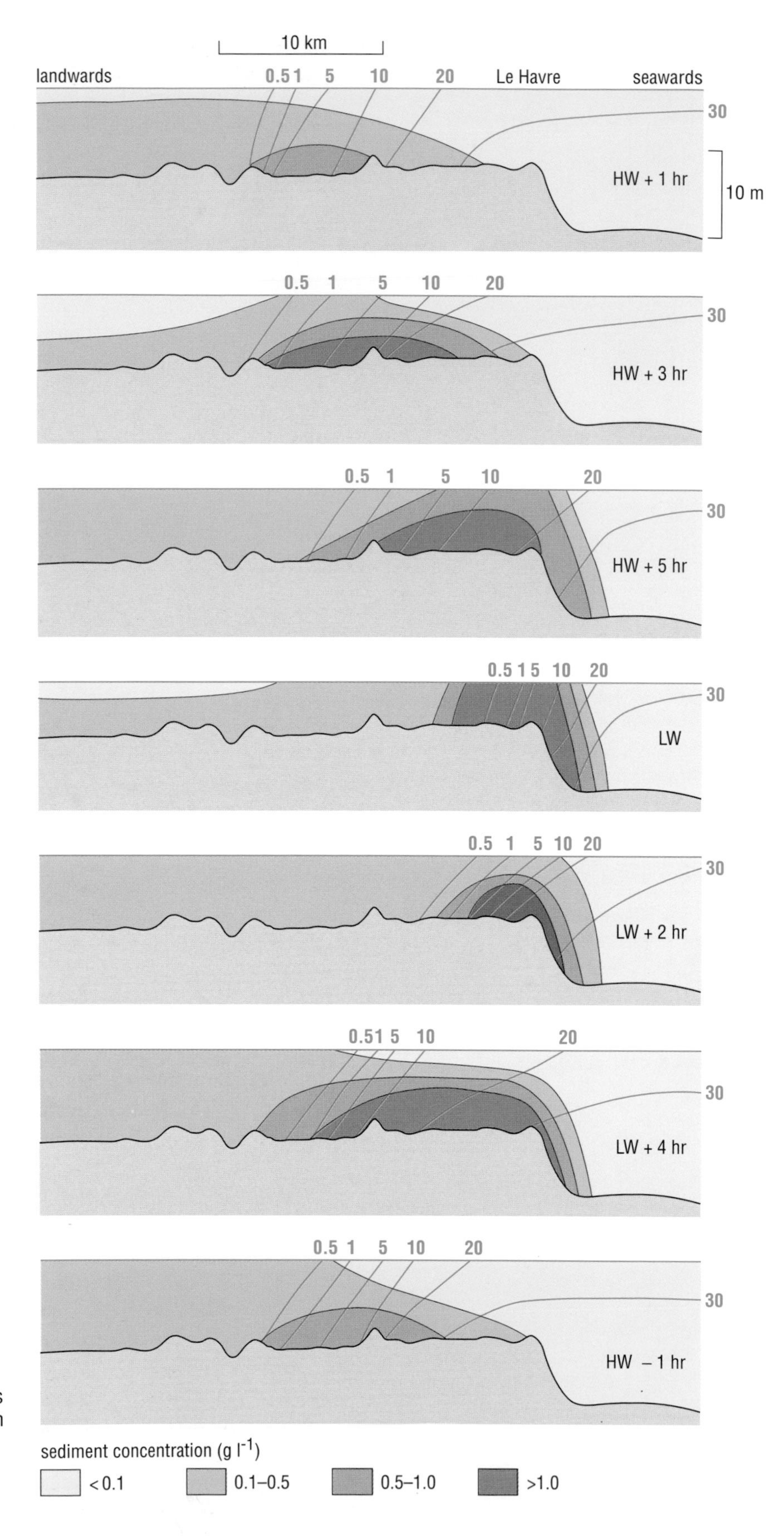

Figure 6.10 The turbidity maximum in the Seine estuary at intervals during a spring tidal cycle, when the river discharge was 780 $m^3 s^{-1}$. HW = high water, LW = low water, HW ± 1 hr (etc.) = high water plus/minus one hour (etc.). Curved blue lines are isohalines with salinity values in parts per thousand. Note that salinities at the core of the turbidity maximum range from 1 (part per thousand) at low tide to 20 at mid-tide, decreasing to around 10 at high tide. (Vertical scale greatly exaggerated.)

The position and size of the turbidity maximum also vary with the river discharge. When discharge is high the turbidity maximum is pushed downstream and diminishes, because more sediment is transported directly out to sea. When discharge is low, on the other hand, suspended sediment is brought further into the estuary on the flood tide, and the turbidity maximum intensifies.

QUESTION 6.5 Figure 6.11 illustrates these two conditions.

(a) Which of the two cross-sections represents (i) high and (ii) low river discharge?

(b) Would you expect an estuary like the Seine to be better stratified during high or during low river discharge?

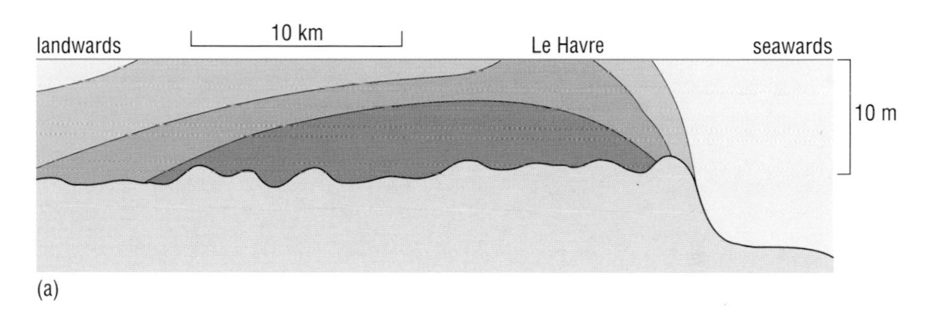

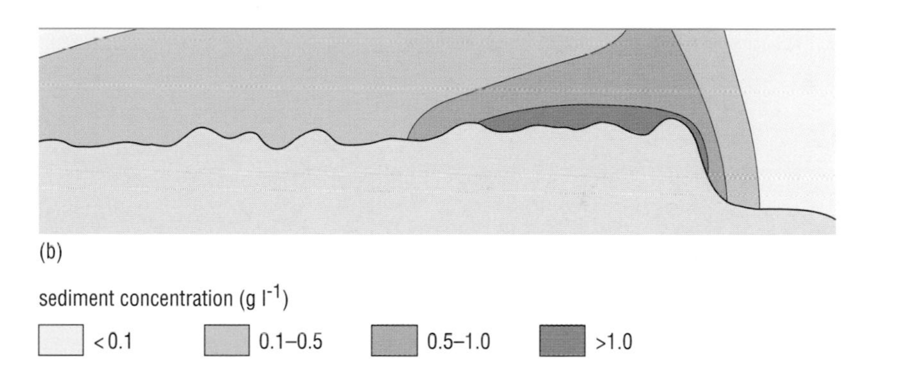

Figure 6.11 The turbidity maximum in the Seine estuary with two different river discharge rates: 200 $m^3 s^{-1}$ and 800 $m^3 s^{-1}$, both at high spring tide. For (a) and (b), see Question 6.5. (Vertical scale greatly exaggerated.)

Fluid mud: In a number of estuaries, very high concentrations of suspended material occur near the bed, beneath the turbidity maximum. Concentrations may exceed 100 $g\,l^{-1}$, much higher than those normally found in the turbidity maximum itself – rarely more than 10 $g\,l^{-1}$, as we have seen. The suspensions are sufficiently coherent to flow, and they can form 'pools' in local depressions in the estuary channel bed. Accumulations of fluid mud can extend over distances of 1–10 km, and move backwards and forwards on the tides, with the turbidity maximum.

As the tidal amplitude and tidal currents decrease after spring tides, less and less material can be re-eroded and suspended, and more of the suspended load settles from the turbidity maximum to form the layer of fluid mud close to the bed. This effect is enhanced by the relatively long periods of slack water around high water during neap tides compared with high water during spring tides. During neap tides, the fluid mud tends to become a little compacted, so that when the tidal range and tidal currents increase again not all of the sediment is re-eroded and some is left permanently deposited.

6.2.4 ESTUARIES IN LOW LATITUDES

In tropical and equatorial regions, mud flats are commonly colonized by mangrove trees whose aerial root systems trap the muds (Figure 6.12). Mangrove swamps, rather than salt-marshes, dominate the zone around the high tide level in such regions.

The actual processes of estuarine circulation and sedimentation at low latitudes differ from those described in previous Sections *only* where there is high evaporation, leading to increased salinity in estuarine surface waters. In such situations, the dense **hypersaline** water sinks and flows *seawards* along the bed of the estuary. For continuity to be maintained, this seawards flow must be replaced by *landwards* flow of seawater *at the surface*. In other words, the 'normal' pattern of residual circulation (Figure 6.4(a–c)) is inverted, to produce a **negative estuarine circulation** (Figure 6.13), which can occur wherever surface salinities are increased by evaporation. Indeed, the outflow of Mediterranean Water at Gibraltar and its replacement by surface inflow of water from the Atlantic is a good example of this type of circulation (though on a much larger scale).

Estuaries of this kind occur mainly in countries with arid climates, and because the low rainfall means that rates of chemical weathering (which produces clays) are generally low, the sediment supply is usually also low, with a high proportion of sandy material. The sands are deposited from bedload where river flow slackens at the head of the estuary, and any fine-grained riverborne sediment that has not flocculated in the hypersaline water and settled from suspension, is carried seawards by the residual hypersaline flow just above the bed (Figure 6.13). Tropical storms in arid regions can cause rivers to flood catastrophically, and when that happens, the whole estuary is likely to be flushed out. Then, as the floods abate, it is possible that a 'normal' pattern of estuarine circulation may be set up, subsequently becoming replaced by negative circulation once more, when the river flow subsides and evaporation intensifies again.

Figure 6.12 The aerial roots of mangroves help to trap muds and bind sediments.

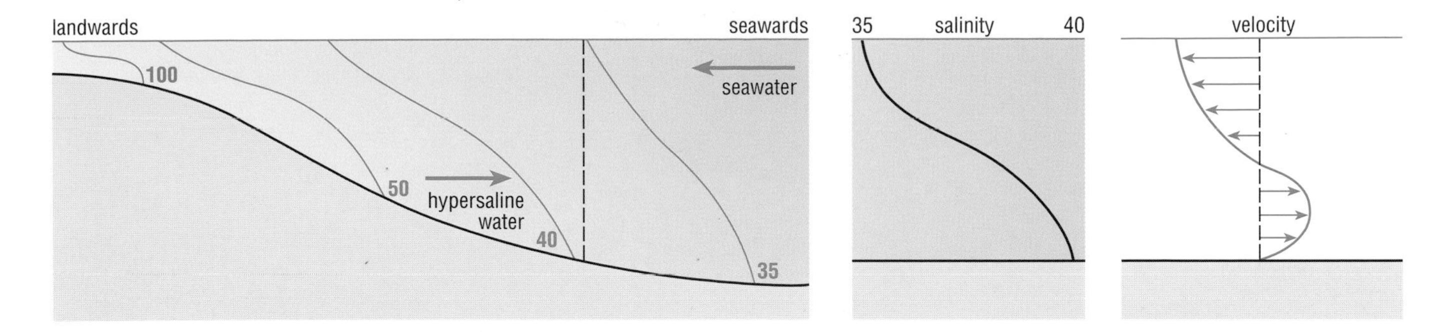

Figure 6.13 Diagrammatic representations of water circulation, salinity distribution and velocity gradients for negative estuarine circulation. Horizontal arrows on the longitudinal section show landwards flow of seawater at the surface and seawards flow of (strongly) hypersaline water near the bed. The broken vertical line shows the position of the salinity and velocity profiles. Note the reversed velocity profile compared with Figure 6.4(a–c), also the increase of salinity with depth

QUESTION 6.6 In an estuary with negative circulation, would you normally expect to find (a) a null point or a turbidity maximum, (b) mixing between the upper and lower layers?

6.2.5 THE DYNAMIC BALANCE OF ESTUARIES

Calculations based on measurements of sediment concentration and river discharge suggest that in many – perhaps most – estuaries, the turbidity maximum contains more sediment than is supplied by rivers, which means that large quantities of sediment are brought into the estuary from the sea. As we have seen, however, during periods of high river discharge much sediment is transported directly to the sea and the turbidity maximum is greatly diminished (Figure 6.11, Question 6.5). In fact, the erosive power of rivers rises very rapidly with increased flow rates, and during major storms, riverborne sediment concentrations can increase by more than an order of magnitude. At such times, the quantity of sediment discharged into the sea can exceed that discharged in a *decade* of normal flow. In other words, sediment discharge from estuaries mostly occurs over very short periods during extreme events. For most of the rest of the time, little or no sediment is being discharged from an estuary to the sea, and if erosion and deposition are in balance, the rate of sediment accumulation will be equal to the rate of supply of sediment from the river (the fluvial flux). Commonly, however, there is *net* accumulation of sediment in the estuary, and as we have seen, the sediment is supplied mostly from the sea. For example, calculations of sediment fluxes in the Humber estuary (Figure 6.5), based on measurements of current speeds, suspended sediment concentrations and rates of sediment deposition (accretion, Figure 6.14, overleaf), suggest that sediment deposition at the estuary mouth is about 9×10^5 tonnes per year. This is more than 40 times greater than the fluvial flux from the Humber catchment area (which is about 2×10^4 tonnes per year), and can only be achieved by net sediment transport *into the estuary* from the sea. Thus, since all estuaries are supplied with sediment not only by rivers but also from the sea, it is likely that all but the largest estuaries are being progressively infilled, as they adjust to the most recent post-glacial rise in sea-level.

This generally landward movement and accumulation of sediment in estuaries can be explained by reference to the tidal asymmetry discussed in Section 2.4.3. As a tidal wave propagates into an estuary, the wave crest (rising tide) travels faster than the wave trough (falling tide), because the speed of propagation depends upon water depth (Equation 1.4).

Figure 6.14 A steep-sided channel in the mud flats of the Humber estuary clearly reveals the successive layers of accreting sediment.

Because of the geometry of estuaries (Figure 6.2), as the tide rises, a large volume of water must flow through the relatively restricted cross-sectional area of the main tidal channel, so it must flow with high speed. At this stage, coarse sand (and even gravel) may be transported into the estuary as bedload.

What will happen as the tide rises further and the water spills over the main channel and starts to flood the intertidal flats?

The water speed will be rapidly reduced because the flow is no longer constrained to move through a small cross-sectional area. Conversely, on the ebb tide, the water initially flows slowly over the extensive areas of intertidal flats, then speeds up as the tide falls further and flow is constrained once more to flow in the channels. There is accordingly an asymmetry in the tidal current velocity between high and low tides. Figure 6.15 illustrates the general form of the resulting velocity curve, with maximum flood and ebb tidal currents either side of low water. Although Figure 6.15 shows the tidal current to be slower on the ebb than on the flood, it is generally augmented by the river flow, and in general it also lasts longer than the flood tidal current.

The settling lag of fine-grained sediments (Section 4.3.2) encourages deposition of muds on the intertidal flats as the current slows down near high tide. Since these are cohesive sediments, they are not easily resuspended, especially at the initially slow speeds of tidal currents when the tide starts to fall again. The result is accretion of the intertidal flats (cf. Figure 6.14).

In the case of large rivers such as the Amazon and the Congo, however, the net flux of sediment through the estuary is seawards. Large amounts of sediment are supplied to the continental shelf from such rivers, and eventually to the deep sea via the submarine canyons eroded by turbidity currents (cf. Section 3.1.1). They are deposited to form extensive 'deep-sea deltas' (**submarine fans**), building up the continental rise. Rates of sediment accumulation on the fans can be very high: as much as 25 m per thousand years on the Amazon fan, for example.

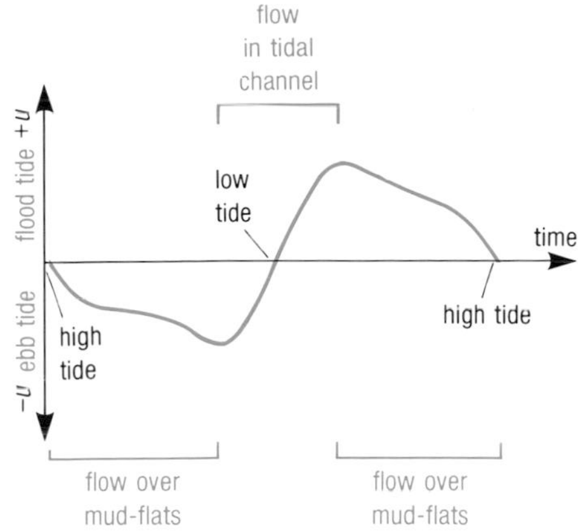

Figure 6.15 The asymmetry of tidal current velocities in estuaries that results from covering and uncovering of intertidal flats during a tidal cycle.

Higher sediment deposition as the sea water carries more sediments up stream.

Estuaries generally make good natural ports and harbours, but as many are slowly silting up, attempts are commonly made to keep them open for shipping as long as possible, e.g. by building 'training walls' to semi-canalize the main channel, and/or by dredging.

QUESTION 6.7 Silting up of the lower Savannah River Estuary (a partially mixed estuary in Florida) was threatening navigation as far upstream as the port of Savannah itself. To combat this, the navigation channel was deepened by dredging, which caused a concomitant increase in salt water penetration up the estuary. Can you suggest what happened as a result, and why?

Dams built across large rivers can have considerable effects on their estuaries, especially where there is high seasonal rainfall, which would normally lead to flooding of the river and flushing of the sediment.

What would happen at the mouth of such an estuary, if a dam were built across the river inland?

Sediment would continue to be transported into the estuary from the sea, and deposited, but seasonal flooding would be inhibited by the dam upstream, there would be no annual removal of sediment to the sea, and the estuary would silt up more rapidly. For example, barely a decade after a large dam was built across the River Porali in Pakistan, the estuary at its mouth (some 80 km north of Karachi) had shrunk to less than half its original width of about 8 km, because of sediment accumulation. The mangrove swamps of the tidal mud flats are also silting up, which impairs their efficacy as nursery areas for young fish. The local fisheries are in decline as a consequence, while waterlogging and saline waters are adversely affecting nearby agricultural land.

6.3 LAGOONS, TIDAL FLATS AND BARRIER ISLANDS

As noted at the beginning of this Chapter, on low-lying coastlines, estuaries can take the form of shallow *lagoons*, commonly behind sand or gravel spits formed by longshore transport (Section 5.2.3). To mention only a few examples, lagoons occur along the Malabar coast of south-west India, the eastern coast of the USA, the Texas Gulf coast, and parts of England's southern and eastern coasts. At low latitudes, of course, lagoons also occur in the shelter of coral reefs (though these have nothing to do with estuaries).

Being shallow, lagoons tend to be well mixed (mainly by winds rather than by currents), and they vary from brackish to hypersaline, depending upon the balance between evaporation, precipitation and river flow. In tropical areas, lagoons can be hypersaline during dry seasons, but may become almost entirely fresh during rainy seasons. Lagoons generally have only narrow connections to the open sea, and although their water levels rise and fall with the tide, tidal currents within them are generally weak, increasing towards the narrow inlets. The lack of significant wave and tidal current action and the generally (but not invariably) low relief of the surrounding land, means that sediments entering lagoons tend mostly to be fine grained, and are deposited on tidal flats, as in the upper reaches of estuaries elsewhere (Figure 6.2). Sediments in lagoons associated with coral reefs consist mostly of calcium carbonate sands and muds.

Coastal lagoons vary considerably in shape and size, according to the balance between tidal range and freshwater supply (whether from rivers or rainfall) on the one hand, and between wave action and sediment supply on the other. Most are simply more or less elongate bodies of open water, but some larger lagoons have developed as extensive and complex areas of low-lying islands and meandering interconnected channels, which may be wide and deep enough to accommodate shipping and harbour works (Figure 6.16).

Where a sand or gravel spit is breached by wave or river erosion, a part may be isolated and a *barrier island* results, typically elongated parallel to the shore. Most barrier islands, however, probably originated during sea-level rise following the last glaciation. Longshore bars (cf. Figure 5.1) were formed by wave action on the sediments left behind by retreating glaciers. They migrated landwards as the sea advanced over the continental shelves, and many were built up above sea-level, especially where additional sediment was supplied by rivers and/or by longshore transport. Good examples are the Friesian Islands off northern Holland and Germany, which shelter the lagoons and extensive tidal flats of the Wadden Sea from the waves of the North Sea (Figure 6.17(a)). The tidal flats in the eastern half of the Wadden Sea merge with those of the estuaries of the major rivers.

(a)

(b)

Figure 6.16 Two views of lagoons near Faro, Portugal.
(a) Low-lying islands with channels, tidal flats and salt-marshes.
(b) Larger channels with harbour works.

(a)

(b)

Figure 6.17 (a) Part of the south-eastern North Sea, showing the West Friesian Islands sheltering the western Wadden Sea, in which large areas of tidal flats are formed. (b) Part of the extensive tidal flats in the Ems–Dollard estuary (far right in (a)).

QUESTION 6.8 Does the configuration of the West Friesian Islands (Figure 6.17(a)) suggest that the longshore drift which probably contributed to their formation was to the east or the west?

Tidal flats can develop wherever wave action is weak, the tidal range is moderate to large, there are high concentrations of suspended sediment in the coastal waters, and the offshore slope is gentle. They are not necessarily confined to estuaries, lagoons and other sheltered embayments that are regularly filled and emptied by the tides. In some circumstances, they can even occur on coasts facing the open sea, e.g. in Surinam on the north-east coast of South America. They are relatively rare in such settings, however, as open coastlines tend to be subjected to significant wave action, and sandy beaches are more likely to develop there (Chapter 5).

You read in Section 6.1 that the sediments of high tidal flats can be colonized by algae which help to bind them. In the warm climates of low latitudes, extensive mats of blue–green algae can form in the intertidal zone, especially where evaporation leads to hypersaline conditions. Some species of these algae secrete calcium carbonate, and successive layers of algal mats can accumulate to form algal mounds known as *stromatolites* (Figure 6.18), which eventually harden into limestone and are well preserved in the fossil record.

Figure 6.18 Section through algal carbonate mats of a stromatolite mound.

Along arid coastlines, such as that of the United Arab Emirates bordering the Persian Gulf, the seaward accretion of sediments leaves the older areas of algal mat stranded above sea-level. They are subject to intense surface evaporation, particularly after they have been flooded by seawater during occasional storms and extra-high tides, so that salts are precipitated within the algal mats. This kind of environment is called a *sabkha*, the Arabic word for salt-flat. In other low latitude coastal regions, carbonate sediments can accumulate wherever terrigenous sediment supplies are low (which includes coral reefs, cf. Chapter 3), and sediments of the intertidal region may be dominated by carbonate muds.

6.4 SUMMARY OF CHAPTER 6

1 Estuaries are tidal inlets at the mouths of rivers where mixing of freshwater and seawater occurs. They are ephemeral features on geological time-scales, and most are now slowly being infilled with sediment. They are characterized by channels and intertidal flats. There is a progression of sediment grain size towards the estuary shore: from sands in the channels, through sands and silts (with some muds) on the main intertidal flats, to muds on the high tidal flats, which are only submerged when tidal currents are weak at slack water. Accretion of tidal mud flats is promoted by the cohesive nature of muddy sediments, by settling lag, and by colonization of the mud flats by algae and eventually by land plants, leading to the formation of salt-marshes at mid- to high latitudes and mangrove swamps at low latitudes.

2 Fine sediment is deposited through aggregation into larger particles with higher settling velocities. Aggregation occurs mainly by flocculation in saline water, aided by turbulence in the water column, and also by biological processes (formation of faecal pellets and 'fluffy' aggregates of organic material). Cation exchange reactions take place between seawater and clay minerals, which can also adsorb heavy metals from solution in contaminated waters.

3 Estuaries range from strongly stratified to well-mixed, depending upon the relative magnitudes of tidal currents and river flow in the main channels. Salt-wedge (well-stratified) estuaries develop in virtually tideless seas, and are dominated by seaward flow of freshwater at the surface, with only minor landward movement (residual flow) of salt water at the bed. Current shear at the halocline leads to entrainment of salt water up into the freshwater layer. Partially mixed (moderately stratified) estuaries develop where there is a moderate tidal range. Greater mixing of fresh and salt water occurs because of turbulence, both at the bed and at the freshwater/seawater interface, and there is significant movement of water both seawards at the surface and landwards at the bed. Well-mixed (unstratified) estuaries develop where the tidal range is high. There is very little variation in salinity with depth, though in wide estuaries (especially if they are well-mixed) there can be lateral salinity gradients because river and tidal flows are on opposite sides of the estuary (as a result of the Coriolis effect) and there is a horizontal residual circulation. Even so, the mean velocity is seawards at all depths.

4 An estuary can exhibit different degrees of stratification and mixing between spring and neap tides and/or as a consequence of changes in river flow: high river discharge promotes stratification, low discharge promotes mixing. The upstream limit of the landward movement of salt water near the bed is called the null point. It occurs at salinities of between about 0.1 and 5 (parts per thousand), depending upon circumstances, and moves up and down the main channel with the tides.

5 The plume of brackish water that flows from the estuary mouth can affect offshore waters over considerable areas, and regions of freshwater influence (ROFIs) can extend up to hundreds of kilometres from the estuary mouth, depending upon the magnitude of the river discharge. Seawater is mixed and entrained into the plume at the base and along the margins, where there are convergent fronts. Where tidal ranges are large, tidal intrusion fronts form on the rising tide at the mouths of some smaller estuaries; and in some well-mixed estuaries longitudinal fronts are observed, the result of transverse surface water movements caused by cross-estuary gradients of salinity (and hence density).

6 A turbidity maximum develops near the null point, because sediment is carried into it both by the river flow and by the landward flow of salt water near the bed, aided by flocculation near the null point. The turbidity maximum also moves up and down the river with the tides, and is the source of most of the muds deposited on the high tidal flats. It tends to be most intense at mid-tides, when erosion by tidal currents is greatest. It is also enhanced during spring tides and/or at times of low river discharge, but is reduced during neap tides and/or at times of high river discharge. In some estuaries, high concentrations of fluid mud may form near the bed during neap tides, to be subsequently dispersed by the spring tides. Most estuaries are net accumulators of sediment since they are supplied with material from both the river and the sea. The landward movement of sediment is aided by the asymmetry of tidal flows in estuaries.

7 Negative estuarine circulation can develop in arid regions, where very high evaporation rates at the head of the estuary lead to sinking of dense hypersaline water, and a landward flow of seawater of normal salinity at the surface to replace it. In such estuaries, sands may be deposited from the bedload at the head of the estuary, while the fine sediments are carried seawards in suspension by the hypersaline flow at the bed.

8 Lagoons commonly form in the shelter of sand or gravel spits formed by longshore transport. Most are shallow and well mixed, and have only narrow outlets to the sea, so that tidal influences and wave activity are relatively weak. Breaching of spits by wave action, or isolation of longshore bars by rising sea-level, can lead to the formation of barrier islands, behind which wave action is limited. If the tidal range is moderate to large, tidal flats (similar to those occurring in estuaries) can form behind barrier islands. In low latitudes, colonization of tidal flat sediments by carbonate-secreting algae leads to accumulations of layers of algal mats. Along arid coastlines, where evaporation is high, salt flats (*sabkhas*) develop, and where terrigenous sediment input is negligible, carbonate muds can accumulate.

Now try the following questions to consolidate your understanding of this Chapter.

QUESTION 6.9

(a) Whereabouts in the estuarine continuum of Figure 6.4 would you place the Seine, according to Figure 6.10?

(b) Does the core of the turbidity maximum in Figure 6.10 lie at the null point, or upstream or downstream of it?

(c) What are (i) the maximum, and (ii) the minimum concentrations of suspended sediment in the core of the turbidity maximum in Figure 6.10, and at what stage of the tide does each occur?

QUESTION 6.10 Examine Figure 6.19.

(a) Suggest an explanation for the form of the surface isohalines in the main estuary.

(b) How confident would you be in concluding that the circulation pattern in this estuary is at the well-mixed end of the continuum in Figure 6.4?

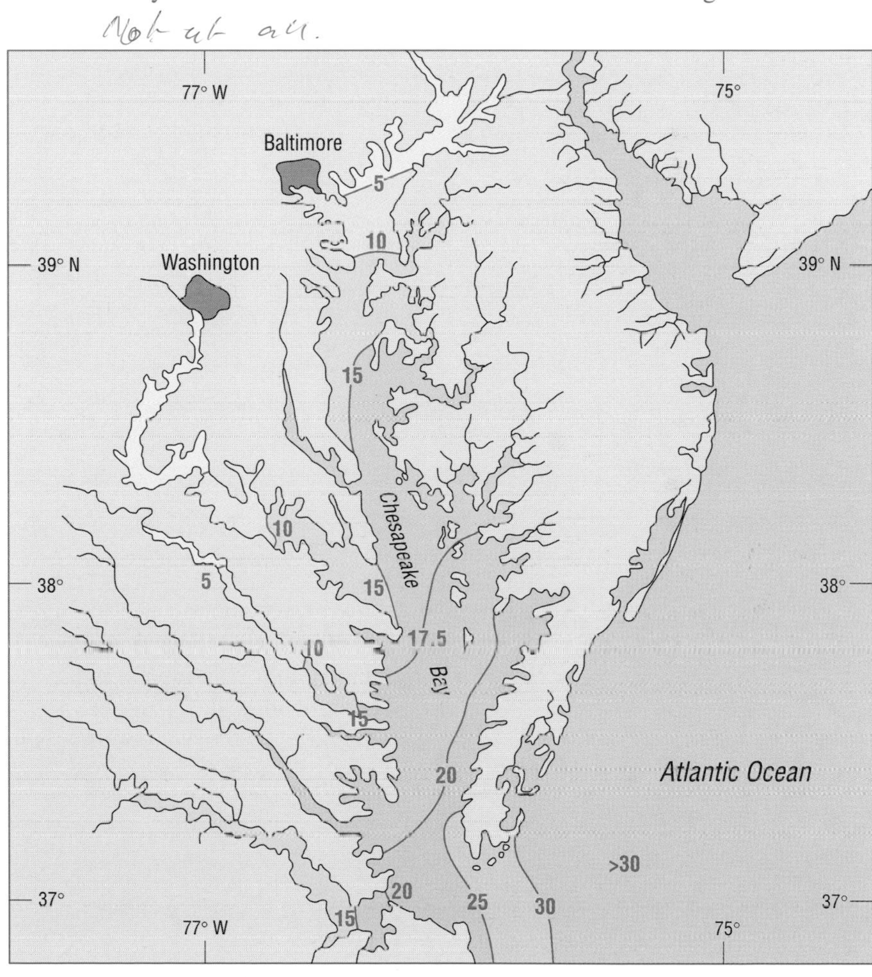

Figure 6.19 Map of the distribution of surface salinity (in parts per thousand) in Chesapeake Bay on the eastern seaboard of the USA. (For use with Question 6.10.)

QUESTION 6.11 Examine Figure 6.20.

(a) At approximately which station would you place the null point?

(b) Do the isohalines slope down landward or seawards?

(c) Approximately what is the salinity in the core of the turbidity maximum?

(d) How can you tell that the water column is well mixed to seawards of about station 17?

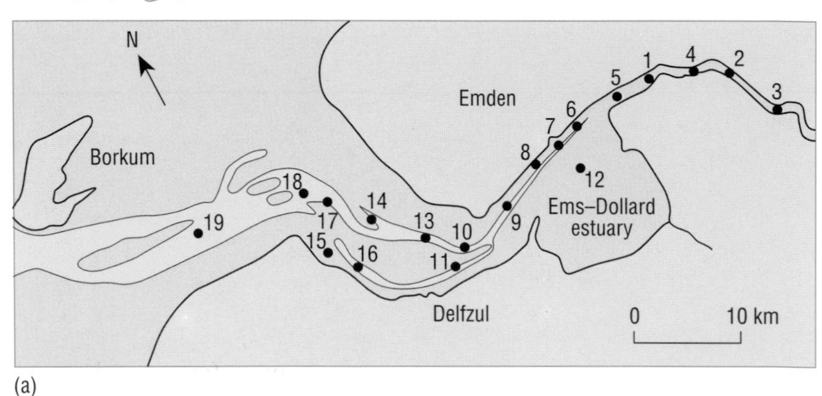

(a)

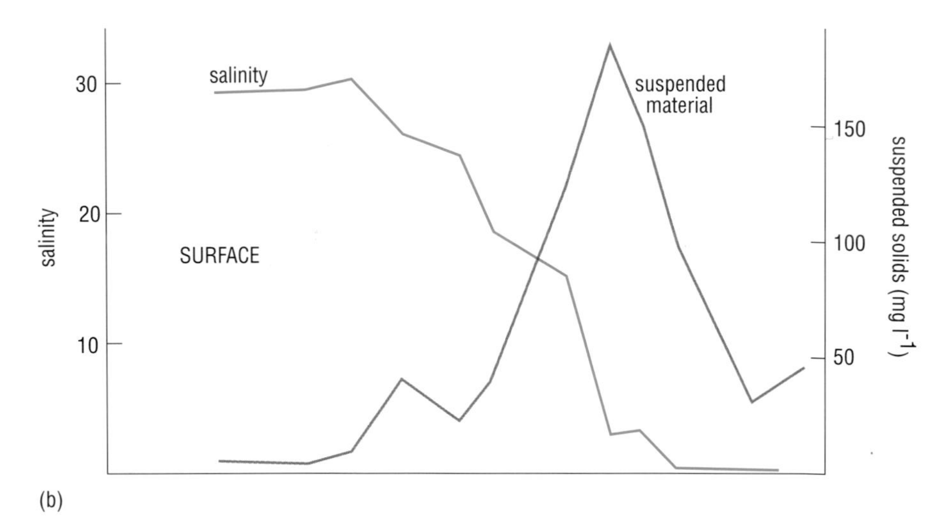

(b)

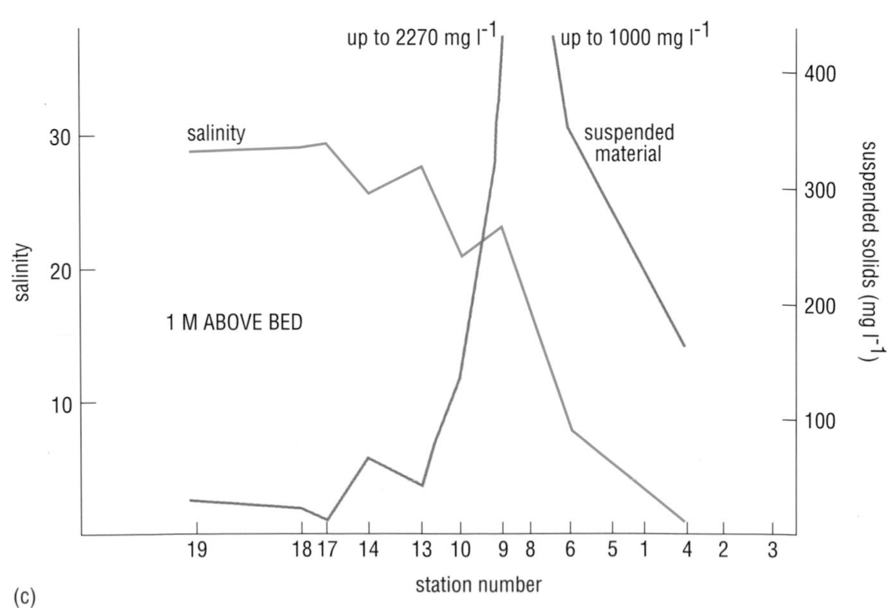

(c)

Figure 6.20 Distribution of salinity and suspended sediment concentrations in the Ems–Dollard estuary, on the German–Dutch border (cf. Figure 6.17).
(a) Map of area (main channel shown blue).
(b) Surface data. (Blue = salinity; brown = suspended material.)
(c) Data for 1 m above bed. Numbers are stations along the estuary; data are not given for all stations.

CHAPTER 7 DELTAS

'... then sands begin
To hem his watery march, and dam his streams,
And split his currents; ...
... till at last
The long'd-for dash of waves is heard ...'

From *Sohrab and Rustum* by Matthew Arnold.

In some ways, it is artificial to deal with deltas in a separate Chapter from estuaries, because the processes of water flow and sediment deposition are similar in both. The principal difference is that whereas estuaries are places where sediments accumulate, the discharge from rivers with deltas is so large that sediment is not deposited until it reaches the river mouth or beyond. If the rate of sediment supply exceeds the rate of sediment dispersal by waves and tidal currents, a coastal accumulation of river-borne sediment forms a **delta**, typically (but not invariably, see below) extending seawards of the river mouth.

Rivers transporting sediment loads sufficiently large to form a delta usually have extensive catchment areas and are fed by many tributaries supplying both water and sediment. Precipitation and erosion within catchment areas (depending on climate, local geology and relief) determine water supply and sediment discharge and hence whether or not a delta is likely to develop. Deltas also develop in freshwater lakes and inland seas – in fact, in any body of water where river flow is slowed on entering a larger body of water.

The term 'delta' was first used by the Greek historian, Herodotus, around 450 BC to describe the triangular accumulation of sediments at the mouth of the Nile (Figure 7.1(a)), resembling the Greek capital letter delta, Δ. As you will see later though, only some deltas develop this particular shape; in fact, there is a wide variety of delta types (Figure 7.1, overleaf), depending on the relative influences of river flow, wave action and tidal currents, and not all deltas form significant extensions to the coastline. Where the river discharges into deep water and/or where there is a significant tidal range, the river mouth tends to resemble a large and rather complicated estuary, with islands and interlinked channels. The mouth of the Ganges–Brahmaputra (Figure 7.1(d)) is a good example, and illustrates how it is not always possible to draw a clear distinction between what is an estuary and what is a delta. Thus, we have treated the mouth of the Amazon (Figure 6.7) as an estuary, but the complex of islands and channels there suggests it might be as valid to classify it as a delta (the Amazon does in fact have a delta, but it takes the form of a submarine fan (Section 6.2.5) rather than an extension to the shoreline). Another important feature that deltas have in common with estuaries is that, irrespective of the variety of deltaic forms that can develop (Figure 7.1), all deltas have *regions of freshwater influence* (ROFIs, Section 6.2.2), extending seawards from channel mouths.

(a)

(b)

Figure 7.1 (a) The Nile delta, photographed from the *Gemini IV* spacecraft.
(b) The Po delta in the Adriatic in October 1984, showing an extensive sediment plume. Venice lies at the top (northern end) of the picture, and the field of view is about 100 km from top to bottom.

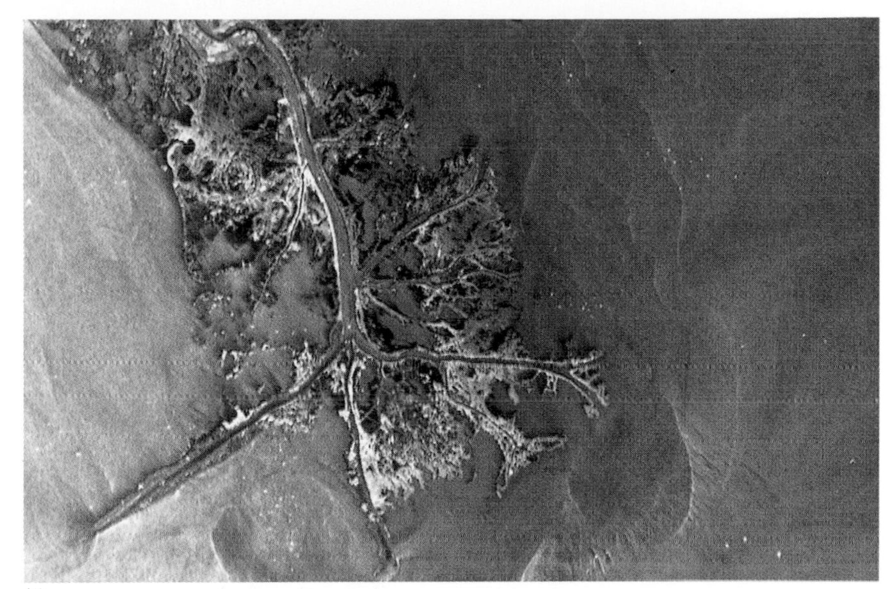

(c)

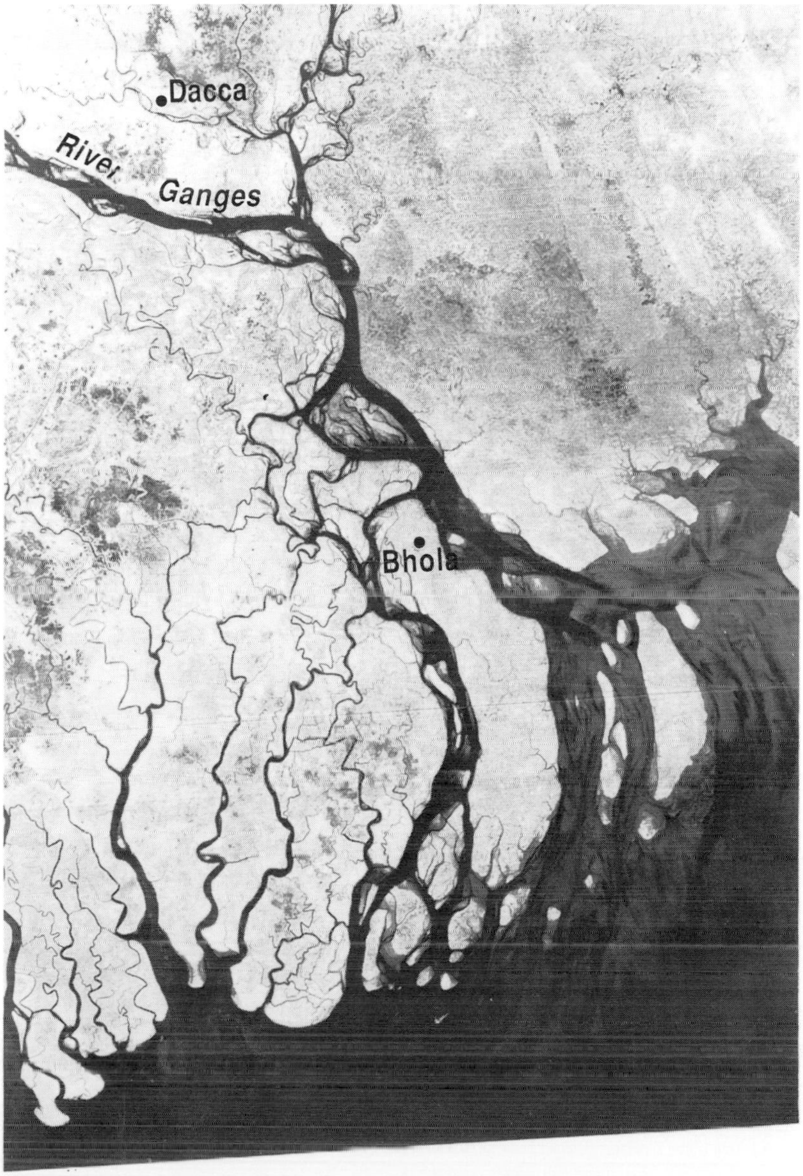

Dacca

River Ganges

Bhola

(d)

(c) A *Seasat* synthetic aperture radar (SAR) image of the Mississippi delta in the Gulf of Mexico, showing the type of delta formation sometimes called 'bird's foot'. The long, thin curving features seen in the waters to the east of the delta are where surface waves interacting with the outflow of water from the Mississippi have steepened and broken. The largest distributary channels are of the order of 1 km across.
(d) Mosaic of two *Landsat* images showing the ragged outline of the eastern active Ganges–Brahmaputra delta, with funnel-shaped distributary mouths and linear sand ridges formed by tidal currents. This view is approximately 180 km from side to side.

7.1 DELTA MORPHOLOGY

Major delta complexes are hundreds of kilometres across. They commonly consist of an extensive lowland area just above sea-level (the **delta plain**, Figure 7.2(a)), crossed by a network of active and abandoned channels. The raised banks of the channels are called *levées*, and are separated by either vegetated or shallow-water (wetland) areas. The numerous channels or *distributaries* range in width from a few tens or hundreds of metres to several kilometres. Deltas are highly dynamic systems, and when a channel becomes blocked with sediment, the flow splits to find new routes round the obstruction, so forming new channels. The proliferation of distributaries by channel-splitting can lead to a delta advancing over a wide front – the Niger delta, for example, has 10 main distributaries and an active width of nearly 300 km.

Deltas of even quite modest size can influence large areas of continental shelf, particularly if, periodically, deltaic diversions occur. This happens when a major breach is made in a channel levée on the delta plain, so that a new set of distributaries develops. Thus, in the past 5000 years there have been seven areas of deposition in the Mississippi delta, and although the active delta is only about 50 km across, the *whole* delta front is over 300 km wide – comparable in size to that of the Niger, which, however, is active over the whole of its front.

Seawards of the delta plain lies the **delta front** (Figure 7.2) which comprises the shoreline and the offshore part of the delta just below sea-level, where the fluvial bedload is deposited and where the sediments consist mainly of sands. The deeper offshore zone is the **prodelta**, which receives much of the silt and clay transported seawards in suspension and deposited in layers that are gently inclined (i.e. that dip) seawards. In most delta systems, the prodelta merges imperceptibly into a normal sedimentary marine environment on the continental shelf. However, where the input of terrigenous sediment is very high and the shelf is narrow (e.g. the Ganges–Brahmaputra delta), turbidity currents from time to time pour down the continental slope to feed vast submarine fans and build up the continental rise – just as they do off the estuaries of large rivers such as the Amazon and Congo (Section 6.2.5). In the shallower waters of shelf seas, high sediment input results in very rapid advance of the delta front and prodelta. For example, the delta of the Yellow (Hwang Ho) River is advancing at about 1 km yr^{-1} where it enters the Bohai Sea.

Most of the world's major deltas form extensive wetlands of high biological productivity and fertility which, among other things, makes them important conservation areas. They are regions where thick sequences of sediment and vegetation rapidly accumulate (Figure 7.2(b)), and many large deltas are also regions of active crustal subsidence (cf. Section 3.1.2), the result of isostatic adjustments in response to loading by the great mass of shallow-water sediment deposited on the deltas. For example, the Mahakan delta in eastern Borneo is part of a large sedimentary basin where sediment thicknesses exceed 15 000 m in places. Over 10 000 m of sediment underlie the Mississippi delta (Figure 7.1(c)), and there are more than 3000 m of

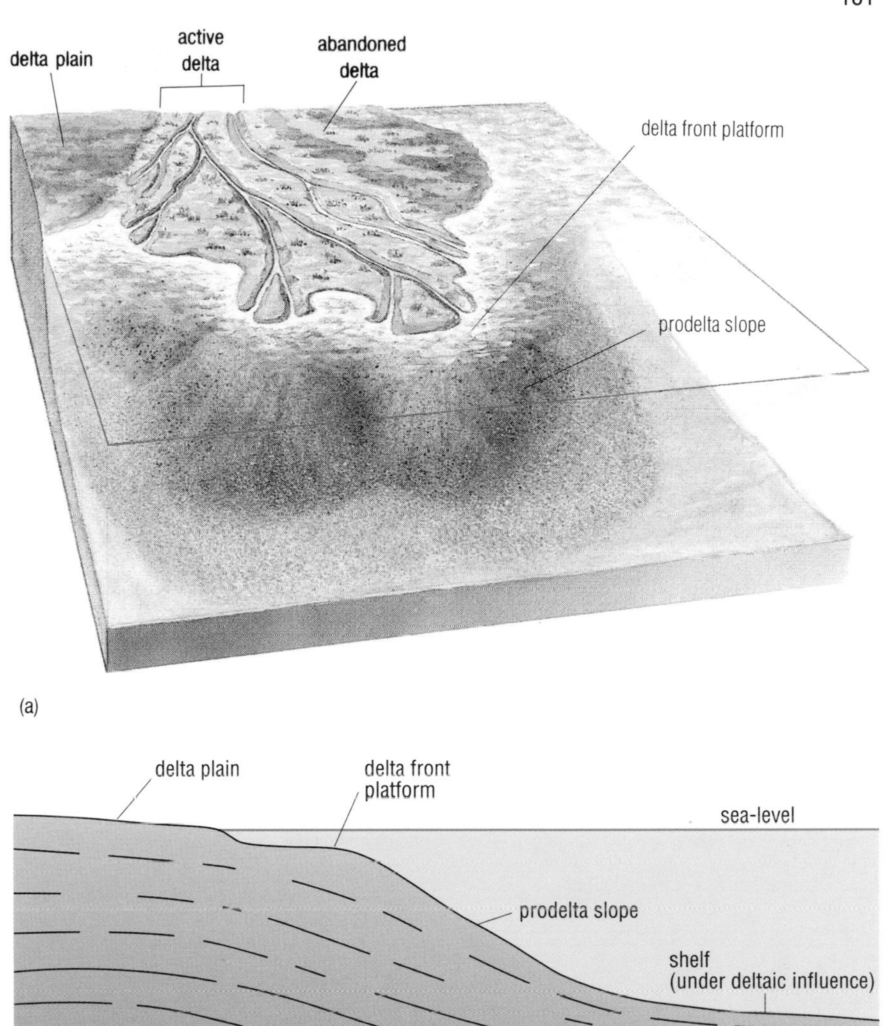

Figure 7.2 (a) Diagrammatic perspective view of a delta, showing the principal components. (b) Highly simplified cross-section through a delta (greatly exaggerated vertical scale), to show the pattern of deposition of sediments in gently seawards-inclined (dipping) layers.

sediments beneath the Nile delta. Deltas are thus geologically more persistent features than estuaries, and in many parts of the world, their thick sediment accumulations are important as source and reservoir rocks for deposits of oil, gas and coal.

Subsidence in deltaic regions results not only from loading by the accumulation of successive layers of sediment (Figure 7.2(b)), but also from compaction of those sediments. Since the lowland plains of deltas remain close to sea-level, rates of sediment deposition and of subsidence must remain broadly in balance. The Mississippi delta is estimated to have subsided by up to 150 m during the past 18 000 years, and the Nile delta by about 40 m in the past 8500 years. These subsidence figures imply sediment accumulation rates of several millimetres per year, but at the end of the last glacial maximum 12 000 years ago (Section 3.1.2), rates of subsidence must have been greater than this to keep pace with rising sea-level.

7.2 MIXING AND SEDIMENT DEPOSITION AT DISTRIBUTARY MOUTHS

Although we have drawn a distinction between estuaries and deltas on the basis of sediment supply and deposition, the processes of mixing between seawater and river water that occur in association with deltas are much the same as those described for estuaries in Chapter 6. Indeed, each individual channel is in effect a small estuary. Differences in the relative influences of river flow, tidal currents and wave action lead to differences in the type and degree of mixing and of sediment transport and deposition, and hence to differences in the way in which the sediment is distributed to give each delta its characteristic shape.

As with estuaries, therefore, we can recognize a continuum, except that in this case there is an additional variable: the effects of wave action.

Why is wave action a more important control on sediment deposition for deltas than for estuaries?

Most deltas form extensions to the coastline and are therefore more exposed to the influence of ocean waves and longshore drift (Section 5.2.3); whereas estuaries are by definition inlets of the sea and so are relatively sheltered from waves generated offshore.

7.2.1 THE DELTAIC CONTINUUM

Figure 7.3 summarizes the deltaic continuum and shows that only a minority of deltas fall neatly into any of the three basic categories, as represented by the apexes of the triangular diagram. Usually, more than one process influences the delta shape, and there is a range of delta morphologies intermediate between the river-dominated, tide-dominated and wave-dominated categories.

Where the tidal range is small, the water column offshore from a delta is stratified, just as it is in estuaries under the same tidal conditions. The **river-dominated delta** is analogous to the salt-wedge estuary (Section 6.2.1), the principal difference being that stratification, mixing and sedimentation take place mostly seaward rather than landward of the distributary mouths. The best-known and best-described example of a river-dominated delta is the one formed at the mouth of the Mississippi in the Gulf of Mexico (Figure 7.1(c)), where the tidal range is less than 0.5 m. Further west, rivers discharging into the Gulf of Mexico from Texas carry less sediment than the Mississippi and form estuaries instead of deltas, as noted at the end of Section 6.2.1.

River water flows from the distributary mouths and spreads out over the denser seawater as a thin layer, an almost two-dimensional plume. The regions of freshwater influence (ROFIs) associated with river-dominated deltas can extend for considerable distances offshore (e.g. over 10 km from the mouth of the Po, over 50 km from the mouth of the Mississippi). Mixing and entrainment of salt water occur at the base of the plume as well as along the sides, where there are convergent fronts, just as in the case of estuarine ROFIs; and the plumes are often turbid because of high suspended sediment concentrations (Figure 7.1(b)). Figure 7.4 (overleaf) presents some data from the Po delta.

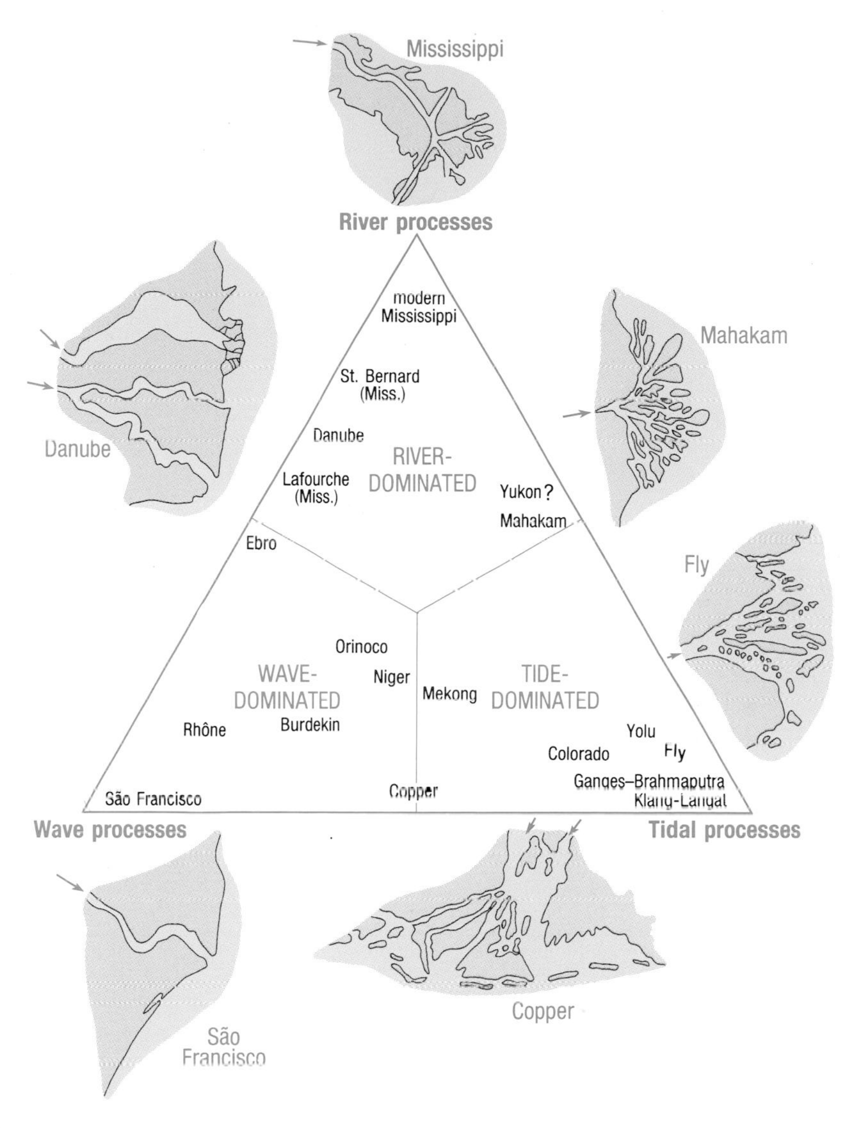

Figure 7.3 The deltaic continuum, based on the relative intensities of river, wave and tidal processes (sketches of deltas not to scale).

QUESTION 7.1

(a) How do the salinity data in Figure 7.4 show that these plumes must be very thin in relation to their lateral extent?

(b) How can you tell that the influence of the plume has almost disappeared by the time it reaches the most easterly station?

In answering Question 7.1, you may have noticed two other features. First, surface salinities within the plume tend to be higher near the margins, where there is intense mixing with the surface seawater. Secondly, the longshore current in the Adriatic here is deflecting the low-salinity influence of the plume (i.e. its ROFI) well to the south of the main river mouth.

In river-dominated deltas, just as in salt-wedge estuaries, the landward flow of saline water beneath the plume (as required by continuity, to balance entrainment of seawater in the plume) results in the landward transport and accumulation of coarser sediment to form transverse sand-bars by deposition near the null points of the distributary channels.

Whereabouts would you expect this to happen, i.e. where would you place the null point of any particular channel in this situation?

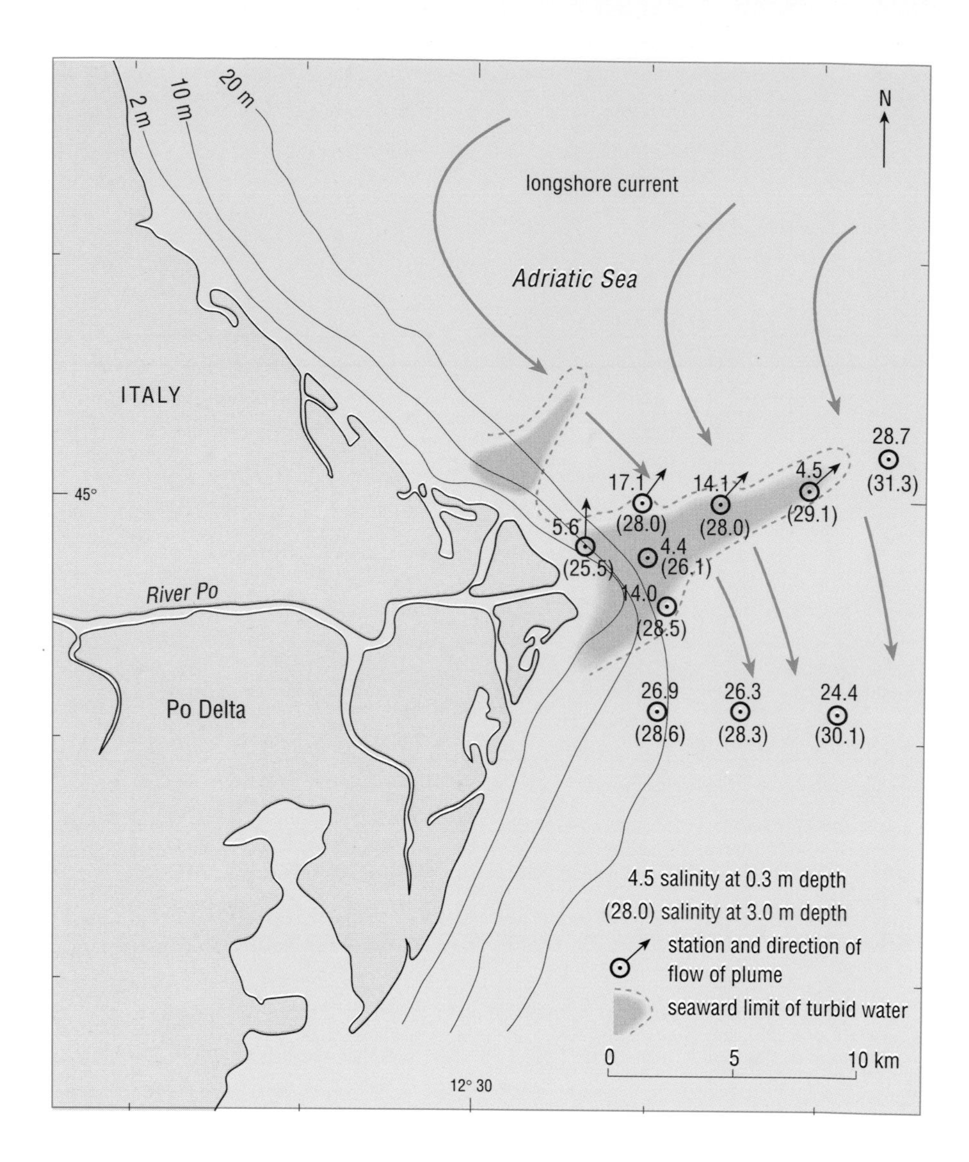

Figure 7.4 The mouth of the Po, showing the effect of the river plume on the salinity and turbidity of the water in this part of the Adriatic. Salinity values at 0.3 m depth (and at 3.0 m depth) are in parts per thousand.

Because the river water begins to slow down as soon as it spreads out on leaving the distributary channel, and entrainment of salt water begins, the null point is likely to be close to the channel mouth. The sand-bar formed at the top of the delta front slope (Figure 7.2) is built of clean sand, because of the influence of waves and currents which winnow away the finer material. Where a sand-bar grows above sea-level, an existing channel may be split to create two channels.

As the plume continues to lose speed through mixing and entrainment along its margins, sediment may be deposited to form linear subaqueous levées, provided wave action is not too severe. As the delta advances seawards, these levées can grow to form the banks of lengthening distributary channels; and if they do not diverge too much as they extend seawards, the deltaic distributaries and their deposits tend to be long and finger-like (the 'bird's-foot' pattern of Figure 7.1(c)).

The sediment load in the lower reaches of most large rivers is predominantly fine-grained, and much is carried out to sea in suspension. For example, only 2 per cent of the sediment being transported by the Mississippi is sand, the bulk being silt and mud in about equal proportions. Even in rivers that have a relatively large bedload, sand generally makes up less than 40 per cent of the total sediment load. This is usually deposited close to the distributary mouth, raising the level of the sea-bed, so that it is quite common for the river to discharge into shallow water, where it can disrupt the stratification. Some of the fine-grained suspended sediment is trapped by estuarine processes (Section 6.2.3) within the distributary channels, but most of it is progressively deposited from suspension offshore, and sediment grain size generally decreases down the prodelta slope (Figure 7.2).

QUESTION 7.2 As sediment-laden freshwater mixes with seawater, what process will aid the rapid deposition of the fine-grained suspended sediment on the prodelta slope?

Like that of many large rivers, the discharge from the Mississippi shows considerable seasonal variations. During most of the year, the total flow is around 18 400 $m^3 s^{-1}$, there is density stratification and a well-defined salt wedge develops beneath the freshwater plume and may even penetrate upstream along the distributary channels (and promote trapping of suspended sediment near the null point). However, for between one and three months a year, the river is in flood, which increases the discharge more than three-fold. The river outflow is then so powerful that it forces the salt water out seawards of the delta bar crest. The density stratification is weakened, even destroyed, by vigorous turbulent mixing, and the associated increase in sediment discharge – of both bedload and suspended load – results in rapid seawards advance of the deltaic deposits.

In deltas where rivers bring large sediment loads into a sea with moderate tidal range (2–4 m), the effects of tidal currents will be greater – and by analogy with partially mixed estuaries (Figure 6.4(b)), mixing between river water and seawater will be more vigorous and the halocline less well defined. Such deltas lie part-way between the 'river processes' apex and the 'tidal processes' apex of Figure 7.3. Sediments deposited on the delta front are continually reworked by tidal currents, so delta front bars tend to be ephemeral (though they are again composed mainly of sands, as finer sediment is winnowed out).

Tide-dominated deltas (at the 'tidal processes' apex of the deltaic continuum in Figure 7.3) occur where the tidal range is large (> 4 m). Where the water offshore is relatively shallow, and/or at times of relatively low river discharge, the water column of a tide-dominated delta resembles that in a well-mixed estuary (Figure 6.4(c)): turbulent mixing extends to the sea-bed and there is no density stratification. The residual flow is seawards at all depths, but superimposed on this are landward flows during flood tides, and seaward flows during ebb tides. Similarly, although net sediment transport is seawards, there is also some landward and seaward movement of sediment on the flood and ebb tides (respectively). The relative importance of fluvial and marine processes will vary during the spring–neap cycle and with seasonal changes in river discharge.

In tide-dominated deltas, sediments transported by river flow and deposited near the distributary mouths are rapidly reworked by tidal currents into a series of linear subaqueous ridges within the distributary mouth, and further seawards. The ridges may be several kilometres long, a few hundred metres wide and over 20 m high. As the delta builds gradually seawards, the sand ridges become exposed above sea-level and are colonized by vegetation to form linear islands. A series of linear islands and sand ridges projects for over 90 km offshore in the Ganges–Brahmaputra delta (Figure 7.1(d)). A similar linear pattern of sediment movement and distribution can be seen from the sediment concentrations in the well-mixed Humber estuary (Figure 6.5), albeit on a much smaller scale. However, the rate of sediment supply from the Humber is too low to permit the formation of linear sand ridges – which is why it remains an estuary!

The third apex of the deltaic continuum (Figure 7.3) is approached where wave energy becomes great enough to affect the distribution of deltaic sediments. If wave energy at the coast is sufficiently high, a **wave-dominated delta** can occur, irrespective of whether the tidal range is large or small. In such situations, the river flow moving seawards behaves as a current flowing counter to the direction of wave propagation, so that wave speed and wavelength decrease, and wave height increases (Section 1.6.1). The effect is very much the same as when waves propagate against an ebbing tide in an estuary (cf. Question 6.4(b)). As a result of these changes, waves approaching the river mouth are liable to break earlier in deeper water than they would normally, which promotes mixing of seawater and freshwater and disrupts the density stratification. Because part of the wave crest impinges on the freshwater plume, it is retarded relative to those parts on either side. The waves are refracted around the plume in a manner analogous to that described in Section 1.6.2, and the effect can be seen in Figure 7.1(c). Refraction concentrates the wave power on the freshwater plume and further enhances the mixing.

The vigorous mixing of seawater with river water leads to rapid deceleration of the freshwater flow, and equally rapid deposition of sediment. For example, observations of flow rates at the mouth of the Shoalhaven River on the south-east coast of Australia show that when the river is in flood, flow speeds are greater than $2\,\mathrm{m\,s^{-1}}$. However, breaking waves immediately mix the freshwater and salt water, and within a distance of about half a kilometre the speed of flow falls to around $0.3\,\mathrm{m\,s^{-1}}$. Only the very fine sediment escapes deposition and is carried seawards to be deposited further offshore. The coarser sediments are deposited in the zone of mixing as a crescentic bar; this is reworked rapidly by waves and the bedload is moved back landwards by wave action and may form a series of swash bars (Section 5.1).

The shorelines of wave-dominated deltas are characterized by straight, sandy beaches and there is usually only a slight protuberance where a distributary mouth meets the sea (see São Francisco delta, Figure 7.3). There are fewer distributaries than in river- and tide-dominated deltas, and they are occasionally blocked by reworked sand being returned landwards, so the river must enter the sea by a new outlet. As the delta grows seawards, the delta plain consists of a series of abandoned beaches stranded above sea-level.

7.3 THE EFFECTS OF HUMAN ACTIVITIES ON DELTAS

Many deltas are subject to interference from human activities, and there are few large deltas that can nowadays be studied in their pristine state. Delta plains are commonly very fertile regions and are extensively exploited for agricultural development, as are the river basins inland of the deltas themselves. Modern farming practices commonly result in large inputs of artificial fertilizers, which are washed off into the distributaries and contribute to enhanced biological production offshore. In some circumstances, this can lead to improved fisheries, but often the excess primary production remains unconsumed and simply decays, so that the water becomes deoxygenated, with adverse effects on marine life. This was the cause of areas of oxygen-poor waters that began to be observed off the mouth of the Mississippi during the 1970s. The influx of nutrients from the river increased in succeeding years, and by the late 1990s these oxygen-poor areas had grown and coalesced into a broad zone in the northern part of the Gulf of Mexico, about 50 km across and extending for nearly 400 km to the west of the Mississippi delta (see also Figure 8.5).

Wetlands in the Mississippi delta region are being lost at a rate of about 100 km^2 annually, and in places the shore is retreating by as much as 25 m per year. Since the early years of this century, the rate of subsidence has exceeded the rate of sediment deposition, because of the effects of human intervention in the deltaic system. Numerous artificial levées or 'training walls' were constructed to prevent channel splitting and to control the annual spring floods, and substantial areas of the delta plain were also reclaimed for building land. In addition, new channels were dug and existing channels deepened to make navigation easier – and these channels themselves became widened as their banks were further eroded by the wakes from passing vessels. All these changes have had two main consequences. First, sediment that would normally be deposited on the delta plain each year is prevented from doing so and most of it is transported out into the Gulf of Mexico. Secondly, both new and deepened channels have allowed seawater to penetrate further into the wetlands, destroying the vegetation (including trees) which cannot tolerate salinities of more than a couple of parts per thousand. The plants die, leaving open water and allowing the sea to spread further inland during high tides and especially during storms. In addition, the artificial levées and the canalization failed in their purpose as flood control measures when in 1993 the Mississippi – swollen by torrential rains so that its normal annual discharge of about 580 km^3 was almost doubled – overflowed its banks and caused billions of dollars worth of damage. The flooding was exacerbated because once-extensive wetland areas across which floodwaters could have spread, and which would have acted rather like a sponge, soaking up the excess water, had been reclaimed and covered with impermeable tarmac and concrete.

The large rivers that supply the deltas are nowadays often dammed, greatly reducing the supply of sediment to the delta front. In the case of the Nile, for example, construction of the Aswan dam has so depleted the sediment loads that the delta front is now actively eroding rather than advancing, while subsidence of the delta plain continues because of compaction of the thick sedimentary pile beneath it. Subsidence of the Ganges–Brahmaputra delta has been exacerbated in recent years both by increased extraction of groundwater for agriculture and domestic use; and by diversion of Ganges water upstream for irrigation, thus reducing river discharge at the delta. A series of dams along the Danube has also resulted in loss of sediments to its delta, while the Po delta was well supplied with sand until its river too was dammed in the 1980s, significantly reducing the sediment discharge. Large rivers such as the Po and the Danube tend also to have industrial complexes along them, and the wastes discharged from such installations further contribute to pollution of coastal waters.

Most damaging of all, however, must be exploitation of hydrocarbons from within the thick deltaic sedimentary sequences, such as those of the Niger delta. Construction work associated with such operations can severely disrupt natural patterns of vegetation and of water flow across the delta plains; water pollution is bound to be considerable, no matter how much care is taken, while extraction of the oil and gas itself accelerates the subsidence that occurs naturally through sediment loading and compaction.

Delta plains are (almost by definition) close to sea-level and are especially vulnerable to flooding, especially from storm surges, which can lead to enormous loss of life among the agrarian communities living on them (cf. Section 2.4.2). The likelihood of severe flooding increases almost annually, because human activity has caused subsidence to outpace sedimentation, and sea-level continues to rise because of global warming (which is also increasing the frequency and severity of tropical cyclones and associated storm surges).

7.4 SUMMARY OF CHAPTER 7

1 Deltas are coastal accumulations of river-borne sediments which accrete when sediment discharge is too large to be dispersed by tidal currents and wave action. The main components of the delta are: the delta plain above sea-level; the shallow water and shoreline region of the delta front, dominated by sands; and the deeper water prodelta, dominated by silts and clays. The shapes of most deltas are controlled by the interaction of fluvial, tidal and wave processes, and a deltaic continuum can be identified on the basis of the relative importance of these three processes, analogous to the estuarine continuum of Chapter 6.

2 Where tidal range is small and wave action is weak, the river discharge plume spreads seawards as a thin layer over the denser seawater, establishing a density stratification. Mixing and entrainment of salt water into the base of the plume generates a landward flow in the salt wedge beneath it. Bedload deposition of sediment forms a delta bar and near-parallel subaqueous levées. When the river discharge is high, vigorous turbulent mixing disrupts the stratification, and increased sediment transport results in rapid advance of the delta front.

3 With increasing tidal range, there is stronger turbulent mixing along the sides and base of the plume, and the water column resembles that of partially mixed estuaries. Where the tidal range is large (>4 m), strong tidal currents inhibit development of density stratification, and the water column at distributary mouths is well-mixed. Sediment movement occurs both up and down the distributary channels, with formation of sediment ridges within distributary mouths, parallel to the direction of tidal ebb and flow, producing a ragged outline to the delta.

4 Where wave energy is high, the outflowing freshwater from the delta distributaries behaves as a counter-current, slowing down and steepening the approaching waves and causing them to break in deeper water than usual. Waves are also refracted so that wave energy is concentrated on the freshwater plume. Both processes lead to vigorous mixing of freshwater and salt water, rapid deceleration of the freshwater flow and deposition of sediments. Wave action reworks the sediments and moves the coarser sediments landwards to form swash bars and beaches, creating a straight shoreline with only minor protuberances at the distributary mouth(s).

5 Human interference with deltas, such as damming of rivers inland, agricultural activity or hydrocarbon exploitation on the delta plain, can disrupt the natural patterns of water flow and of sediment transport and deposition. Sediment supplies to the delta front may be reduced, and as subsidence continues there is likely to be erosion and retreat of the delta front, instead of deposition and advance. The discharge of agricultural and industrial wastes can significantly contaminate the coastal waters.

Now try the following question to consolidate your understanding of this Chapter.

QUESTION 7.3

(a) Explain whereabouts on Figure 7.3 you would plot the Nile delta (Figure 7.1(a)).

(b) Explain whereabouts on Figure 7.3 you would plot the Po delta (Figure 7.1(b)).

a Between wave & river dominated.

b. mainly River dominated but some wave action.

CHAPTER 8 SHELF SEAS

> 'All the rivers run into the sea, yet the sea is not full.'
>
> From *Ecclesiastes,* 1.7.

Shelf seas can conveniently be considered to extend from just below the low tide level at the coastline out to the shelf break. As summarized at the start of Chapter 3, most are bounded on one side by land, on the other by open ocean, but some, such as the North Sea (Figure 8.1), are semi-enclosed. As you also read in Chapter 3, most continental shelf areas are underlain by considerable thicknesses of sediment. During periods of fluctuating sea-level in the Pleistocene, coastal environments migrated back and forth across the shelves, which are now largely submerged by the rise in sea-level that followed the last glacial maximum. As a result, the sea-bed in shelf seas is generally formed of both modern and relict sediments. The latter were deposited when sea-level was lower and were subsequently reworked by waves and currents, so that finer-grained material was removed, leaving behind residual deposits of sand and gravel (Section 3.1.2). On shelves at high latitudes, the sediments consist of material resulting from physical (mechanical) weathering and ice transport (Chapter 3), and commonly contain >30 per cent sand and gravel.

Carbonate sediments accumulate wherever the supply of terrigenous material is low and other conditions are favourable (Section 3.1). The remains of shelly animals (molluscs, echinoids, etc.) can occur almost anywhere, but reef-building corals and inorganic carbonates are mostly confined to low latitudes and to water shallower than about 25 m (although relict reef sediments have been found in deeper shelf sea waters). The 'cold-water' corals of higher latitudes (e.g. *Lophelia*) do not build reefs as such, but form colonies in water depths ranging from less than 100 m to more than 600 m, where water temperatures may be only around 10 °C (Section 3.1).

Planktonic organisms that secrete skeletons (tests) of calcium carbonate have also contributed to shelf sediments in the geological past. The thick Chalk deposits of southern England and northern France consist of the calcareous remains of **coccolithophores** (also called coccolithophorids), minute algae that lived in an extensive shelf sea covering the region some 80–90 million years ago (Section 3.1.1). Such sediments are not accumulating in present-day shelf seas, but they do constitute a significant proportion of deep-sea sediments.

Shelf seas have an importance that is out of proportion to the relatively small fraction of the total area which they occupy. They are the regions where much of the ocean's tidal and wave energy is dissipated, and they are biologically highly productive, supporting most of the world's major fisheries. They can be important sources of economically important minerals and of hydrocarbons – and many are important maritime thoroughfares. Their proximity to centres of population and industrial development, coupled with their shallow depths and restricted area (relative to the ocean basins), makes them especially vulnerable to anthropogenic pollution via rivers and atmosphere, while coastal engineering works often disrupt the natural processes of land–sea interaction. The growing importance of shelf seas to coastal states thus makes it increasingly necessary to adopt a multidisciplinary 'systems' approach to the study of these regions. One such multidisciplinary research programme was the North Sea Project, which ran from 1987 to 1992, involving observations, measurements and modelling, and covering virtually every aspect of the oceanography of the North Sea.

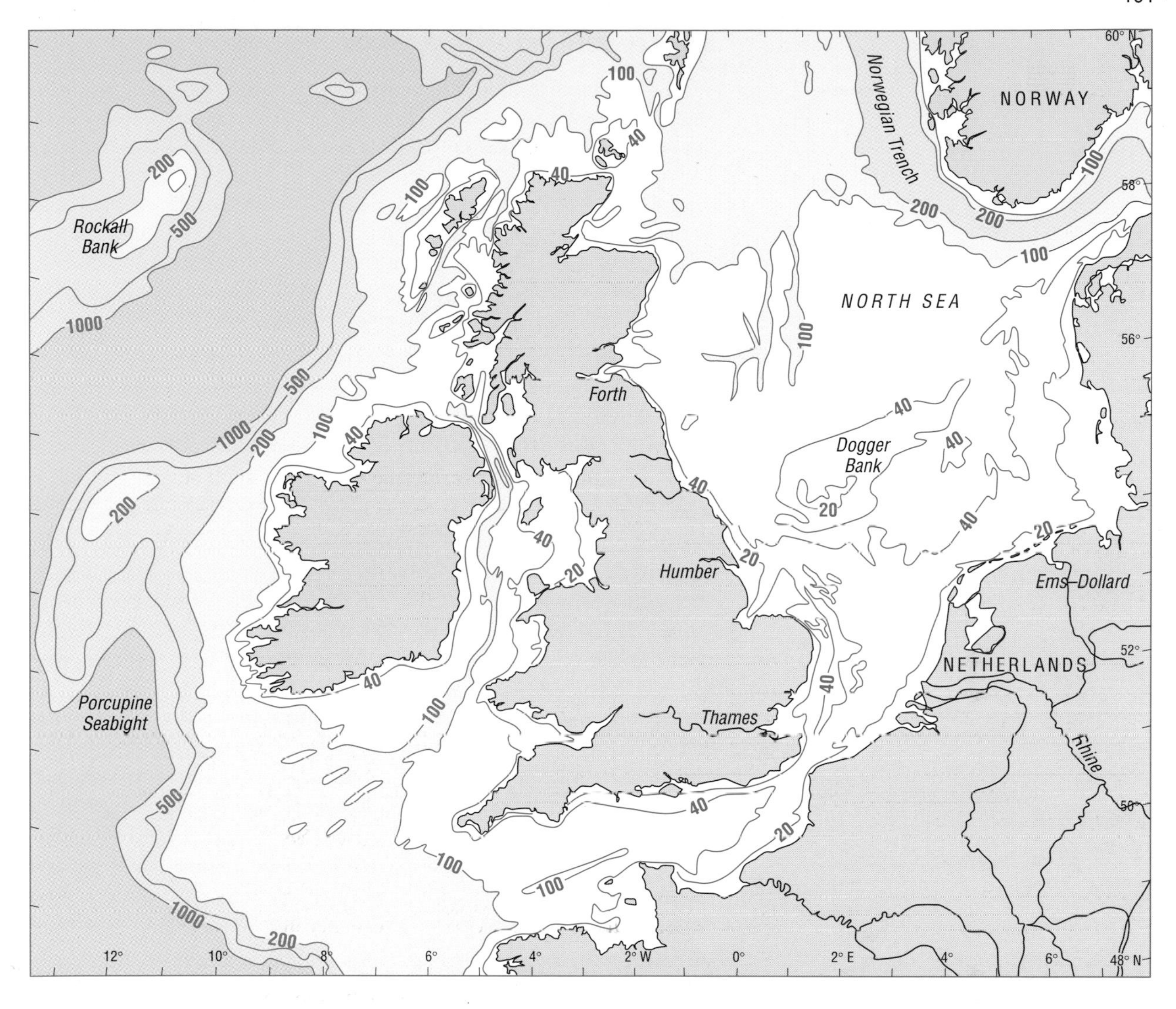

Figure 8.1 Shelf seas around the British Isles include continental shelf areas open to the north-east Atlantic as well as the semi-enclosed North Sea basin. Bathymetric contours in metres.

8.1 SHELF SEA PROCESSES

Deposition of modern sediments on present day continental shelves depends on the rate of sediment supply, on whether it is in suspension or bedload, and on the influence of waves and currents. Where pipelines or oil rigs are placed on the shelf sea-bed, it is important to know the speed and direction of current flow and the magnitude of wave action, and the resulting sediment movements on the sea-bed. Sea-bed roughness and the distribution of bed forms can be determined using acoustic instruments that are sophisticated versions of the echo-sounders used for depth measurement. As outlined in Section 4.4, the combination of waves and currents produces a wide variety of bed forms in shelf seas, ranging in size from centimetres to tens of kilometres, that change and migrate with variations in the wave and current regime. Although tidal currents and storm waves are the most important water movements affecting the sea-bed, oceanic currents, local wind-driven currents and storm surges (Section 2.4.2) can also lead to sediment movement in shelf seas. Tidal currents are considerably faster in shelf seas than in the open ocean, typically reaching speeds of 1 m s^{-1}, strong enough to rework sands and gravels and even cohesive muds; and the greater the tidal range, the greater the current speed (Section 2.4.1).

8.1.1 COASTAL AND OCEAN CURRENTS

Wind-driven currents can also contribute to the residual current field in shelf seas, because onshore winds tend to pile up water near coasts (cf. Section 2.4.2). This leads to a downward slope of the water surface towards the sea, and a seaward-directed horizontal pressure gradient. Any pressure-driven current will tend to be deflected by the Coriolis force (Section 2.3), to the right in the Northern Hemisphere, to the left in the Southern Hemisphere, with the result that there may be **geostrophic** coastal currents, flowing with the coast to the right in the Northern Hemisphere, to the left in the Southern Hemisphere. The effects of onshore winds may be enhanced in wet weather if large amounts of less dense freshwater flow into the sea and further elevate the water surface along the coast, thus increasing the seaward pressure gradient. These coastal currents are on a larger scale, flow further offshore and are generally slower than the longshore currents associated with wave action, described in Section 5.2.3.

Where prevailing winds blow *along* the coast, the combination of wind stress and the effect of the Coriolis force results in a surface current at an angle to the wind direction, and average movement of the wind-driven surface layer as a whole is at *right angles* to the wind direction. This is known as **Ekman transport**. In the Northern Hemisphere, if you stand with your back to a longshore wind and the sea is on your right, there is net *offshore* Ekman transport of surface water. Because of the requirements of continuity, this surface water is replaced by **upwelling** of cooler sub-surface water. As this water is rich in nutrients, it commonly supports high primary production and rich fisheries. Conversely, if the sea is on your left when you stand with your back to the longshore wind (still in the Northern Hemisphere), there is net *onshore* movement and coastal **downwelling** of surface water. Landward Ekman transport of surface water has the same effect as onshore winds (and/or of rain), creating or reinforcing a seaward slope of the sea-surface, and contributing to the coastal geostrophic current described above.

In the Southern Hemisphere, the Coriolis force acts in the opposite sense, so when your back is to the longshore wind, there will be upwelling where the sea is on the left, downwelling where it is on the right

QUESTION 8.1

(a) Look at Figure 8.2 and identify (i) which part portrays a situation likely to lead to upwelling and which to downwelling, and (ii) which hemisphere is represented?

(b) What are the likely directions of sediment movement on the sea-bed in each of these situations?

Major ocean currents can affect the outer parts of some continental shelves, either where large eddies 'spin off' from a major current onto the shelf, or where a major current migrates shorewards. For example, during summer months the Florida Current encroaches over the shelf of eastern North America, and its speed is high enough to erode sediment on the Blake Plateau; during winter, it moves offshore. The Canary Current is strong enough at times to form megaripples on the outer Saharan continental shelf.

As we noted in Section 2.4.1, the pattern of shelf-sea currents (and hence of sediment distribution) is modified by the shape of coastlines and bottom topography, and by the presence of fronts between stratified and mixed

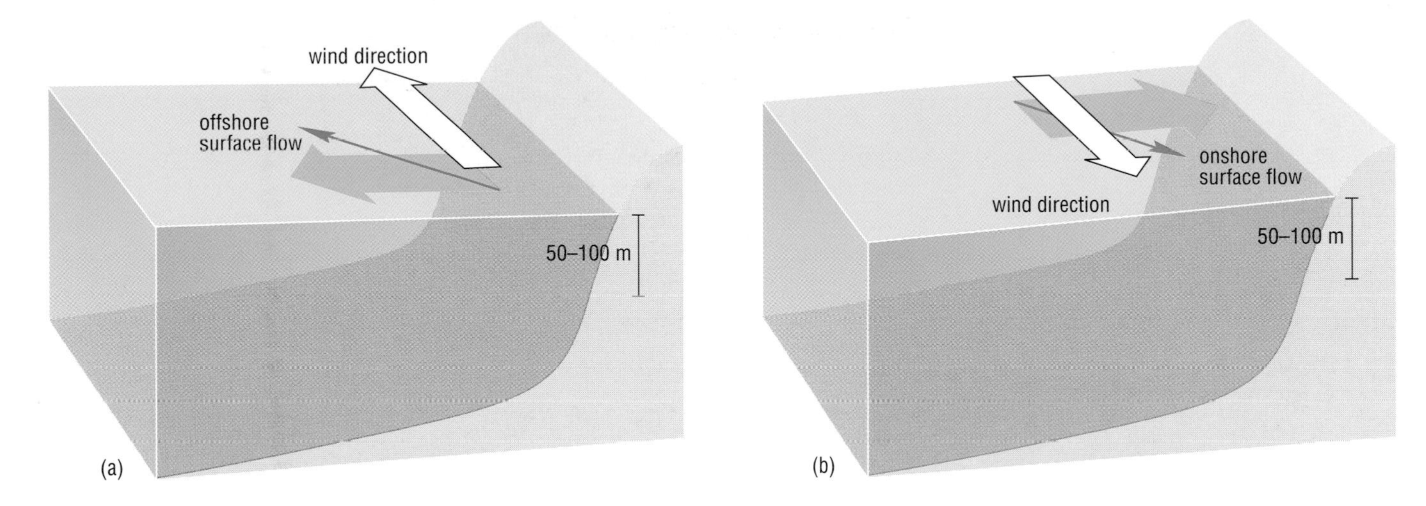

Figure 8.2 Longshore winds result in either (a) offshore or (b) onshore surface water movement (by definition, Ekman transport of the wind-driven layer (broad blue arrows) is at right angles to the wind direction). See also Question 8.1.

waters on the shelf (Figure 2.22). The overall average (residual) current field of a shelf sea is a result of all these factors, and it determines net directions of sediment transport and hence of areas of erosion and deposition (cf. Figures 4.13 and 4.14, see also Figures 8.3 and 8.4).

8.1.2 THE EFFECTS OF WAVES AND OF BIOTURBATION

Waves are more variable than currents and their influence depends greatly on the frequency and intensity of storms; their effect is to rework shelf sediments rather than to supply them. Waves are generally responsible for resuspending and sorting sediment, winnowing out the finer particles to be transported and deposited elsewhere by currents that would otherwise be too weak to move them. Wave action can also lift cohesive muds into suspension, when tidal currents alone cannot shift them. During fair weather, wave action disturbs sediment down to water depths of only 10–20 m, but storm waves can affect sediments at depths as great as 200 m (Section 5.2.2). On the Washington shelf off north-western USA, for example, it has been estimated that waves affect sediments at depths of about 75 m (the mid-shelf area) on about 50 days each year, while sediments at the shelf break (*c.* 170 m depth) are reworked by waves as often as five days per year. As noted in Section 5.2.2, despite their infrequent occurrence, storm events can be the dominant influence on sedimentation processes on shelves. This includes the effect of storm surges (Section 2.4.2), which cause local temporary increases in tidal range and increased current speeds, reinforcing the erosive action of large storm waves.

Burrowing animals can also play an important role in reworking sea-bed sediments, a process known as *bioturbation*. In parts of the North Sea, for example, communities of the mud shrimp *Callianassa subterranea* form numerous sediment mounds that significantly modify the surface of the sea-bed. The mounds have been described as 'volcano-like', and average about 5 cm in height and about 10 cm across at the base. Mud shrimps are estimated to turn over (i.e. rework) of the order of 10 kg of sediment per square metre per year, and a substantial proportion of this sediment is fine material put into suspension, to be redistributed by currents.

QUESTION 8.2 With reference to Section 4.2.3, can you explain why a current of given speed can winnow fine-grained material from poorly sorted gravels and then allow it to be deposited on a smooth silt-covered bed some distance away, without any change of near-bed current velocity?

8.2 SEDIMENT TRANSPORT AND DEPOSITION IN SOME SHELF SEAS

Figure 8.3(a) is a generalized map of the average current field in the North Sea, compiled from measurements and observations made over the last hundred years or so. It represents both residual tidal currents and wind-driven surface and near-surface currents, i.e. it represents the overall residual current field. This long-term average pattern conceals considerable short-term variability, for flow directions can be temporarily changed by strong winds, especially during storms, which will also affect currents associated with density gradients such as those along fronts and the boundaries of ROFIs.

The North Sea receives very little sediment – of the order of $5–10 \times 10^6\,\mathrm{t\,yr^{-1}}$, the bulk of it supplied from erosion of the cliffs of eastern England rather than by rivers (cf. Section 5.3). As Figure 8.3 suggests, the suspended load mostly ends up in the eastern North Sea, especially on the tidal flats and in the estuaries of the Netherlands and northern Germany (cf. Figures 6.17 and 6.20). During spring and summer, when less sediment is resuspended by waves, and the upper part of the water column is stratified, blooms of phytoplankton can locally contribute a significant proportion of the material in suspension in shelf seas (Figure 8.3(b)).

Most of the sand supplied from coastal erosion remains fairly near to coastlines, and is subjected mainly to bedload transport (cf. Figures 4.13 and 4.14). Figure 8.4 (overleaf) is a compilation of *bedload transport paths* around the British Isles. The net sediment transport paths have been deduced from tidal data and simulations, supplemented by other lines of evidence, such as the migration and asymmetry of bed forms (Section 4.4) and systematic changes in sediment grain size on the bed. Bedload transport is determined almost exclusively by tidal currents, whereas the currents that carry suspended sediment have a significant wind-driven component. Hence, Figure 8.4 suggests that off the bulge of East Anglia in eastern England, transport in the bedload is generally north-westward, while Figure 8.3 shows that transport in suspension is generally south-eastward to eastward.

Even where wind-driven currents are weak, net flow directions at the surface and at the bed may be different. For example, the velocity profiles in Figure 2.21 indicate no obvious net movement in either direction at the surface, but near-bed flow velocities are on the negative side of the zero line for five of the seven hours represented, on the positive side for only two. In other words, residual tidal flows can be in opposite directions near the surface and at the bed; and as illustrated in Figure 4.13, a difference of only a few centimetres per second in the speed of ebb and flood tidal currents can be enough to determine the direction of long-term net sediment transport at the bed.

QUESTION 8.3 From your understanding of Chapter 4, explain why small differences of current speed in opposite directions can result in significant amounts of bedload sediment being moved in one direction.

The sediment transport paths in Figure 8.3 originate at zones of divergence known as *bedload partings*. These tend to be located in relatively narrow straits where tidal currents are particularly strong (Section 2.4.1). There is a notable exception in the Straits of Dover, where sediment transport paths meet at a bedload convergence. The transport paths are mostly along the coast, in the direction of the tidal streams (across the co-tidal lines, Figure 2.14). Transport also occurs into estuaries and embayments, e.g. the Solway Firth, Liverpool

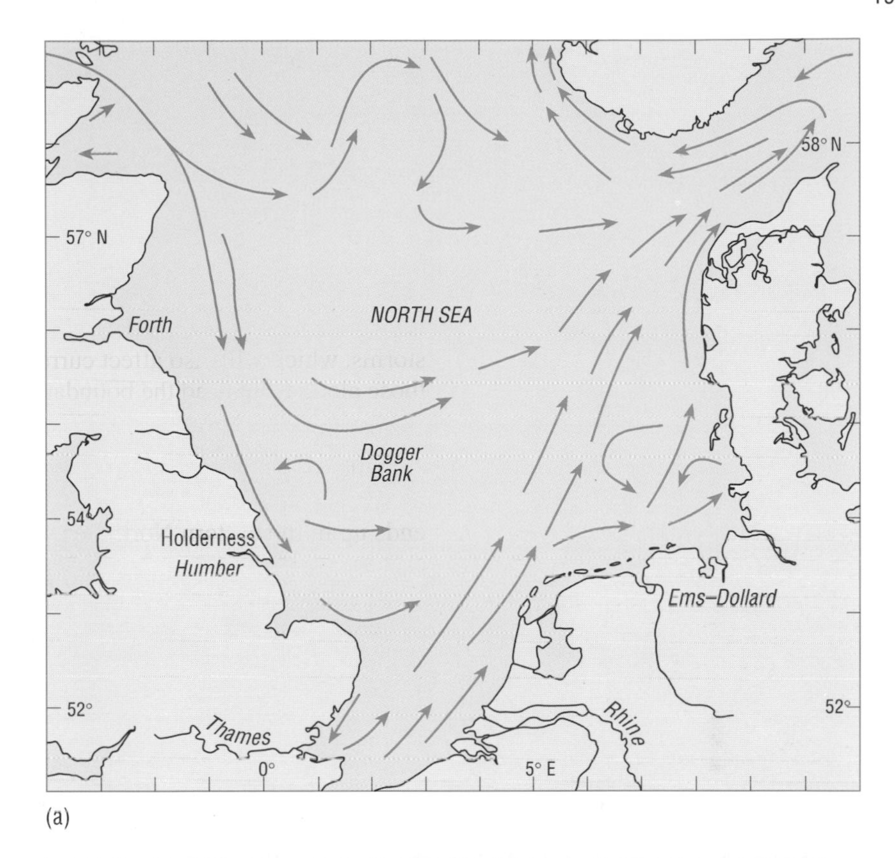

(a)

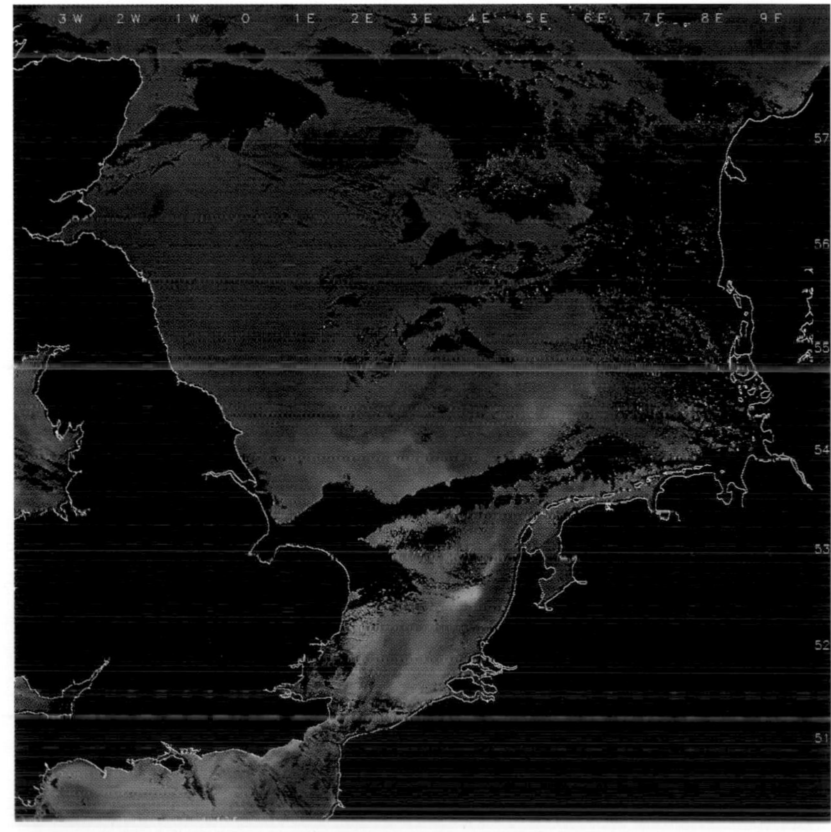

(b)

Figure 8.3 (a) Map of the long-term average current field in the North Sea, effectively showing the average paths followed by fine sediment carried in suspension (for bedload transport, see Figure 8.4).
(b) Ocean colour image from the *SeaWIFS* satellite showing suspended sediment (and plankton) in the southern North Sea in October 1997. Concentrations are high along the UK coast and in the Thames estuary, in the Wadden Sea and the Ems–Dollard estuary; to some extent, the patterns in such images reflect the transport paths for suspended sediment in the North Sea (cf. (a)). (The image is a composite of individual wavebands to produce a near true-colour picture.)

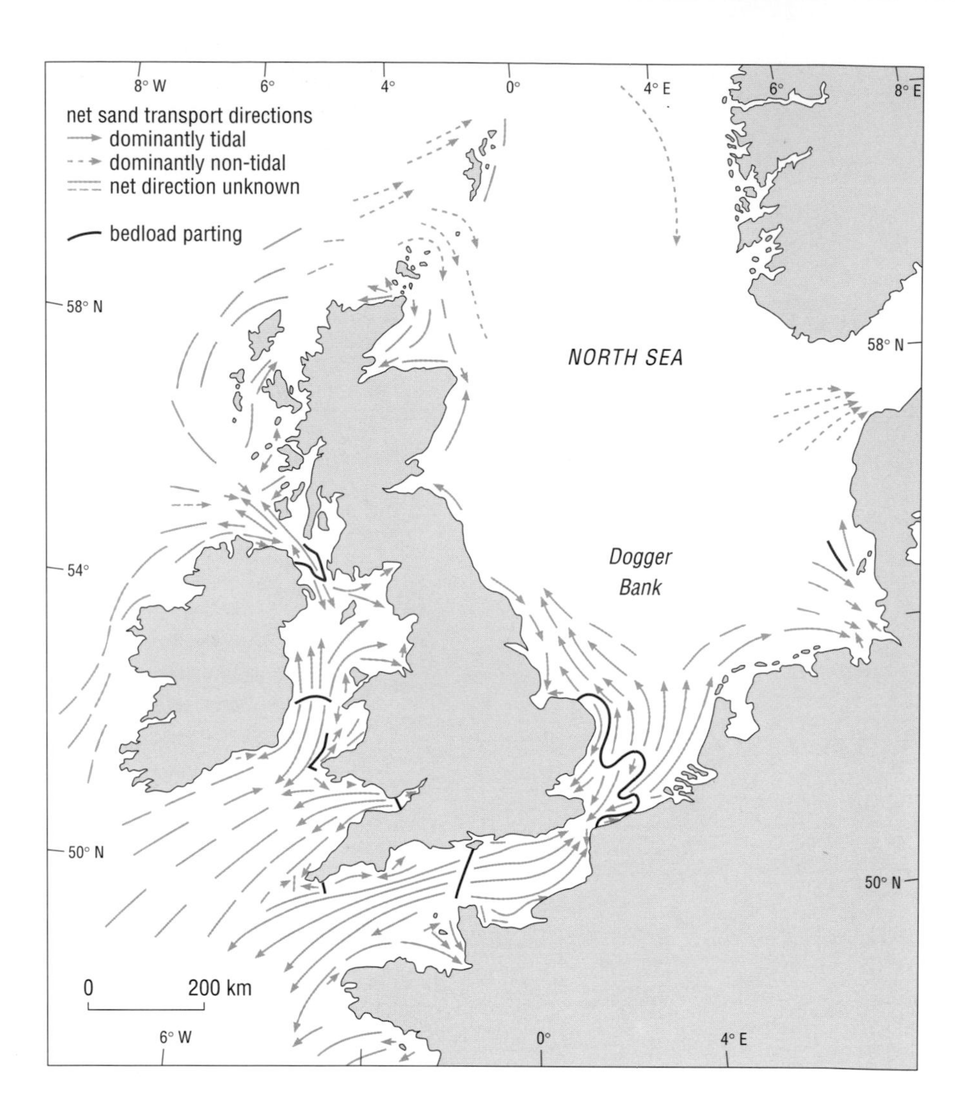

Figure 8.4 Directions of net sand (bedload) transport paths around the British Isles.

Bay, the Humber (Section 6.2.5), and the Wash. Rates of sediment transport along these paths vary with time, changing with the spring–neap cycle, which itself varies on time-scales of decades to centuries (cf. Section 2.3.1).

Most of the floor of the North Sea is covered by relict sediments, consisting mainly of sand and gravel deposited by rivers from melting glaciers, during past periods of low sea-level (Section 3.1.2). In fact, the area of the southern North Sea was land about 10 000 years ago, and the Dogger Bank (Figure 8.4) was an island only about 8000 years ago. The North Sea is relatively shallow south of the Dogger Bank (*c*. 50 m or less), and tidal currents are strong enough to rework the sands and form sand-waves and sand-banks (Section 4.4). In the northern North Sea (north of the Dogger Bank), water depths are of the order of 100 m, the surface speeds of tidal currents are generally less than 2 knots (*c*. 1 m s^{-1}), and there is unlikely to be significant redistribution of sediments on the bed, except during storms. In general, sands and gravels occur in the shallower parts and mud in the deeper parts, winnowed from topographic highs by storm waves and currents.

The Yellow Sea (between China and Korea), which is similar in size to the North Sea, is supplied with vast amounts of sediment (over 2000×10^6 t yr^{-1}) by the Yangtze and Hwang Ho (Yellow) Rivers, two of the world's largest (cf. Figure 3.4). The distribution of sediments in the Yellow Sea is a function both of this enormous input and of the anti-clockwise residual circulation in the basin. The eastern (Korean) side is dominated by sands, because currents are stronger there (and form sand-banks in places); whereas weaker currents on the western (Chinese) side permit the accumulation of muds and sandy muds (less than 62.5 μm, Table 4.1).

Figure 8.5 shows the sediment distribution patterns on the continental shelf in the north-western part of the Gulf of Mexico (west of the Mississippi delta, Figure 7.1(c)). Most of the sediment in this region is supplied by the Mississippi, 98 per cent being silt and clay (Section 7.2.1); so most of the modern deposits are muds, except for a narrow strip of sands along the shore. Further west, where sediment supplies are generally small, and rivers end in estuaries, most of the shelf is covered with relict sands and muds and there is negligible sedimentation, except for an area north of the Rio Grande, where muds are deposited. The reason for this distribution is that around the Mississippi delta and north of the Rio Grande, waves cannot resuspend all the sediment that is supplied, so mud accumulates. Elsewhere on the shelf, however, wave action is more vigorous and the supply of sediment is less, so rates of deposition are low. The prevailing westward current flows from the Mississippi delta are also responsible for the spread of oxygen-deficient waters into the northern part of the Gulf of Mexico (described in Section 7.3).

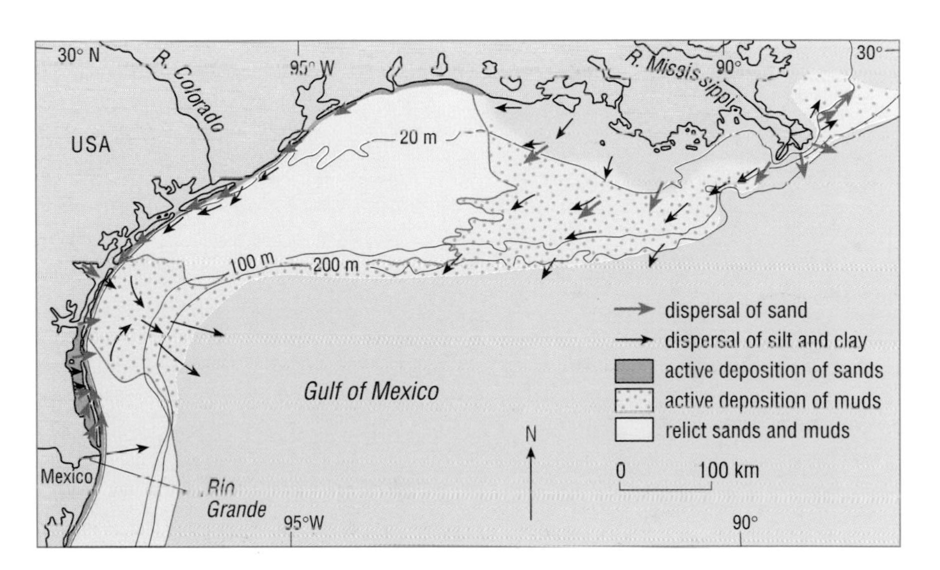

Figure 8.5 Sediment transport paths and areas of relict sediments and modern sediment deposition on the continental shelf in the north-western Gulf of Mexico. Bathymetric contours in metres.

Sea-bed resources in shelf seas

Three main groups of resources can be identified among the mobile or unconsolidated sediments of shelf seas (Figure 8.6, overleaf):

1 *Aggregates* comprise mainly residual deposits of sands and gravels (principally reworked relict sediments of glacial origin), used by the construction industry. They can normally be recovered economically from water depths of less than about 50 m, but many potential sites cannot be dredged. This is partly because dredging can upset the dynamic equilibrium of the coastal zone (Section 5.3), partly because it can damage fisheries, and partly because of the presence of underwater cables and oil and gas pipelines. Shell deposits and coral rubble provide a source of calcium carbonate for cement manufacture and agricultural purposes, particularly in those parts of the world where there is a deficiency of terrestrial limestone.

2 *Placer deposits* consist of concentrations of dense, resistant and economically valuable minerals that have been eroded from terrestrial rocks, transported by rivers to the sea and concentrated by wave and/or current reworking, which removes finer and less dense particles. Many placers were originally deposited as river sediments or were concentrated in beach sands when sea-level was much lower than it is now. They are exploited by offshore dredging (e.g. deposits of the tin mineral, cassiterite, have been extracted off Thailand and Malaysia for over a hundred years). In some parts of the world, modern placer deposits occur on present-day beaches – e.g. the titanium minerals ilmenite and rutile are extracted from beach sands in Australia and California.

3 *Phosphorites* are widespread on continental shelves close to the shelf break, and they contain economic concentrations of the calcium phosphate minerals used in fertilizers. Most deposits appear to be relict sediments originally formed in regions of upwelling (Section 8.1.1). Enhanced biological production resulted in accumulation of phosphorus-rich material which was then slowly converted to phosphate by chemical reactions within the sediments.

Although not strictly a sea-bed resource, springs of *freshwater* issue from the sea-bed in some shelf seas. There is some evidence that in former times these springs were used by crews of sailing vessels to replenish their water supplies. It has been suggested that the springs may also provide a chemical stimulus in the breeding cycle of demersal fishes (e.g. plaice, Section 2.4.1) in the North Sea. The springs can only occur where there is geological continuity of water-bearing rock strata (aquifers) between land and sea, as illustrated in Figure 8.7. Rainwater falling on land percolates into the rocks and is constrained by the geological structure to accumulate as groundwater in the basin-shaped aquifer, formed of permeable sediments, being trapped there by impermeable strata above and below. If the aquifer is 'filled' nearly to the land surface, groundwater in the aquifer will be under sufficient pressure ('artesian pressure') for the water to be forced up to the sea-bed, to emerge as springs through cracks or fissures (faults) in the overlying impermeable strata.

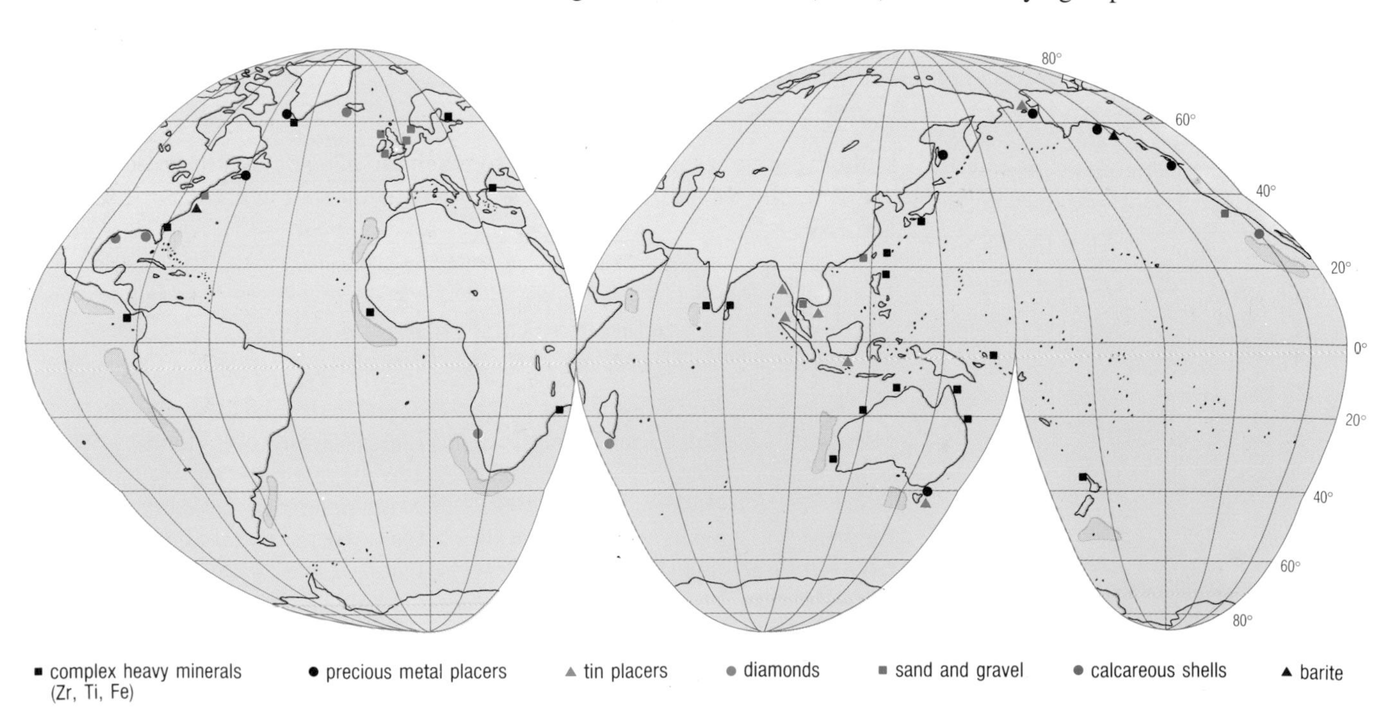

Figure 8.6 Approximate locations of identified deposits of aggregates, placers and phosphorites.

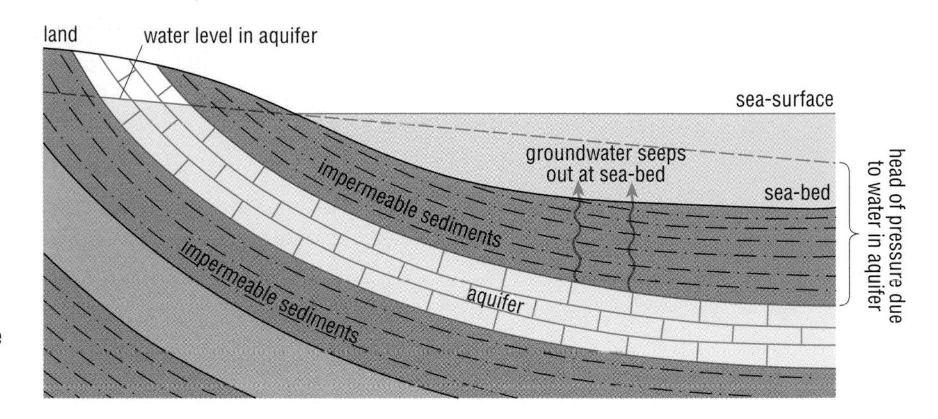

Figure 8.7 Diagrammatic cross-section illustrating how groundwater can seep out at the sea-bed as freshwater springs. For explanation see the text.

The most important commercial resources of shelf seas are of course fossil fuels, especially *oil and gas*, but these occur at depth within the underlying sediments, not at the sea-bed, and we do not consider them further here.

8.3 THE SHELF SEA SYSTEM FROM SHORE TO SHELF BREAK

We have considered beaches, estuaries, tidal flats, lagoons and deltas as separate environments in preceding Chapters, but such divisions are artificial. For example, whether a beach develops rather than a tidal flat is a question of whether waves or tidal currents have the greater effect on coastal sediment transport. Coastlines influenced more or less equally by waves and tidal currents develop features characteristic of both beaches and tidal flats. The transition from beach (or tidal flat) to shelf sea is somewhat arbitrarily placed at the depth of water below which sediment is not disturbed by waves during normal fair weather conditions. Estuaries are river mouths into which the sea penetrates, but where sediment discharge is large, the sediments accumulate seawards of the river mouth to form a delta, the shape of which largely depends upon the interplay between waves and tidal currents.

In this Volume, we have been able to touch on only some of the processes that determine water movements and water column structure in shelf seas, and shallow marine environments generally, and hence influence sediment transport and deposition. A number of these processes are summarized in Figure 8.8, which is intended to help consolidate several of the concepts, principles and processes introduced in previous Chapters. Moving seawards from the coastline, we can identify the following features, labelled 1 to 5 on Figure 8.8:

1 Regions of freshwater influence (ROFIs) occur chiefly off the mouths of estuaries and deltas. In stratified ROFIs, there may be a tidal 'phase difference' between the top and bottom of the water column. Changes in tidal current speed and direction at the surface may precede or lag behind those near the bed, in some cases by as much as an hour or so (i.e. *c*. 30–40° of the tidal current ellipse, Figure 2.19).

QUESTION 8.4 What conditions are likely (a) to promote, and (b) to break down, the stratification in ROFIs?

2 The water column over shallower parts of the shelf is generally well mixed by waves and tidal currents (where there is no ROFI, this influence extends to the shore). The effects of the quarter-diurnal M_4 tidal constituent (Section 2.4.1) are greatest here, and result in strong tidal stirring and sediment resuspension, with consequent high turbidity.

3 A *tidal mixing front* (a seasonal front between tidally mixed waters and stratified waters) is shown at the mid-shelf in Figure 8.8 – an obvious example of such a front is the Flamborough Head front (Figure 2.22(b)). Another important tidal mixing front in British waters lies between southern Ireland and south-west England, in the southern Irish (Celtic) Sea. The position of such fronts is determined partly by water depth and partly by the tidal current regime.

An empirical relationship that can be used to predict whether or not a front is likely to develop at a particular location is:

$$P = H/u_S{}^3 \quad (8.1)$$

where H is mean *water depth* (i.e. not wave height), and u_S is the mean maximum surface tidal current speed (i.e. mean of maximum current speeds on both floods and ebbs, ignoring direction).

The quantity P in Equation 8.1 can for our purposes be considered to be a measure of the probability that turbulent mixing caused by tidal currents interacting with the sea-bed will extend to the surface (cf. Section 2.4.1). The critical value of P for fronts (or more properly frontal zones) to develop in shelf waters is 70–80 (with units of $m^{-2}\,s^3$). Smaller values than this indicate a well-mixed water column, while larger values indicate stratified conditions. To take a simple example: for a mean surface tidal current speed of $1\,m\,s^{-1}$, we would expect a water column 50 m deep to be well mixed ($P = 50$), whereas a water column 100 m deep would be well stratified ($P = 100$).

We can also think of P as representing the relative importance of the buoyancy forces that promote stratification (especially seasonal warming of surface waters), and the forces that encourage mixing (especially tidal currents, cf. Question 8.4). More formally, it represents the rate of decay/dissipation of tidal energy per unit mass of water (an important parameter in theories of turbulence).

Equation 8.1 can have only rather general predictive value, not least because stratification varies seasonally, being intensified by summer warming and by rainfall, and destroyed by winter winds and cooling.

With reference to the '~14.5-day cycle' in the position of the tidal mixing front between stratified and tidally mixed waters in Figure 8.8, would you expect the front to move further offshore during spring or during neap tides?

The front will move further offshore during spring tides, when tidal currents are stronger and turbulent mixing can extend to the surface through a greater depth of water, i.e. in Equation 8.1 the value of P for a given depth will decrease.

The ratio of water depth to the cube of mean current speed (Equation 8.1) must increase sharply at the shelf break. In consequence, where the continental shelf is narrow, the tidal mixing front is likely to be a *shelf-break front*. By contrast, where the continental shelf is wide, fronts may develop both at the shelf break and in the mid-shelf region.

4 Seawards of (3) in Figure 8.8, the water is too deep for tidal stirring to extend to the surface (cf. Section 2.4.1), and in summer the water column becomes stratified. In such circumstances, surface waters become depleted in nutrients as dead organic matter sinks below the **seasonal thermocline**, where the rate of change of temperature with depth is greatest, and which, being a pycnocline (Section 1.1.1), forms a barrier to upward mixing. However, in winter a combination of winds and surface cooling causes downward mixing from the surface to break down the stratification and 'link up' with tidal mixing from below, so that the whole water column is mixed – and the tidal mixing front (3 on Figure 8.8) disappears.

5 Tidally generated internal waves (with periods comparable to those of the tides) are shown in Figure 8.8 above the shelf break, along the thermocline/pycnocline. Reflection or breaking of these internal waves at the shelf break causes mixing of the water column and brings nutrients to the surface. Biological production can therefore be high in this region, especially if a shelf break front has developed and/or if there is upwelling of deeper nutrient-rich water.

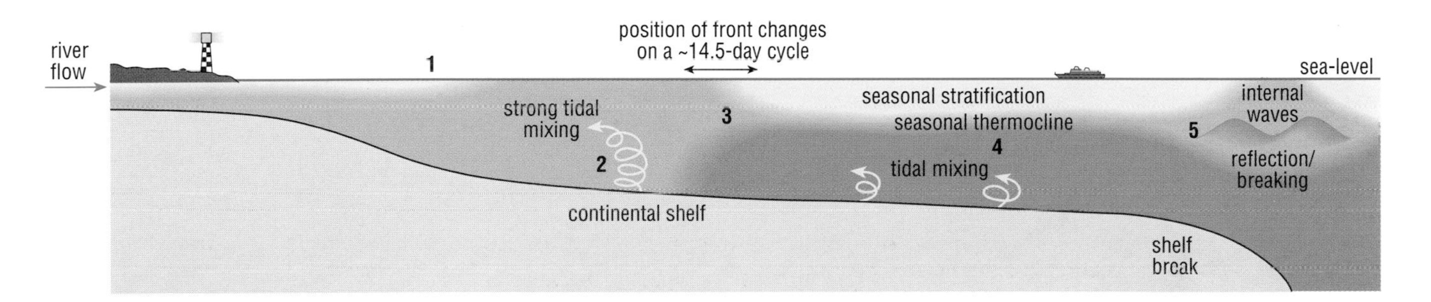

Figure 8.8 Generalized schematic section across a shelf sea in summer, from the mouth of an estuary (or delta) to the shelf break, with greatly exaggerated vertical scale. For numbers see text. Generally, lighter blue = less dense water, darker blue = more dense water.

8.4 SUMMARY OF CHAPTER 8

1 Shelf seas extend from just below low tide level out to the shelf break, and shelves are underlain by considerable thicknesses of sediment. Both modern and relict (glacial) sediments form the sea-bed and are being reworked by waves and currents. The principal factors determining sediment distribution in shelf seas are the residual current field and the rate of supply of sediments from rivers and coastal erosion.

2 Coastal geostrophic currents develop where offshore pressure gradients are set up in response to winds and/or heavy rain. Longshore winds lead to Ekman transport of surface water offshore or onshore, and hence to upwelling or downwelling respectively, influencing both biological production and sediment transport.

3 Sediment transport in shelf seas is determined by the residual current field, which may not be the same at the surface and the bed. In general, the stronger the currents at the sea-bed, the coarser the sediments, and reworking by waves tends to move finer sediments from shallow to deep water. Waves have a large effect during storms, and can aid sediment transport by lifting sediment into suspension to be carried away by currents. Bioturbation by burrowing animals can locally also resuspend significant amounts of sediment. Bedload partings, where sediments are transported in opposite directions, occur at zones of divergence in near-bed currents. Sea-bed resources in shelf seas include aggregates (sands and gravels, and shell deposits), placers (heavy mineral concentrations) and phosphorites, all resulting mainly from reworking or chemical alteration of relict sediments. Freshwater (artesian) springs may occur at the sea-bed if the underlying geological structure is suitable.

4 A rough zonation can be discerned in shelf seas, between shore and shelf break. The 'zones' may not always be developed in all shelf seas and their relative widths and positions can vary with tidal range and season. Regions of freshwater influence (ROFIs) off estuaries or deltas, may be stratified or mixed depending on wind and tidal state. A tidal mixing front separates the shallower and vertically mixed marine water column seaward of the ROFI from the deeper and seasonally stratified waters of the mid-outer shelf, and its position varies according to water depth, tidal range and season. Near the shelf break, there may be localized disruption of the seasonal stratification by tidally generated internal waves.

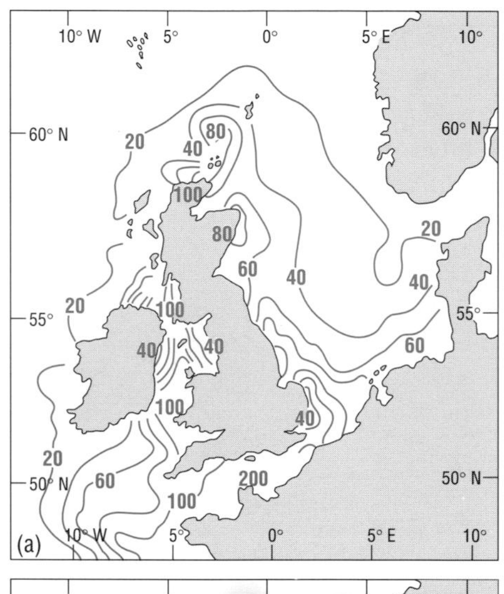

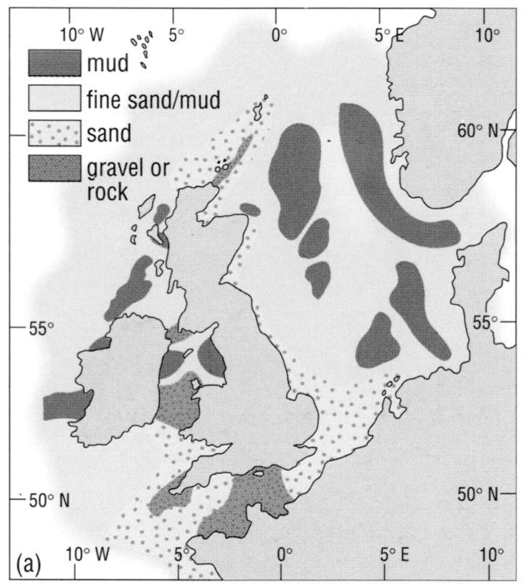

Figure 8.9 (a) Mean near-surface tidal current speeds around the British Isles (speeds in cm s^{-1}). (b) Distribution of shelf sediments around the British Isles.

Now try the following questions to consolidate your understanding of this Chapter.

QUESTION 8.5

(a) Would you be more likely to find muds or sands forming the bed to seaward of the tidal mixing front (3 on Figure 8.8)?

(b) What would you expect to happen to the tidal mixing front in a storm?

QUESTION 8.6 Figure 8.9(a, b) shows simplified maps of mean near-surface tidal current speeds and sediment types in the shelf seas around Britain. Comment on the relationship between average tidal current speed and the nature of the sea-bed.

QUESTION 8.7 In the upper right of Figure 8.5, some of the (brown) arrows indicating dispersal of sand are shown in a large area of mud deposition. From your understanding of the sediment load of the Mississippi, suggest an explanation of why this is not labelled as an area of sand deposition.

QUESTION 8.8 To what extent is it justifiable to consider marine placer deposits to be both residual and relict sediments?

SUGGESTED FURTHER READING

ALLEN, J. R. L. (1985) *Principles of Physical Sedimentology*, Allen and Unwin. A moderately mathematical introduction to the physics of fluid flow and sediment movement, illustrated with simple laboratory experiments.

CARTWRIGHT, D.E. (1999) *Tides: A Scientific History*, Cambridge University Press. A fascinating history of the development of ideas about tides. Some chapters contain quite complex mathematics, but much of the book is purely descriptive.

DYER, K. R. (1998) *Coastal and Estuarine Sediment Dynamics* (2nd edn), John Wiley and Sons. An advanced textbook which integrates marine sedimentology with experimental results and theory from oceanography. The approach is quantitative and moderately mathematical, and the problems of applying theories to the real marine environment are brought out.

GROEN, P. (1976) *The Waters of the Sea*, Van Nostrand. Useful background material including a good general coverage of waves and tides.

KINSMAN, B. (1965) *Wind Waves: Their Generation and Propagation on the Ocean Surface*, Prentice-Hall. A more detailed survey of the dynamics of wind waves, with a more mathematical treatment than that given in this Volume.

LEEDER, M. R. (1982) *Sedimentology: Process and Product*, Allen and Unwin. Provides a more geological approach to physical sedimentology and covers a wide range of topics, including the origins of sediment grains, continental and deep-sea environments and sediment diagenesis, in addition to the physics of sediment movement.

MINISTRY OF AGRICULTURE, FISHERIES AND FOOD (MAFF) (1981) *Atlas of the Seas around the British Isles*, HMSO. Complete set of maps and descriptions of waves, tides, tidal currents, sediment distribution and transportation, and bed forms, as well as temperature and salinity structure and a host of other topics including fisheries, pollution, marine transport and communications.

PETHICK, J. (1984) *An Introduction to Coastal Geomorphology*, Edward Arnold. Do not be misled by the use of 'geomorphology' in this title. This is a simple introduction to waves, tides, coastal sediments, the littoral zone and estuaries which is easily accessible to the mathematically less confident reader.

PUGH, D. T. (1987) *Tides, Surges and Mean Sea-Level*, Wiley. A comprehensive explanation of tides, surges and sea-level changes and their effect upon shallow water and coastal environments. The book is written for marine and coastal engineers, hydrographers, sedimentologists and similar professions, and has a strong practical bias.

PYE, K. (ed.) (1994) *Sediment Transport and Depositional Processes*, Blackwell Scientific. A thorough summary of sediment transport and depositional processes in a variety of environments. Some chapters are at a fairly advanced level.

ANSWERS AND COMMENTS TO QUESTIONS

CHAPTER 1

Question 1.1 Five seconds. The frequency is $0.2\ s^{-1}$, i.e during one second '0.2 of a wave' passes a fixed point. To find out how long it takes for the whole wave to pass (the period), we need to divide 1 by $0.2\ s^{-1}$:

$T = 1/0.2\ s^{-1} = 5\ s.$

Question 1.2 The less steep of the two. Since steepness = H/L, and H is the same for both waves, the less steep wave will have the greater wavelength, and hence travel faster.

Question 1.3 It decreases. Figure 1.4 shows that the higher the average frequency of the wave field, the smaller the area under the curve. The $\sim 20\ m\ s^{-1}$ (40-knot) spectrum contains much more energy than either of the other two spectra. Most of this energy is related to the low frequency (long period) waves that a 40-knot wind would generate.

Question 1.4 Sixteen waves in 64 seconds = a period of 64/16 seconds = 4 s. The frequency is thus the reciprocal of 4 s, i.e. 1/4 s = $0.25\ s^{-1}$.

Question 1.5 First, convert frequency f to period T. $T = 1/f = 1/0.05 = 20$ s.

(a) $-a$ (a trough at P), because 30 s = $1\frac{1}{2} \times 20$ s.

(b) $+a$ (a peak at P), because 80 s = 4×20 s.

(c) 0, because 85 s = $4\frac{1}{4} \times 20$ s (η changes from $+a$ to $-a$ in 10 seconds, so five seconds after a peak ($+a$), the displacement is zero).

For (d), (e) and (f): the distance between P and Q is half a wavelength. Note that if the displacement at P is zero and is diminishing, then the displacement at Q is zero and is increasing (and *vice versa*). Hence:

(d) 0

(e) $+a$

(f) 0

Question 1.6 If $k = 2\pi/L$, and $\sigma = 2\pi/T$, then $L = 2\pi/k$, and $T = 2\pi/\sigma$. Substituting into $c = L/T$, we have:

$$c = \frac{2\pi / k}{2\pi / \sigma} = \frac{1/k}{1/\sigma} = \frac{\sigma}{k}$$

In basic units, angular frequency/wave number is $s^{-1}/m^{-1} = m\ s^{-1}$, i.e. speed.

Question 1.7 (a) If d is greater than $0.5L$, then $2d$ is greater than L, and the expression $2\pi d/L$, becomes greater than π. The tanh of numbers greater than π approximates to 1. So $\tanh(2\pi d/L) \approx 1$ and Equation 1.2 approximates to:

$$c = \sqrt{\frac{gL}{2\pi}}$$

(b) If d/L is very small, then $2\pi d/L$ is also very small, and hence tanh $(2\pi d/L)$ approximates to $2\pi d/L$. So Equation 1.2 becomes:

$$c = \sqrt{\frac{gL2\pi d}{2\pi L}}$$

$$= \sqrt{gd}$$

Question 1.8 From Equation 1.1, $c = L/T$.

From Equation 1.3, $c = \sqrt{gL/2\pi}$

So $\sqrt{gL/2\pi} = L/T$, and (squaring both sides)
$gL/2\pi = L^2/T^2$.

From which $L/T^2 = g/2\pi$ and
$L = gT^2/2\pi$.

Question 1.9 (a) $g = 9.8\ \text{m s}^{-2}$, and $\pi = 3.14$, so $g/2\pi = 1.56\ \text{m s}^{-2}$.

(b) In Equation 1.7: $c = \sqrt{1.56\ \text{m s}^{-2} \times \text{m}}$ which gives units of $\sqrt{\text{m}^2\ \text{s}^{-2}} = \text{m s}^{-1}$.

In Equation 1.8: $L = 1.56\ \text{m s}^{-2} \times \text{s}^2$, which gives units of m (s^{-2} and s^2 cancel out).

In Equation 1.9: $c = 1.56\ \text{m s}^{-2} \times \text{s}$, which gives units of m s^{-1} ($\text{s}^{-2} \times \text{s} = \text{s}^{-1}$).

Question 1.10 (a) $31.2\ \text{m s}^{-2}$. You may have done this the hard way by $L = 1.56 \times 20 \times 20 = 624$, followed by $c = 624/20 = 31.2\ \text{m s}^{-2}$. Better still, you may have used Equation 1.9, and done the sum in one step.

(b) The deep water speed of the wave will be $22.1\ \text{m s}^{-1}$. From Equation 1.7,

$$c = \sqrt{1.56 \times 312} = \sqrt{486.7}$$
$$= 22.1\ \text{m s}^{-1}.$$

(c) The answer is $10.8\ \text{m s}^{-1}$ in *both* cases. If the depth is less than 1/20 of the wavelength, all waves will travel at the same depth-determined speed, i.e. the depth is the only controlling factor. So from Equation 1.4, we get:

$$c = \sqrt{gd}$$

$$= \sqrt{9.8 \times 12} = \sqrt{117.6}$$

$$= 10.8\ \text{m s}^{-1}$$

Question 1.11 No. It would be quadrupled, because the energy of a wave varies with the square of the wave height (Equation 1.11), and hence with the square of the wave amplitude.

Question 1.12 (a) If amplitude is 1.3 m, then wave height is 2.6 m. The values of the constants g and ρ, and also the above value for wave height, can be plugged into Equation 1.11, giving:

$$E = \tfrac{1}{8} \times 1.03 \times 10^3 \times 9.8 \times 2.6^2$$

$$= 8.5 \times 10^3\ \text{J m}^{-2}$$

(The units work out as: $\text{kg m}^{-3} \times \text{m s}^{-1} \times \text{m}^2 = \text{kg s}^{-2}$;
$\text{J} = \text{kg m}^2\ \text{s}^{-2}$, so J m^{-2} is $\text{kg m}^2\ \text{s}^{-2}\ \text{m}^{-2} = \text{kg s}^{-2}$.)

(b) The wave power per unit length is the product of the wave energy per unit area and the group speed. We know the wave energy from (a) above, and can calculate the group speed from the height and steepness as follows:

steepness (0.04) = height (2.6 m) / wavelength.

So wavelength = 2.6/0.04 = 65 m.

From Equation 1.3, wave speed, $c = \sqrt{gl / 2\pi}$

So $c = \sqrt{1.56L}$ (Eqn 1.7)

$= \sqrt{1.56 \times 65}$

$\approx 10\ \text{m s}^{-1}$

From which group speed, $c_g = 10/2 = 5\ \text{m s}^{-1}$.

So wave power = $8.5 \times 10^3\ \text{J m}^{-2} \times 5\ \text{m s}^{-1} = 42.5\ \text{kW m}^{-1}$.

Question 1.13 Spectrum (a) of Figure 1.12 shows the wave energy distributed amongst a wide range of frequencies. The peak is rather poorly defined, and hence must represent the storm-generating area. Spectrum (b), on the other hand, has a much narrower range of frequencies, and a clearly defined peak which has shifted to lower frequencies. It thus represents the regular swell waves at a point well away from the storm. The total energy of spectrum (b), as represented by the coloured area under the curve, is smaller than in (a), because the waves have lost some of their energy in transit, as outlined in the text.

Question 1.14 The wave refraction diagram (Figure 1.17(b)) illustrates how the offshore Hudson Canyon is effective in defocusing storm waves as they approach the Long Branch coastal section from the east–south-east, and in refracting them onto other beaches. Fishermen can leave their boats on the beach at Long Branch during all seasons of the year, despite its apparent exposure to the full force of Atlantic storms. The wave rays are, if anything, focused as they enter the mouth of the Hudson River, so that the energy of storm waves would certainly not be diminished there, and might even be increased. People leaving their boats in this apparently sheltered region could thus be courting disaster.

Question 1.15 If the change in the beach slope was sufficient, you might expect to see collapsing breakers, and if it got really steep, surging breakers as well.

Question 1.16 You have no information on the length of the ship, but if you calculate the wavelength corresponding to a period of 30 s, using Equation 1.8, you get $L = 1.56T^2 = 1.56 \times 900 = 1404$ m. You might conclude that the sailor is trying to tell you that his ship was about 700 m long (nearly half a mile). The longest of today's supertankers are only about 320 m. However, return to the main text and read on …

Question 1.17 *Exeter* was 175 m in length, so if the seaman's story were true, the wavelengths concerned were 350 m. The ship was travelling at $11.8\ \text{m s}^{-1}$, so in 30 seconds it would have travelled 30×11.8 m = 354 m. In 30 seconds, an overtaking wave would have travelled one wavelength plus the distance the ship had travelled, i.e. 354 + 350 = 704 m, and the wave speed would be $704/30 = 23.5\ \text{m s}^{-1}$.

From Equations 1.3 and 1.7, we can find the wave speed corresponding to a wavelength of 350 m, i.e.

$$c = \sqrt{gL/2\pi} = \sqrt{1.56L}$$

$$= \sqrt{1.56 \times 350} = 23.4\ \text{m s}^{-1}$$

which means the sailor's tale is at least consistent with simple wave theory. Full marks if you suspected something of this sort while attempting Question 1.16.

Question 1.18 Because the wavelength is very long compared with an ocean depth of 5500 m over the abyssal plains, the tsunami must be treated as a shallow-water wave (Equation 1.4):

$$c = \sqrt{9.8 \times 5500} = 232\ \text{m s}^{-1},$$

which is more than 800 kilometres per hour!

Question 1.19 In Equation 1.18, $l = 90$ m, $d = 10$ m.
So $T = 4 \times 90 / \sqrt{9.8 \times 10} = 36.36$ seconds. Because the resonant period of the harbour is close to 36 s, waves of period 18 s (half of 36 s) would set up a standing wave in the harbour.

Question 1.20 Table 1.1 gives the relationships between wind force and significant wave heights.

(a) (i) ± 8 cm, because $H_{1/3}$ is about 0.15 m, which is below 8 m.

(ii)± 0.3 m, because $H_{1/3}$ is below 3 m.

(b) (i) ± 8 cm, because $H_{1/3}$ is about 1.5 m, which is below 8 m.

(ii)± 0.3 m, because $H_{1/3}$ is below 3 m.

(c) (i) ± 12 cm (1 cm for each metre of $H_{1/3}$, which is about 12 m).

(ii)± 1.2 m (10% of $H_{1/3}$).

Note that the extent of the cloud cover is irrelevant, because clouds are transparent to the radar pulses used here, and will not affect the signals.

Question 1.21 (a) From Equation 1.9, $c = 1.56T$.
So wave speed $c = 1.56 \times 10 = 15.6\ \text{m s}^{-1}$,
and group speed $c_g = c/2 = 7.8\ \text{m s}^{-1}$.

(b) From Equation 1.8, wavelength $L = 1.56 \times T^2 = 156$ m.
So steepness, $H/L = 1/156 = 0.0064$.

(c) Wave energy is $\frac{1}{8}(\rho g H^2)$ (Eqn 1.11)

$= \frac{1}{8} \times 1.03 \times 10^3 \times 9.8 \times 1$

$= 1.26 \times 10^3\ \text{J m}^{-2}$.

From Section 1.4.1, wave power is the product of the wave energy per unit area and the group speed, so:

wave power $= 1.26 \times 10^3\ \text{J m}^{-2} \times 7.8\ \text{m s}^{-1} = 9.8 \times 10^3\ \text{J m}^{-1}\,\text{s}^{-1}$

$= 9.8\ \text{kW m}^{-1}$.

(d) The wave power per metre of wave crest will be the same as the answer in part (c) above, because wave power remains constant (Section 1.5). You may have found this out the hard way by calculating the wave height and wave energy in water 2.5 m deep (Equations 1.12 and 1.13, and wave power $= Ec_g$, Section 1.4.1).

Question 1.22 (a) Because swell waves and short waves have little interaction (Section 1.4.3), they can be treated separately. From Equation 1.7, the short waves will have a speed of:

$$\sqrt{1.56L} = \sqrt{1.56 \times 6} = 3.06 \text{ m s}^{-1}.$$

Hence, group speed is 1.53 m s^{-1}, and these waves will not propagate against a current of 3 knots (1.54 m s^{-1}), because this current exceeds half the group speed of the waves. In the narrow inlet, these smaller waves will increase in height until they become unstable and break. Because $c = 1.56T$ (Eqn 1.9), the swell waves will travel at a speed of 34.3 m s^{-1}. Therefore $c_g = 17.15$ m s^{-1}, and because the counter-current is less than half the group speed, these waves will propagate against the current, but will show an increase in height and a reduction of wavelength.

The wavelength of the swell waves can be found from the period, using Equation 1.8 ($L = 1.56T^2$), i.e. $L_0 = 1.56 \times 22^2 = 755$ m. If we now apply Equation 1.16, the wavelength in the narrow inlet can be found. The period of 22 s $= L_{current}/(34.3 - 1.5)$. So $L_{current} = 22 \times 32.8 = 722$ m.

(b) Beyond the inlet, only the swell waves will be encountered, although the swell waves emerging from the inlet will not have the same set of characteristics as they had before entry. They will have lost height and steepness in passing through the inlet, but now they are no longer propagating against a current. The period of 22 s and the 'new' wavelength of 722 m are now not consistent with the relationship between T and L summarized in Equation 1.8 ($L = 1.56T^2$). Once beyond the influence of the current, the waves will tend to re-establish those relationships between c, L and T which are typical of sinusoidal deep-water waves,

Question 1.23 (a) Wave speed = apparent speed of overtaking waves, plus the ship's speed.

Apparent speed of waves = 146 m/6.3 s = 23.17 m s^{-1} (= 45 knots).

Added to the ship's speed of 5.14 m s^{-1} (10 knots), this gives an actual wave speed of 28.31 m s^{-1} (55 knots).

Note: The speed of the waves calculated in Question 1.23(a) of 28.31 m s^{-1} (55 knots) is exactly the same wave speed which was reported by the *Ramapo*, and there is no reason to doubt it. The discrepancy between the period consistent with such a wave speed, and that reported, may reflect the fact that the observed waves were not simple sinusoidal waves. As waves get steeper, so their shape becomes trochoidal (Figure 1.7) rather than sinusoidal, their wavelengths and periods become shorter, and the simplifying assumptions made in this Chapter no longer apply.

(b) From Equation 1.7, $c = \sqrt{1.56L}$. So:

$L = c^2/1.56$ (i.e. wavelength = (wave speed)^2/1.56)

$= 28.31^2/1.56 = 514$ m.

Wave height = 34 m (given) so steepness = 34/514 = 0.066.

(c) $c = 1.56T$ (Eqn 1.9), so $T = c/1.56 = 0.64c$ and, from part (a), wave speed = 28.31 m s^{-1}. So the period = 0.64 × 28.31 = 18.1 s. This answer is some 22% more than the period of 14.8 s reported by the *Ramapo*.

(d) A wave of period 14.8 s would have a wavelength of $1.56 \times 14.8^2 =$ 342 m. The steepness of such a wave is 34/342 = 0.1. This is a very steep wave indeed.

Question 1.24 (a) After a long calm spell, only low waves of long period would be arriving. Such waves would become surging breakers on a beach of intermediate slope.

(b) Winds of this force would generate steep waves, which on a beach of intermediate slope would become spilling breakers.

CHAPTER 2

Question 2.1 (a) The magnitudes of the gravitational and centrifugal force and hence of the resultant tide-producing force would be the same as at D and J on Figure 2.3. The tide-producing force would be directed into the Earth (into the plane of the page as you look at Figure 2.3).

(b) This is the only point where the gravitational force exerted by the Moon on the Earth is exactly equal to, and acting in exactly the opposite direction from, the centrifugal force. The resultant tide-producing force at that point is zero.

Question 2.2 (a) The waves would be required to travel 40 000 km in 24 hr 50 min, i.e. in 24.83 hours. That is equivalent to 1611 km hr^{-1}, or 448 m s^{-1}.

(b) Using Equation 1.4, if $c = \sqrt{gd}$, then $d = c^2/g$.

So, depth required = $448^2/9.8$ = 20 480 m, i.e the ocean would have to be more than 20 km deep.

Question 2.3 (a) Nil. About seven days (i.e. one-quarter of 27.3 days) after the scenario shown on Figure 2.9, the Moon will be overhead at the Equator (Figure 2.5), which occurs at positions 2 and 4 in Figure 2.8. There will then be no diurnal inequality in the lunar tide anywhere on the globe.

(b) Diurnal inequality will again be at a maximum, and the tidal bulges will be as in Figure 2.9 once more, although the Moon will be on the opposite side of the Earth (corresponding to position 3 on Figure 2.8).

Question 2.4 The tidal contribution from the Sun will be greatest around 3 January, when the Earth is at perihelion (which happens to be quite close to the time of the winter solstice, 21 December).

Question 2.5 (a) 14.75 days (half of 29.5).

(b) Neap tides. Spring tides coincide with syzygy, so 14.75 days after that there will be another spring tide, and 7.4 days more will bring the cycle to neap tide (14.75 + 7.4 = 22.15).

(c) 3–4 days, i.e. half-way between the spring tide associated with the new Moon and the neap tide which will occur 7.4 days after the new Moon.

(d) In this simplest case, Figure 2.12(a) corresponds to a solar eclipse (Moon directly between Sun and Earth) and Figure 2.12(c) to a lunar eclipse (Earth's shadow on Moon).

Question 2.6 (a) (1) *The Wash:* Just after low tide (i.e. nearly 6 hours to go to high tide).

(2) *Firth of Forth:* About $1\frac{1}{2}$ lunar hours to go to high tide (i.e. about $4\frac{1}{2}$ hours after low tide).

If you had difficulty with (a) and (b), note that if high tide is at '0' on Figure 2.14, it has not yet reached the Firth of Forth, and that area is expecting a high tide in two hours' time. Similarly, the high tide will take five-and-a-half hours to reach the Wash.

(b) The tidal range of the Wash (over 6 m) exceeds that of the Firth of Forth (more than 4 m, but less than 5 m).

Question 2.7 (a) The amphidromic systems in the South Atlantic, mid-Pacific and round Madagascar are all in the Southern Hemisphere and rotate anticlockwise rather than clockwise as the theory would suggest; whereas the system in the North Pacific is in the Northern Hemisphere and rotates clockwise, not anticlockwise. It may be worth noting that the mid-Pacific and North Pacific amphidromic points are also quite close to islands (the Tuamoto and Hawaiian groups respectively).

(b) The amphidrome centred in the north-west Indian Ocean would seem to be sufficiently enclosed to fulfil the criteria illustrated in Figure 2.16. Being in the Northern Hemisphere, it should rotate anticlockwise, but it is in fact a clockwise system. The reason for this may be that there are in fact two systems in the northern Indian Ocean, both centred nearly on the Equator where the Coriolis force is zero.

Question 2.8 (a) As you might expect, a high value of F corresponds to a dominant diurnal tidal component (average period of 24 hr 50 min). A low value of F indicates dominance of a semi-diurnal component.

(b) Yes. Syzygy occurs every 14.75 days, and as noted in the text following Question 2.5 it affects tides all over the world simultaneously.

Question 2.9 (a) Using Equation 1.18, with $l = 270$ km, and $d = 60$ m, we get:

resonant period of basin = $4l/\sqrt{gd} = 4 \times 270 \times 10^3/\sqrt{9.8 \times 60} = 44\,538$ s = 12.37 hours.

This is very close to the semi-diurnal period of 12.42 hours, i.e. the resonant period of the basin is the same as that of the semi-diurnal tides. The reason for using Equation 1.18 is that the Bay of Fundy is a basin open at one end. Equation 1.17 is for standing waves in closed basins.

(b) From the calculation in (a) and Figure 1.20(c), the wavelength of the standing wave would be 4×270 km, which works out to 1080 km, i.e. the wavelength is of the order of 10^3 km.

(c) We would expect the tidal range to be small near the entrance to the Bay of Fundy, because Figure 1.20(c) suggests this is where the node of the standing wave should be – and by definition, *vertical* displacements of the water surface there should be minimal, cf. Figure 1.20(d).

Question 2.10 If a 10 m column of water (= 1000 cm) corresponds to one atmosphere of pressure (1000 mbar), then 1 cm of water corresponds to 1 mbar. A reduction in atmospheric pressure of 50 mbar would therefore result in a sea-level *rise* of 50 cm, i.e. 0.5 m.

Question 2.11 The tide-producing force at P (TPF_P) = the Moon's gravitational attraction (F_{gP}) at P minus the centrifugal force at P.

$$F_{gP} = \frac{GM_1M_2}{(R - a\cos\psi)^2} \qquad \text{(Eqn 2.3)}$$

and the centrifugal force at P is the same as at all other points on Earth, and therefore equal to F_g at the Earth's centre, i.e. GM_1M_2/R^2 (from Eqn 2.1). So by analogy with the reasoning leading to Equation 2.2:

$$TPF_P = \frac{GM_1M_2}{(R - a\cos\psi)^2} - \frac{GM_1M_2}{R^2}$$

Do not worry if you could not manage the subsequent algebra, but the expression simplifies to:

$$TPF_P = GM_1M_2\frac{a\cos\psi(2R - a\cos\psi)}{R^2(R - a\cos\psi)^2}$$

This can be further simplified to the approximate relationship:

$$TPF_P \approx \frac{GM_1M_2 2a\cos\psi}{R^3}$$

Question 2.12 None of the statements is true.

(a) False. The term 'syzygy' includes *both* conjunction and opposition. If the Moon is in opposition, it is also in syzygy. However, if the Moon is in syzygy, it is not necessarily in opposition (i.e. it might be in conjunction).

(b) False. Spring tides would occur (see Figure 2.12(a)).

(c) False. Spring tides occur every 14.75 days throughout the year.

(d) False. The lowest sea-levels occur at low *spring* tides. When the Moon is in quadrature, neap tides occur, and these have a smaller range than spring tides.

Question 2.13 The actual observed tides are constrained by the shallow depth of the oceans, and also by inertia, friction and shape of the ocean basins. So there is a time-lag between the application of the tide-generating forces and the oceans' responses. In addition, tidal currents are subject to the Coriolis force.

Question 2.14 (a) There would be a decrease in the tidal range. The difference between perihelion and aphelion in terms of Earth–Sun distance is only about 4% (Section 2.2), and the Sun has slightly less than half the tide-producing influence of the Moon. The decrease in tidal range as the Earth–Sun distance increased would therefore be small (say a centimetre or so).

(b) As noted in Section 2.4, changes in the Moon's declination, which cause equatorial and tropic tides (Section 2.1.1), have less effect at higher latitudes. As Figure 2.18(a) shows, there is little diurnal inequality in the tidal range at Immingham (F-value = 0.1), although close inspection of the tidal curve does reveal very slight diurnal inequalities around days 6–8 and 21–23.

(c) No effect on tidal *range*, but by analogy to the answer to Question 2.10 a depression of the sea-surface of about 30 cm would result, whatever the state of the tide at the time.

CHAPTER 3

Question 3.1 (a) To the east of the Andes, the Amazon and its tributaries flow across the vast drainage basin of the wet equatorial and tropical zones of South America; chemical weathering dominates, river discharge into the Atlantic is high, and the water carries much suspended sediment. West of the Andes, there is little river run-off because the western coast of South America between the Equator and 30° S is arid. Physical weathering dominates and, although much weathered debris is produced in the Andes, it cannot easily be transported by rivers westwards into the Pacific.

(b) Sediment discharge is high in the western Pacific partly because this region is characterized by humid equatorial/tropical conditions; and partly because the rivers flowing off mountainous regions have steep gradients. The result is a combination of intense physical and chemical weathering and high river discharge, leading to the transport of large quantities of sediment to the western Pacific Ocean.

Question 3.2 You should have selected (b), (c) and (e). Relict sediments (a) are ubiquitous on the continental shelves (because sea-level fell world-wide and rivers once flowed to the sea across what are now continental margins), and volcanoes (d) occur at all latitudes. Although recent glacial sediments (b) are restricted to high latitudes, glacial material also occurs in relict sediments at mid-latitudes in the Northern Hemisphere, deposited when ice-sheets extended much further south during the Quaternary. Wind-blown dust (c) mostly originates from subtropical deserts, though it tends to be fairly widely distributed by global wind patterns. Carbonate sediments (e) are most abundant at low latitudes, along coastlines where terrigenous sediments are scarce, though they sometimes occur at higher latitudes, usually in the form of shelly debris.

CHAPTER 4

Question 4.1 It is because this part of the diagram applies to cohesive sediments, and the proportion of cohesive clay minerals is greatest in sediments with the smallest particle sizes.

Question 4.2 (a) In laminar flow, turbulence is suppressed and eddy viscosity is effectively zero, so Equation 4.2 simplifies to:

$$\tau_0 = \mu \times \frac{d\bar{u}}{dz}$$

(b) In turbulent flow, eddy viscosity exceeds molecular viscosity by several orders of magnitude, so molecular viscosity can effectively be neglected, and Equation 4.2 becomes:

$$\tau_0 = \eta \times \frac{d\bar{u}}{dz}$$

Question 4.3 (a) According to Equation 4.5, $\tau_0 = \rho u_*^2$, so

$$\tau_0 = 10^3\ \text{kg m}^{-3} \times (0.083\ \text{m s}^{-1})^2$$

$$= 1000\ \text{kg m}^{-3} \times 0.0069\ \text{m}^2\,\text{s}^{-2}$$

$$= 6.9\ \text{kg m}^{-1}\,\text{s}^{-2} \text{ or } 6.9\ \text{N m}^{-2}$$

(b) The time-averaged current speed at 1 m above the bed is 1.2 m s^{-1} which is roughly an order of magnitude greater than the (derived) shear velocity.

(c) Yes, because the shear velocity profile is a straight line. Expressed mathematically:

On Figure 4.8, for the interval 1 m to 10^{-1} m, d log$z$ = $\log_{10} 1.0 - \log_{10} 0.1$ = $0 - (-1) = 1$, which is the same as over the interval 10^{-1} to 10^{-2}. As d$\bar{u}$ over the interval 1 m to 10^{-1} m is the same as that over the interval 10^{-1} m to 10^{-2} m (1.2 m s^{-1} – 0.72 m s^{-1} = 0.48 m^{-1}), it follows that the shear velocity d$\bar{u}$/d log$z$ must also be the same.

Question 4.4 (a) (i) If the bed roughness remained constant, then the roughness length, z_0, would stay the same; to accommodate an increase in speed, the slope would become flatter; in other words, the two slopes would diverge from z_0 (Figure A1).

(ii) In this case, the roughness length would increase, so the intercept z_0 would occur higher up the depth axis. As the current speed well away from the bed (>1 m) is unchanged, the slope would again become flatter. In other words, the two slopes would converge (Figure A2).

(b) In both cases, a flattening of the slope implies a steeper velocity gradient. As $\tau_0 \propto d\bar{u}/dz$ (Equation 4.2), this also means an increase in τ_0. Consequently, both an increase in current velocity and an increase in bed roughness will lead to an increase in shear stress and hence in erosion at the bed.

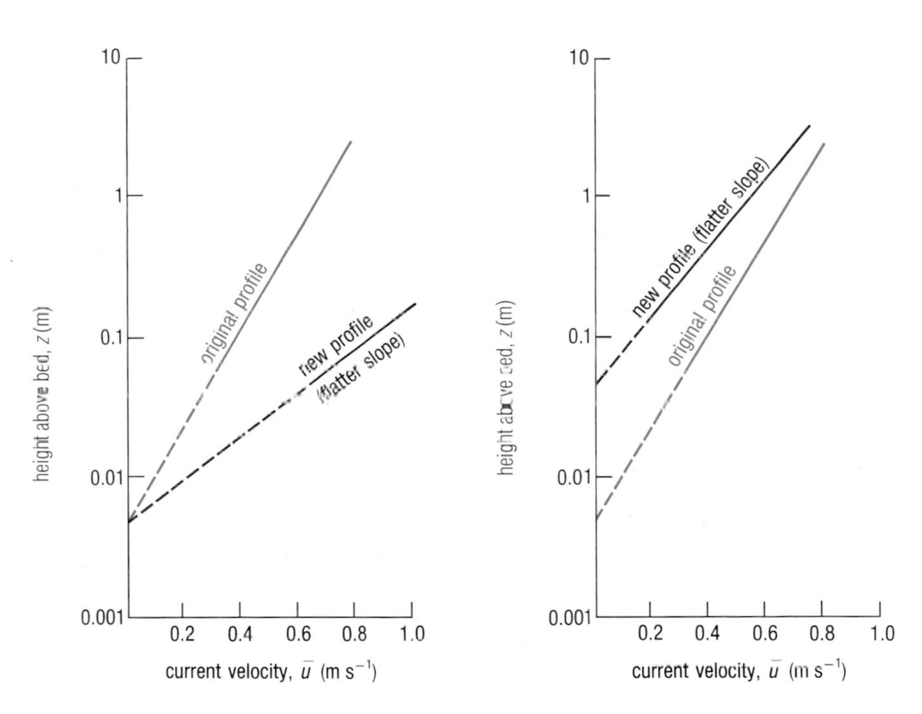

Figures A1 and A2 Logarithmic velocity profiles to illustrate the answers to Question 4.4(a).

Question 4.5 Both equations show that the shear stress is proportional to density. Sediment-laden water is denser than sediment-free water, and will exert a higher shear stress at the bed (τ_0), so its erosive power should be greater.

Question 4.6 (a) According to Figure 4.12, the minimum shear velocity needed to erode 1 mm particles is ~2.3×10^{-2} m s^{-1} (remember the scales are logarithmic). So, substituting in Equation 4.5 to give the critical shear stress, that gives:

$$\tau_0 = \rho u^2_*$$

$$\tau_0 = 10^3 \text{ kg m}^{-3} \times (2.3 \times 10^{-2})^2 \text{ m}^2 \text{ s}^{-2}$$

$$= 5.3 \times 10^3 \times 10^{-4} \text{ kg m}^{-3} \text{ m}^2 \text{ s}^{-2}$$

$$= 0.53 \text{ kg m}^{-1} \text{ s}^{-2}$$

$$= 0.53 \text{ N m}^{-2}$$

(b) According to Figure 4.12, the minimum shear velocity required to keep 1 mm particles in suspension is close to 2×10^{-1} m s^{-1}. From Equation 4.5, that corresponds to a shear stress, τ_0, of $10^3 \times (2 \times 10^{-1})^2$

$$= 4 \times 10^3 \times 10^{-2} = 40 \text{ N m}^{-2}$$

The reason why this is nearly two orders of magnitude greater than the shear stress required to set the particles in motion as bedload (see (a)) is that shear stress is proportional to the *square* of the shear velocity – so even a quite small change in shear velocity results in a large change in shear stress.

Question 4.7 (a) The difference in maximum velocity between the two tidal currents is about 0.03 m s^{-1}, or 3 cm s^{-1}, about 4% difference.

(b) The 'SSW' shaded area in Figure 4.13(b) is about one-third as large again as the 'NNE' shaded area. The shaded areas are proportional to the total amount of sediment being transported, and so there must be significant net south–south-westerly transport of sediment. This also means that the relatively small difference in current velocity has led to an approximately 30% increase in sediment transport in one direction.

(c) The dashed lines represent the threshold velocity (in Figure 4.13(a), of the actual average current, $\bar{u}_1$ and in Figure 4.13(b), of $\bar{u}_1{}^3$) required to move sediment grains of 0.3 mm diameter. In Figure 4.13(b), the shaded areas represent those parts of the tidal cycle where this threshold is exceeded and so sediment grains coarser than 0.3 mm are being moved, certainly in the bedload, and probably in suspension for part of the time. In the unshaded areas, where the curve falls below the threshold velocity, only grains smaller than 0.3 mm are being moved, and the grain sizes being moved decrease as the velocity minima are approached.

Question 4.8 The coarser sediments (0.075 to 0.1 mm) are concentrated in the bottom two or three metres of the water column (Figure 4.15(b)) and, despite the somewhat slower currents at these levels due to friction with the bed (Figure 4.15(a)), the sediment fluxes are greatest here (Figure 4.15(c)). In contrast, the finer-grained sediments are much more evenly spread through the water column, as you might expect, because the small particles are much more easily kept in suspension by turbulent flow. Concentrations rise slightly in the bottom two metres and fall off in the top two metres. When concentrations are combined with the velocity distributions (Figure 4.15(a)), the minimum fluxes are at the top and bottom of the water column, a pattern very different from the one observed for the coarser suspended sediments.

Question 4.9 As the settling velocity of grains 0.1 mm in diameter is 10 mm s^{-1} and most of this sediment is concentrated within the lowermost three metres (or 3×10^3 mm) of the water column (Figure 4.15(b)), it would take about ($3 \times 10^3/10$) seconds, or about 5 minutes, for all of this sediment to reach the bed.

Question 4.10 (a) Over the depth interval 10^{-1} to 1 m, the average current velocity changes from about 0.4 m s^{-1} to 0.7 m s^{-1}, so d log$z$ is 1 and d$\bar{u}$ is 0.3 m s^{-1}. From Equation 4.3:

$$\bar{u}_* = \frac{1}{5.75} \times \frac{d\bar{u}}{d \log z} = \frac{0.3}{5.75 \times 1} = 0.052 \text{ m s}^{-1}$$

From Equation 4.5:

$$\tau_0 = \rho u_*^2$$

$$= 10^3 \times (0.052)^2 \text{ kg m}^{-3} \text{ m}^2 \text{ s}^{-2} = 2.7 \text{ kg m}^{-1} \text{ s}^{-2}$$

$$= 2.7 \text{ N m}^{-2}$$

The current velocity at one metre above the bed is 0.7 m s^{-1}, an order of magnitude greater than the calculated shear velocity.

(b) As this is a tidal current, it will not remain steady. Within a matter of hours at most it will either accelerate or decelerate. This means that the values for u_* and τ_0 you calculated are valid only for a short time.

(c) The shear velocity calculated in (a) is 5.2×10^{-2} m s^{-1}. From Figure 4.12, the range of grain sizes that could be transported only as bedload is about 0.3 mm to 4.4 mm. This includes medium sands to fine gravel (Table 4.1).

(d) The maximum size of particles transported fully in suspension by a current with this shear velocity is about 0.07 mm (70 μm), at the lower end of the very fine sand range (Table 4.1).

Question 4.11 The steeper slopes of asymmetrical current-produced bed forms should mostly face in the direction indicated by the arrows in Figure 4.14.

Question 4.12 As the size of ripples generally increases with current speed, the megaripples formed at higher current speeds, while the smaller ripples may either have formed as the current waned, or formed later as a separate event.

CHAPTER 5

Question 5.1 (a) No, the zones must migrate landwards on the rising tide, seawards on the falling tide, but their relative positions will of course stay the same (so will their relative widths, unless there is a great change in beach slope).

(b) No, they will be narrower off steep than off shallow beaches (Figure 5.1).

(c) Yes, the plunge point will move landwards on the rising ride, seawards on the falling tide – though you need to bear in mind that it is an average position, not all waves break there, see Section 5.1.1.

Question 5.2 From Figure 1.18, for waves of given height and steepness, (a) spilling and plunging breakers are more likely on shallow sloping beaches, (b) collapsing and surging breakers are more likely on steep beaches.

Question 5.3 Although it may not be immediately obvious at first sight, Figure 5.4 is NOT a collection of beach profiles!

(a) The curve for the largest grains (3.44 mm, gravel-sized, Table 4.1) indicates that beach slope is more or less constant at around 20° for a wave steepness up to about 0.02, Steeper waves result in progressive decrease in the angle of slope. For coarse beach sediments, wave steepness appears to be less important than grain size in controlling beach slope until quite high values of wave steepness are reached. A similar, but not so marked, trend is seen in the curve for coarse sand-sized sediment (0.97 mm). For the medium- and fine-grained sand-sized sediment, there is a continuous decrease in beach slope with increasing wave steepness. In summary, irrespective of grain size, the steeper the waves, the smaller the beach slope.

(b) (i) Just over 7° and (ii) about 18°, i.e. for waves of given steepness, the coarser the sediments, the steeper the slope.

Question 5.4 If wave height is doubled, so is the speed of water particles near the bed (Equation 5.2). We should expect shear stress to quadruple, because, as noted at the end of Section 4.2.4, the shear stress due to waves is proportional to the square of the horizontal orbital velocity, u_m. In general, the frictional force exerted by a fluid flowing over a surface is proportional to the *square* of its speed; Section 4.1.1.

Question 5.5 If you are interested, the calculations using Equation 5.1, referred to at the start of this question, are as follows, using the approximation that $\sinh x \approx x$ where $x \leq 1$, which is valid here since the sinh for $2\pi d/L \approx 0.6$:

$$u_m = \frac{\pi H}{T \sinh\ (2\pi d / L)}$$

$$u_m = (3.14 \times 3)/(10 \times 6.28/10) \approx\ 1.5\ \mathrm{m\ s^{-1}}\ \text{for 10 s period waves}$$

$$u_m = (3.14 \times 3)/(15 \times 6.28/10) \approx\ 1\ \mathrm{m\ s^{-1}}\ \text{for 15 s period waves}$$

(a) From Figure 5.7, particles up to about 20 mm and 10 mm diameter are moved by the 10 s and the 15 s waves, respectively.

(b) (i) Equation 5.1 shows that u_m is inversely proportional to T, i.e. shorter period waves have greater orbital velocities, and therefore move larger particles. (ii) In general, shorter period waves also have shorter wavelengths, so if H is the same but L is smaller, then H/L must be larger (waves are steeper); and since we are told that d/L is the same for both sets of waves ($d = L/10$), then water depth, d, must be less if L is smaller.

(c) No. From Equations 5.2a and b, if wave height decreases, u_m must decrease because u_m is proportional to H (see also Figure 5.8).

(d) No. If water depth increases, u_m must also decrease because from Equation 5.2b, $u_m \propto \dfrac{1}{\sqrt{d}}$ (see also Figure 5.8).

Question 5.6 Other things being equal, sediment grain sizes in the intertidal zone should become smaller in a seaward direction, something you may have seen for yourself on a beach around low tide.

Question 5.7 (a) If we insert the values given in the question into Equation 5.3, then we have:

$$P_l = 0.5\,(\tfrac{1}{8} \times 10^3 \times 9.8 \times 1^2) \times 0.5 \times 0.87 \approx 266\ \text{W m}^{-1}.$$

From Equation 5.3 the units for wave power are derived as follows: $P_l = (\text{m s}^{-1}) \times (\text{kg m}^{-3}) \times (\text{m s}^{-2}) \times (\text{m}^2) = \text{kg m s}^{-3}$, which is the same as $\text{kg m}^2\,\text{s}^{-2}\,\text{s}^{-1}\,\text{m}^{-1}$. In other words, joules per second per metre (1 joule = $\text{kg m}^2\,\text{s}^{-2}$), or watts per metre.

(b) If the wave crest approaches parallel to the shoreline, then the angle between the wave crest and the shoreline must be zero. So, sin α must also be zero and the value of P_l will be zero as well, so there is no *longshore* wave power. All the wave power must be applied perpendicular to the shoreline.

Question 5.8 (a) From Figure 5.7, the particle size would be about 0.3 mm.

(b) From Figure 5.8, in water 100 m deep these waves would have been about 4 m high, given that they moved particles of 0.3 mm diameter.

Question 5.9 This is a good example of the consequences of disturbing the dynamic equilibrium of a coastline. Interrupting the longshore drift caused accretion to the north, erosion to the south. Here is an account of the actual outcome reported in 1972 by the Jacksonville City Engineer, Oscar G. Rawls:

> 'The jetties trapped the southward-migrating sand and held it north of the north jetty. Sand accreted there. That accretion meant, however, that there was almost equal starvation of the beaches south of the jetties ... and we have an eroding beach to the south of the jetties extending for several miles.'

CHAPTER 6

Question 6.1 Given the strong tidal currents, you would expect to see asymmetrical current-formed ripples and even megaripples, their form depending on whether the tide is on the ebb or the flood. There may also be symmetrical wave-formed ripples.

Question 6.2 (a) A settling velocity of $5 \times 10^{-6}\ \text{m s}^{-1}$ means that it would take about 2×10^5 seconds for particles of this size to settle through a water column of 1 m depth. That is something like two-and-a-half days, i.e. time for four or five high tides to have occurred.

(b) If they were flaky clay particles, it would take even longer, because the shape would slow the descent. In theory, it should not be possible for clays to settle out of suspension in the course of a tidal cycle.

Question 6.3 In a well mixed estuary, the salinity is more or less constant with depth at any particular place and time, but will increase on the rising (flooding) tide (as the whole estuarine water mass moves upstream), and decrease on the falling (ebbing) tide (as the water mass moves back downstream). In other words, the isohalines in Figure 6.4(c) move bodily upstream and downstream.

Question 6.4 (a) Given the large tidal range and the relatively broad aspect of the estuary, it would probably be classified as well mixed (and the water column is indeed likely to be well mixed at high tide over the wide expanse of intertidal flats to the right of the main channel (looking landwards)). Some stratification could well develop in the middle and upper estuary, however, especially during neap tides and periods of high river discharge.

(b) On the ebb, tidal currents in the channel run counter to the incoming waves, slowing them down, steepening them, and causing them to break further offshore (cf. Section 1.6.1).

(In answering Question 6.4, you may have concluded that the main channel in Figure 6.6 is on the right-hand side (looking seaward) because of the Coriolis effect leading to a horizontal residual circulation as illustrated in Figure 6.4(d). However, the tidal part of this estuary is not much more than about 6 km long, which is probably not sufficient to allow the Coriolis effect to influence the course of the tidal channel significantly. The estuary has its present form mainly because of late 18th century drainage and canalization works upstream of the present head of the estuary.)

Question 6.5 (a) Figure 6.11(b) represents (i) high river discharge ($800\,\mathrm{m}^3\,\mathrm{s}^{-1}$), as the turbidity maximum is near the estuary mouth and relatively small, compared with Figure 6.11(a), which represents (ii) low river discharge ($200\,\mathrm{m}^3\,\mathrm{s}^{-1}$), as the turbidity maximum is larger and extends further upstream.

(b) As outlined in Section 6.2.1, we would expect the Seine to be better stratified when river discharge is high (Figure 6.11(b)); this would increase the thickness of the low salinity (low density) upper layer, and would tend to 'flatten' the slope of the isohalines (cf. Figure 6.4(b)).

(These effects might be offset to some extent by greater suspended sediment concentrations in the turbidity maximum when river discharge is low. This would slightly increase the density of the lower layers and contribute to stratification at such times – but the density contrasts due to salinity differences are likely in general to be a more important determinant of stratification.)

Question 6.6 (a) Neither, under normal conditions. The pattern of flow (Figure 6.13(b)) is such that there is no convergence of flows at the bed, and therefore no null point; and since the sediments are mostly sands, there can be no turbidity maximum either.

(b) Yes, of course. The flows are in opposite directions from those in 'normal' estuaries, but there must be mixing and exchange of water between the upper and lower layers.

Question 6.7 The result was an *increase* in the rate of sedimentation in the estuary, because the increasing landward flow of seawater at the bed would have led to an increase in the sediment load brought into the estuary from the sea.

Question 6.8 The curving form of the island 'chain' in Figure 6.17(a), and the way it tapers off towards the east suggest that this was the prevailing direction of longshore drift – and it is consistent with the present-day residual current circulation in this part of the North Sea (see Chapter 8).

Question 6.9 (a) Sloping isohalines (as opposed to the near-vertical ones of well-mixed estuaries), a tidal range approaching 5 m, and high suspended sediment concentrations in the turbidity maximum show the Seine to be a partially mixed estuary.

(b) Assuming the null point to be at the intersection with the bed of the 1 (part per thousand) salinity isohaline, the core of the turbidity maximum lies well downstream of it for most of the time.

(c) (i) Maximum concentrations in the core are greater than 1 g l^{-1} (and extend throughout the water column) at low water, and (ii) minimum concentrations are 0.5–1 g l^{-1} (near the bed) during slack water around high tide.

Question 6.10 (a) The surface isohalines are not straight but are deflected further up-estuary on the eastern side, further down-estuary on the western side. There is a lateral salinity gradient, consistent with deflection by the Coriolis force being to the right in the Northern Hemisphere (cf. Figure 6.4(d)). However, the lateral salinity gradient could also result, in part at least, from greater supplies of freshwater from tributaries to the west of Chesapeake Bay.

(b) You should not be at all confident, in the absence of information about river discharge, main channel depth and so on (see last part of Section 6.2.1). You can infer that the tidal range is large, because a horizontal circulation appears to be developed, but no more. In fact, the estuary is commonly more or less stratified, because a relatively shallow sill near the estuary mouth isolates the bottom waters, especially in summer, and anoxic conditions can develop in them. At such times, the estuary is certainly *not* well mixed.

Question 6.11 (a) According to Figure 6.20(b), the salt intrusion clearly penetrates upstream as far as about station 4, so the null point must be near there, where bottom salinity is in the order of 1 (part per thousand).

(b) Isohalines must slope down landwards, as in Figure 6.4(b). Comparison of Figures 6.20(b) and (c) shows that at any station between 1 and 17, near-bed salinity is greater than surface salinity. The difference is relatively small so this estuary is approaching the well-mixed end of the continuum (Section 6.2.1), i.e. the slope of the isohalines is quite steep.

(c) The core of the turbidity maximum (2.27 g l^{-1}) probably lies between stations 8 and 9, and the near-bed salinity approaches 20 (parts per thousand) there. At the surface, the turbidity maximum contains much less suspended sediment, and the maximum concentration is reached a little further upstream than it is near the bed.

(d) Seaward of station 17, salinities in both surface and bottom waters are the same, so the water column must be well-mixed. That is what you might expect at the mouth of the estuary, but the salinity data show no evidence of a ROFI, possibly because freshwater flow was low and/or tidal mixing was especially strong when the measurements were made.

CHAPTER 7

Question 7.1 (a) Salinities are very low at the surface (e.g. 4.5 at the station in the middle of the plume about 8 km from land), while at only 3 m depth they are all greater than 25. So by 5 m depth, salinities must be close to that of normal seawater (35). These plumes are thus very thin.

(b) Salinities at the station just beyond the north-eastern end of the plume boundary, *c.* 10 km from land, are very similar near the surface and at 3 m depth: both close to 30 (parts per thousand), i.e. approaching the salinity of normal seawater.

Question 7.2 Flocculation (Section 6.1.1) where river water and seawater mix. Clay particles flocculate to form larger aggregates which settle much more rapidly than individual clay particles.

Question 7.3 Note that in answering this question, you could eliminate the tide-dominated 'sides' of Figure 7.3, because both rivers flow into the Mediterranean, where (as noted at the end of Section 2.4) the tidal range is very small.

(a) The Nile delta shoreline is relatively straight with few distributaries, and the photograph shows how the sediment has been reworked into beaches. The dominant influence appears to be wave action. However, two distributaries discharge into the sea from quite elongate protuberances which suggests that fluvial processes also play a part in shaping the delta. The Nile should therefore plot in the wave-dominated area of Figure 7.3, but part-way towards the river-dominated apex.

(b) The Po delta has features in common with the Nile delta.
There is some indication of sediment movement landwards to form beaches, suggesting wave action may be important. However, the delta protrudes into the sea far more than the Nile delta, and the delta plain is crossed by a larger number of distributaries, so fluvial processes are more significant. The Po delta should therefore plot in the river-dominated area, part-way towards the wave-dominated apex.

CHAPTER 8

Question 8.1 (a) (i) Upwelling is in Figure 8.2(a), with offshore Ekman transport; downwelling in Figure 8.2(b), with onshore Ekman transport.
(ii) Transport of surface water is to the *left* of the wind direction, so this must be the Southern Hemisphere.

(b) While the wind keeps blowing, sediment transport at the sea-bed will be onshore in Figure 8.2(a), offshore in (b); though sediment movement is likely to occur only where the water is shallow.

Question 8.2 Gravels are coarse sediments which disrupt the viscous sublayer (and the roughness length is likely to be large) and allow turbulence to extend to the bed, increasing the potential for sediment movement. Over the silt, a viscous sublayer develops (and the roughness length decreases), and fine particles settling into it can be deposited.

Question 8.3 As bedload transport depends upon the *cube* of the current velocity (Equation 4.6), small differences in current speed can result in large differences in sediment transport. So a current that flows slightly faster in one direction can transport significantly more bedload than a slightly slower current flowing in the opposite direction. (For suspended sediment, the settling rates of most particles are so low that small residual currents can transport suspended sediments for considerable distances, especially as the turbulence associated with both waves and currents helps to keep them in suspension.)

Question 8.4 (a) Stratification will be promoted by conditions which increase the buoyancy of surface water and/or reduce the likelihood of mixing. Hence, any combination of neap tides (weaker tidal currents), enhanced river flow, and calm weather, promotes buoyancy and stratification.

(b) Stratification will be broken down by conditions which reduce the buoyancy (increase the density) of surface waters and/or encourage mixing. Any combination of spring tides, reduced river flow and cold strong winds will therefore break down stratification. During storms, when strong winds are combined with enhanced river flow, stratification is likely to be weak, irrespective of tidal conditions.

Question 8.5 (a) Other things being equal, we would expect to find muds to seawards of the tidal mixing front, where the lower part of the water column is mixed only by tidal currents. The water is deeper there and speeds are generally less – whereas landwards of the front the water is shallower, and both wind and tidal mixing occur.

(b) A large storm could break down the front, by generating large waves that – along with tidal currents – could completely mix the water column as far as the shelf break (Section 8.1.2).

Question 8.6 In general, there is a direct relationship between tidal current speed and the nature of the sea-bed. Exposed rock surfaces or gravel are found where tidal currents exceed $100\,cm\,s^{-1}$ ($1\,m\,s^{-1}$), while coarse and fine sands occur where current speeds are less. Muds seem to occur mainly where current speeds are less than about $40\,cm\,s^{-1}$ ($0.4\,m\,s^{-1}$). Muds resuspended by waves during storms may not be carried very far, and some of the muddy areas in Figure 8.9(b) are in deeper shelf waters, where near-bed tidal current speeds are lower. (The areas of gravel/rock in the Channel and in the Irish Sea coincide roughly with bedload partings, as shown in Figure 8.4.)

Question 8.7 Only about 2 per cent of Mississippi sediment is sand, the rest is mud (Section 8.2), so while it is likely that both sand and mud are being dispersed across the shelf, any sand that is deposited is overwhelmed by the much greater quantities of mud.

Question 8.8 It is justifiable to consider placer deposits as relict sediments because they were deposited from rivers at a time of lower sea-level. They are also residual sediments in the sense that they have been reworked by waves and currents which removed smaller and less dense particles, but left the denser minerals in place.

ACKNOWLEDGEMENTS

The Course Team wish to thank the following: Dr Martin Angel (Southampton Oceanography Centre) and Prof. Kenneth Pye (University of Reading), the external assessors; also Dr Chris Vincent, Maurice Crickmore and Mary Llewellyn who provided useful advice and comments on the first edition.

The structure and content of this Volume and of the Series as a whole owes much to our experience of producing and presenting the first Open University Course in Oceanography (S334), from 1976 to 1987. We are grateful to all those who prepared and maintained that Course, and to the students and tutors who provided valuable feedback and advice on both Courses.

Grateful acknowledgement is also made to the following for material used in this Volume:

Figure 1.2 and Table 2.1 H.J. McLellan (1965) *Elements of Physical Oceanography*, Pergamon; *Figures 1.11(a) and 1.17* B. Kinsman (1965) *Wind Waves*, Prentice-Hall; *Figure 1.11(b)* J. Williams (1962) *Oceanography: An Introduction to the Marine Sciences*, Little Brown & Co.; *Figures 1.16 and 2.13(b)* H.V. Thurmann (1993) *Essentials of Oceanography*, Prentice-Hall; *Figures 1.18(c), 5.1(c), 5.2(b,c), 5.15(d), 6.2(b), 6.3(c) and 6.17(b)* Angela Colling, Open University; *Figure 1.19(a)* C.E. Vincent; *Figures 1.21, 1.22 and 7.1(a–d)* NASA; *Figure 1.23* Ocean Scientific International Ltd.; *Figure 2.5* J.G. Harvey (1976) *Atmosphere and Ocean*, Artemis Press; *Figure 2.15* The Royal Society; *Figure 2.16(a–c)* H.U. Sverdrup *et al.* (1942) *The Oceans*, Prentice-Hall; *Figures 2.18(a) and 2.21* A. Defant (1961) *Physical Oceanography*, Pergamon; *Figures 2.18(b), 4.17, 5.1(e), 5.2(a,d), 5.12, 5.13, 5.16(b), 5.17(b), 6.3(b), 6.6 and 6.16* John Wright, Open University; *Figure 2.22(b)* NERC Satellite Station, Dundee/ NERC CCMS Plymouth Marine Laboratory; *Figure 2.25(a)* B.J. Skinner and K.K. Turekian (1973) *Man and the Ocean*, Prentice-Hall; *Figure 2.25(b)* French Embassy, London; *Figure 3.2(a)* Peter Skelton, Open University; *Figure 3.2(b)* BP Development Ltd; *Figure 3.4* J.D. Milliman and R.H. Meade (1983) in *Journal of Geology*, **91**, University of Chicago Press; *Figure 3.5* Prof. Derek Mottershead, Manchester Metropolitan University; *Figures 4.1 and 4.11* J.R.L. Allen in K. Pye (ed.) (1994) *Sedimentary Transport and Depositional Processes*, Blackwell Scientific; *Figure 4.12* I.N. McCave and the Geological Society; *Figure 4.14* I.N. McCave (1971) in *Marine Geology*, Elsevier; *Figures 4.16 and 4.22(b)* NERC CCMS Proudman Oceanographic Laboratory; *Figure 4.18* K. Dyer (1986) *Coastal and Estuarine Sediment Dynamics*, Wiley; *Figure 4.19(a,b)* J.R.L. Allen (1985) *Principles of Physical Sedimentology*, Allen & Unwin; *Figures 4.20(a), 4.22(a), 4.23 and 8.4* A.H. Stride (ed.) (1982) *Offshore Tidal Sands*, Chapman & Hall; *Figure 5.1(d)* Dee Edwards; *Figure 5.1(f) and 5.10(a)* Evelyn Brown; *Figure 5.3* SEPM (1969) *Field Trip Guide Book*, Coastal Research Group; *Figure 5.4* American Society of Civil Engineers; *Figure 5.5* J. Hardisty in K. Pye *op cit.*; *Figures 5.8 and 7.2(a)* Society for Economic Palaeontologists; *Figure 5.7* Elsevier Science Publishers; *Figures 5.9 and 5.11* P.D. Komar (1976) *Beach Processes and Sedimentation*, Prentice-Hall; *Figures 5.13, 5.16(b) and 5.17(b)* Andy Tindle/John Watson, Open University; *Figure 5.15(b)* J. Pethick (1984) *An Introduction to Coastal Geomorphology*, Edward Arnold; *Figure 5.18* Paul Gray, Middleton-on-Sea;

Figure 6.1(b) Tony Waltham, Geophotos; *Figure 6.3(a)* David Paterson, Gatty Marine Laboratory, University of St. Andrews; *Figure 6.5* Phil Smith, AEC; *Figure 6.8* John Simpson and Bill Rowntree, School of Oceanographic Sciences, University of Wales, Bangor; *Figures 6.10 and 6.11* K.R. Dyer in K. Pye *op cit.* after a thesis by J. Avoine (1981); *Figure 6.12* Jonathan Silvertown, Open University; *Figure 6.14* Mike Fennessy, Institute of Marine Studies, University of Plymouth; *Figure 6.15* Devon Association for the Advancement of Science, Literature and Art; *Figure 6.17* Teunis Lauters/Directoraat-Generaal Rijkwaterstaat; *Figure 6.18* Jim Ogg, Purdue University, USA; *Figure 6.19* J.V. McCormick and J.M. Thiruvathukal (1985) *Elements of Oceanography*, CBS College Publishing; *Figure 7.3* W.E. Galloway (1983) *Terrigenous Clastic Depositional Systems*, Springer-Verlag; *Figure 7.4* J.P. Morgan and R.H. Shaver (eds) (1970) *Deltaic Sedimentation: Modern and Ancient*, SEPM; *Figure 8.5* Tj. H van Andel (1960) in *Recent Sediments, Northwestern Gulf of Mexico*, AAPG; *Figure 8.3(b)* NASA SeaWIFS Project and Orbital Sciences Corp.; *Figure 8.8* John Simpson, School of Oceanographic Sciences, University of Wales, Bangor; *Figure 8.9* Institute of Oceanographic Sciences.

INDEX

Note: page numbers in italic refer to illustrations